FOOD SCIENCE AND TECHNOLOGY

PECTIN

CHEMICAL PROPERTIES, USES AND HEALTH BENEFITS

FOOD SCIENCE AND TECHNOLOGY

Additional books in this series can be found on Nova's website under the Series tab.

Additional e-books in this series can be found on Nova's website under the e-book tab.

FOOD SCIENCE AND TECHNOLOGY

PECTIN

CHEMICAL PROPERTIES, USES AND HEALTH BENEFITS

PHILLIP L. BUSH
EDITOR

nova publishers
New York

For permission to use material from this book please contact us:
Telephone 631-231-7269; Fax 631-231-8175
Web Site: http://www.novapublishers.com

Library of Congress Cataloging-in-Publication Data

ISBN: 978-1-63321-438-5

Library of Congress Control Number: 2014943360

Published by Nova Science Publishers, Inc. † New York

CONTENTS

PREFACE

Pectins are biopolymers with multiple applications because of their structural diversity and complexity. Although pectins from different sources have some common structural characteristics, many aspects of the common structure change according to the species and the physiological stage of the plant. Moreover, the application of pectin is determined by its chemical features, including galacturonic acid content, methoxyl content, degree of esterification and acetyl value. The most traditional raw materials used for the extraction of pectins are either apple pomace or citrus peels that are supplied as by-products of juice production. Both materials contain significant amounts of pectic substances, but with different chemical characteristics that make them suitable for specific applications. Considering that pectin is widely used as a functional ingredient, many researchers have been testing the use of other materials and alternative methods of extraction for industrial exploitation. This book discusses the chemical properties of pectin. It addition, it includes the uses and health benefits that pectin may have.

Chapter 1 - Regenerative Medicine is an interdisciplinary field, which applies the principles of engineering and life sciences to the development of biological substitutes that restore, maintain, or improve tissue function, combining a scaffold/support material with appropriate cells and bioactive molecules. Briefly, a scaffold serves as temporary support for cell growth and to present stimuli directing the growth and formation of a new desired tissue. Depending on the specific application, the required scaffold material and its properties can vary. Within the possible materials to fabricate a scaffold, natural-based polymers are among the most attractive, mainly due to their similarities with the extracellular matrix, their degradability, hydrophilicity, biocompatibility and versatility as well as typically good biological performance. In particular, natural-based hydrogels, due to their capability of retaining water and other biomimetic properties, offer an attractive alternative for numerous biomedical applications, such as bioactive molecules delivery systems, 3D scaffolds and cell immobilization.

Recently, several studies have been focused on the production and characterization of different gels obtained with pectin, a polysaccharide non-conventional in regenerative medicine, which presents the peculiar characteristics for this application, such as tunable physical properties, high water content and ability to homogeneously immobilize cells, genes, proteins, drugs or growth factors.

Additionally, pectin was chemically modified by grafting an oligopeptide containing the RGD sequence, that is known to promote cell adhesion. A partial oxidation of pectin was aimed to obtain faster degradation rates of pectin gels. MC3T3 preosteoblasts and human mesenchymal stem cells were immobilized in RGD-modified pectin microspheres for bone tissue regeneration showing good cell viability, metabolic activity and osteogenic differentiation. In vivo experiments confirmed that RGD-modified pectin microspheres provided a complete adaptation to bone defect and induced bone regeneration avoiding the dispersion of cells after implantation.

Chapter 2 - The term dietary fiber includes a wide group of dietary compounds of plant origin and resistant to digestion by human gastrointestinal enzymes. Traditionally, this group of compounds has been classified according to their chemical structure and physiological activities, as its behavior in water, being insoluble ones the most frequently consumed in Westernized countries. The assessment of the different types of fiber provided from diet implies some difficulties since factors, such as fruit maturation or the intake of peeled fruit, may affect the fiber content of a food and they are impossible to quantify in a food composition table. These handicaps result in few nutritional studies providing a detailed intake of these dietary compounds.

From soluble fibers, pectins, provided by citric fruits and vegetables, have attracted a deal of attention in the last decades, giving the epidemiological evidences linking their intake with protection against cardiovascular disease, type II diabetes, colorectal cancer and gastrointestinal diseases. Also, pectins evade digestion by intestinal enzymes and passes directly into the colon, where they are metabolized by some intestinal bacteria such as *Bifidobacterium* and *Lactobacillus,* which constitute the traditional target of prebiotics, and contributing to the increase of short chain fatty acid production. Thus, pectin confers benefits upon host health by decreasing the risk of some diseases, such as the irritable bowel syndrome, inflammatory bowel disease, cardiovascular disease and cancer.

The contributors to this chapter will provide a brief description about the dietary sources of pectin in humans and review the most relevant literature discussing the prebiotic effect of this dietary compound together with its implications for health by means of the increase in the production of bacterial metabolites.

Chapter 3 - Pectins are biopolymers with multiple applications because of their structural diversity and complexity. Although pectins from different sources have some common structural characteristics, many aspects of the common structure change according to the species and the physiological stage of the plant. Moreover, the application of pectin is determined by its chemical features, including galacturonic acid content, methoxyl content, degree of esterification and acetyl value. The most traditional raw materials used for the extraction of pectins are either apple pomace or citrus peels that are supplied as by-products of juice production. Both materials contain significant amounts of pectic substances, but with different chemical characteristics that make them suitable for specific applications. Considering that pectin is widely used as a functional ingredient, many researchers have been testing the use of other materials and alternative methods of extraction for industrial exploitation. Among them, different waste materials have been tested, such as sugar beet and passion fruit pomace. The yield and quality of extracted pectins are essential for their commercialization and are highly affected by the method used. The usual extraction process is based on the combination of acidic solutions and high temperature. Moreover, it is a very time-consuming process - up to 12 h. Microwave-assisted extraction has been tested and presented good results in the extraction of pectins from passion fruit peel, berries and from watermelon waste fruit rinds. The extraction of pectin from navel orange peel assisted by ultra-high pressure showed that the intrinsic viscosity and viscosity-average molecular weight were much higher than those extracted by traditional heating, microwave and commercial pectin. In order to obtain time-saving, and eco-friendly extraction methods, the use of microbial enzymes has attractive properties that justify more extensive researches. Some studies confirm that the implementation of a biotechnological method would greatly increase the pectin production and contribute to free the processing plants from expensive works to neutralize the acidic components of the traditional technology. Given the importance of this biopolymer regarding the wide application in medicinal and food products, this chapter reviews current issues regarding the prospect for new sources of pectin and the advances in their extraction methods.

Chapter 4 - Pectins are polysaccharides containing D-galacturonic acid and galacturonic acid with methyl ester residues that can be acetylated to some degree. This biopolymer has been used as a gelling agent for the last two centuries and is extensively applied in food and pharmaceutical industries. In this case, pectins with a methylation degree lower than 50%, called low-methoxyl pectin (LMP), form gel in the presence of calcium ions, and hence, may be used as a gelling agent in numerous types of products such as: low-calorie jams and jellies, confectionery jelly products, and other food applications. However, one highlighted use of LMP is for the entrapment, encapsulation or immobilization of enzymes and cells for biotechnological applications. The encapsulation of a lipase in pectin gels cross-linked with calcium ions brought three to four times more enzymatic activity in water miscible organic co-solvents compared with aqueous systems. In another study, α-amylase and glucoamylase enzymes were immobilized to pectin by covalent binding showing greater thermal and pH stability over the free enzyme system with the complete retention of original activities. The immobilized enzymes showed the highest release of glucose compared with free enzymes when applied in starch hydrolysis. Another important use of LMP is in the entrapment of microbial cells for biocatalytic/ bio-transformation and fermentation uses. When the cells of the *Nocardia tartaricans* bacterial strain were immobilized in pectate gel to obtain L-tartrate, higher *cis*-epoxysuccinate hydrolase activity was observed compared with the free cells. An additional study reports the immobiliza-tion of *Saccharomyces cerevisiae* cells in pectin gel for ethanol produc-tion, indicating that no significant changes occurred. Cells maintained their growth capacity, and the beads could be reutilized several times in successive batch fermentations, which is one of the major advantages of cell immobilization. The uses of pectin will be reviewed in this chapter since different high-added-value-compounds can be obtained showing the remarkable relevance of this matrix for biocata-lysts immobilization.

Chapter 5 - Pectin is a natural polysaccharide and its specific enzymatic degradability by colonic microbiota makes it a promising material for designing drug release systems, mainly those intended for targeting drugs to the colon. However, in despite of pectin resistance against proteases and amylases, remaining as aggregates of macromolecules in acid medium, a great challenge to optimize the performance of pectin in such systems lies in its high hydrophilicity that, in several times, results in an undesirable premature release of drugs. Blends of pectin with other polysaccharides and cross-linking reactions are valuable tools to modulate such properties of pectin, particularly reducing its solubility. These approaches have been focus of important researches of our research group and our findings have been published in important scientific journals. Blends of pectin and retrograded starch (RS) allowed the preparation of free films with suitable mechanical properties and reduced dissolution of films in acid media, while their high resistance against enzymatic digestion by pancreatin was demonstrated. The same polymer association was exploited for preparing tablets containing sodium diclofenac (SD), and the presence of pectin reduced significantly the drug dissolution in acid medium. In another study with free films, the blends of pectin-high amylose starch (HAS) cross-linked with sodium trimetaphosphate (STMP) contributed to the reduction of their hydrophilicity. This polymer association was also exploited for preparing hydrophilic matrices from which the drug release rates in acid medium were lowered. In addition, this same cross-linked HAS/pectin blend was employed for preparing microparticles loaded with SD by immersion and the mixtures containing the same proportion of polymers allowed a more effective control of drug release rates. Furthermore, microparticles obtained by physical mixture of polymers showed the lower percentage of drug released in acid medium and this behavior was attributed to the pectin that provides a diffusion layer of high viscosity that reduces the drug release rate. The association of pectin with gellan gum for preparing mucoadhesive beads by ionotropic gelation provided a pH dependent dissolution behavior, allowing reduced drug release rates in acid media. The purpose of this review is to evidence the importance of pectin as a carrier in the design of different drug release systems, aiming the targeting of drugs. Besides, the association of pectin with other polysaccharides and the cross-linking reaction are demonstrated to be reliable strategies to modulate the properties of the systems according to specific therapeutic needs.

Chapter 6 - Homogalacturonan (HG), rhamnogalacturonan-I (RG-I) and rhamnogalacturonan-II (RG-II) are structural pectic polysaccharides (i.e. pectin) found within the cell wall of terrestrial plants, and common sources of dietary fibre. The human genome does not contain any enzymes predicted to be involved in pectin digestion; therefore, in order to extract nutritional value from HG, RG-I, and RG-II humans rely on a consortium of symbiotic intestinal bacteria, commonly referred to as the distal gut microbiota (DGM), to deconstruct and ferment pectins and other complex carbohydrates into host-absorbable products. Currently, intestinal applications for bioactive pectins, such as HG, are under intensive investigation as nutraceuticals, prebiotics, and drug delivery systems. In this light, elucidating the incremental process of HG recognition and deconstruction by intestinal pectinolytic bacteria will provide new insights into the dynamic relationship between diet, human intestinal health, and DGM community structure. This chapter will define the different types of pectin structure, review mechanisms of pectinase function, provide insights into pectinolytic genes present within the genomes of intestinal pectinolytic bacteria, such as *Bacteroides thetaiotaomicron*, and summarize key functions of pectin in the maintenance of intestinal health.

Chapter 7 - Pectic polysaccharides, commonly termed pectins, are a group of natural polymers containing (1→4)-linked α-D-galacturonosyl residues such as homogalacturonan, arabinan-rich rhamnogalacturonan and xylogalacturonan. Pectins are abundant in many fruits and higher plants such as citrus, apples, pears and carrots, and have long been used as food additives or as active/stabilising components of pharmaceutical and cosmetic products. Global annual use of pectins is estimated at around 45 million kilograms, with a market value exceeding €400 million. The positive physiological effect of pectins (mainly as soluble dietary fibers) on humans stimulates manufacturing opportunities for novel pectin fibre ingredients and derived functional foods. However, ensuring that the desirable biological functionality and food processing properties of pectins are retained during product manufacturing remains a challenge. This chapter explores novel approaches for manipulating and optimising the positive role(s) of pectins in food processing and digestion based on pectins' sensitivity to pH, temperature, enzymes and other matrix factors. Several case studies, such as pectins' inhibitory effects on the activity of lipase enzymes and pectin's role in preservation of ascorbic acid and other phytochemicals during food processing, will be used to highlight key concepts. Plant origin, extraction method and degree of methyl esterification, acetylation and amidation of pectin ingredients determine finished food quality attributes as well as bioactive content and activities. The importance of advanced characterisation techniques, such as SEM, FT-IR, HPLC, GC, rheometry, cyclic voltammetry and solid-state NMR spectroscopy, in addition to conventional chemical analysis assays, for evaluating the suitability of a pectin ingredient for a specific functional food application, is demonstrated. The future outlook suggests that pectins will be increasingly utilised as soluble fibers for conventional food systems, and also as encapsulants for bioactive delivery systems. It is expected that 'designer pectins' such as those enriched in uronic acid or oligosaccharide fractions, or having specific methylation or branching degree to confer desirable health-promoting functionality (e.g. preserving antioxidants and enhancing prebiotic effects) will gain increased market share in the food ingredient sector.

Chapter 8 - Pectins of different nanostructure were assayed in their ability to develop film networks able to stabilize L-(+)-ascorbic acid (AA) to hydrolysis in view of antioxidant protection at interfaces, nutritional supplementation or therapy. Compartmentalization into edible films can permit not only to increase the AA stability but also to achieve localized antioxidant activity and controlled release. The AA hydrolysis was specifically studied in the present work. Hence, films were stored at controlled relative humidity (RH) in the absence of air. Films were made with each one of the enzymatically tailored (50, 70 and 80% DM) pectins (Cameron et al., 2008) or commercial high methoxyl pectin (HMP; 72% DM). A random distribution of demethylated blocks is expected to characterize commercial pectins whereas ordered patterns are obtained by enzymatic action. Calcium ions are necessary for crosslinking of low methoxyl pectins. Hence, the ability of Ca-mediated junction zones to stabilize AA into the edible films made with commercial pectins of low (LMP; 40%) or high (HMP; 72%) DM, at the same Ca^{2+} concentration (film systems called Ca-LMP and Ca-HMP, respectively), was also evaluated. Glycerol was used for plasticization. Kinetics of AA loss and subsequent browning development were determined by film storage at constant 57.7% RH and 25ºC, in the dark. Since AA stability was dependent on water availability in the film network, determined by ^{1}H-NMR, it was observed that the pectin nanostructure affected the AA kinetics. Higher AA retention and lower browning rates were achieved in HMP films than in enzymatically tailored pectin films, and the immobilization of water and consequent AA stability increased with the proportion of calcium-crosslinked junction zones present in the film network. As determined through tensile assays, the presence of Ca^{2+} in the film network produced significant decrease in elongation at fracture. This assay also revealed some sensibility of the HMP (commercial 72% DM pectin) to calcium ions. The glass transition temperature values of pectin films decreased ($T_g \approx -88$ to -102ºC) with the moisture content increase, indicating the contribution of water to the network plasticization by glycerol. However, water was mostly confined in the Ca-LMP network ($T_g \approx -93.99$ºC) followed by Ca-HMP ($T_g \approx -88.56$ºC), as inferred from the water availability determined by the ^{1}H-NMR. This was attributed to the water interaction at the Ca^{2+}-junction zones. Random distribution of demethylated blocks in the HG chains in addition to the presence of some disordered (amorphous) regions of RG-I may produce better immobilization of water than more rigid networks like those developed from the tailored pectin macromolecules.

Chapter 9 - Pectins are soluble dietary fibres that constitute a family of complex polysaccharides present in the primary cell wall and middle lamella of plants. The major sources of commercial pectins are citrus peel and apple pomace. Most pectin is produced by the extraction of the raw material with hot aqueous mineral acid at pH~2. In the food industry, pectins are used as gelling agents, thickeners, and stabilizers. New applications are constantly developing and their use as emulsifiers is one of the latest new-comes.

Utilization of by-products of the fruit and vegetable industrialization as source of pectins may contribute to the efficiency of the processes and also to the sustainability of the environment.

In the present work, pectins were extracted through different procedures from three unconventional sources and were characterized in their chemical composition and rheological behaviour. Beetroot pectin isolated through cellulase digestion and alkaline pretreatment, presented 54 % (w/w) of uronic acids (UA) and showed low degree of methylation (DM) and acetylation. The aqueous solution of this pectin presented low viscosity and pseudoplastic behaviour in flow. Gels were formed by addition of Ca^{2+}. On the other hand, butternut pectin, isolated through cellulase digestion, also presented 54% (w/w) of UA and a high DM, and in the presence of high sugar concentrations and at low pH, produced viscous solutions with pseudoplastic behaviour. Pectins were also obtained from Japanese plums with water at different temperatures. Those extracted at room temperature contained 56% (w/w) of UA and low DM as well as pseudoplastic behaviour in water. Pectin fraction extracted with boiling water contained 50% of UA and showed high DM. Although the later procedure increased considerably the yield, the extracted pectin showed significant lower apparent viscosity in water, in spite of its high molecular weight.

The isolating procedures assayed permitted the extraction of pectin enriched fractions from non-conventional sources with interesting yields and diverse rheological characteristics.

Chapter 10 - Pectin is a heteropolysaccharide found in cell walls and middle lamellae of higher plants, its structure is linear or branched complex. It is composed of acidic and neutral sugars molecules and the molecular weight ranges from 50,000 to 180,000 daltons and are negatively charged in neutral pH. Pectins can be characterized by different molecular parameters such as the degree of methoxylation, the molecular weight, and the galacturonan content. The most common commercial sources for pectin are apple pomace and citrus peel, however other novel sources, including sugar beet, potato, sunflower heads and papaya, have been investigated. Commercial pectins have diverse composition, polymer size distribution, acylation pattern, esterification degree, neutral sugar substitution, and this variability can influence the optimal condition of extraction. In plant biology, this polysaccharide plays important roles in plant growth, development, morphogenesis, defense, cell-cell adhesion, wall structure, signaling, cell expansion, wall porosity, pollen tube growth, seed hydration, leaf abscission, and fruit development. Recently, remarkable progress has been made in elucidating pectin structure/function relationships at the molecular level, and this is leading to the design of pectins with specific functionalities in several applications. In human nutrition, pectin represents a prebiotic and soluble dietary fiber that has extended its use as a nutraceutical ingredient for its properties of hypoglycemic, hypolipidemic, immunostimulant and anticancer. Fermentative pectin is resistant to human digestion, but is degraded in the colon by species of *Aerobacillus*, *Lactobacillus*, *Micrococcus* and *Enterococcus*. The bacteria produce pectolytic and fermentative enzymes capable to hydrolyze pectin into short chain fatty acids (acetic acid, butyric acid, propionic acid) and carbon dioxide. In addition, many studies supports the benefit and protective effects of soluble fiber intake against debilitating diseases including obesity, diabetes, coronary heart disease and elevated cholesterol. In food technology, pectin is used as a gelling and stabilizing agent and is also widely used in industries of edible and biodegradable films, adhesives, paper substitutes, foams and plasticizers, medical devices, biomedical implants, and drug delivery. A recent application for modified-pectins have been explored as a specific targets, e.g., for binding to galectin-3 (Gal3), a multifaceted and prometastatic protein whose expression is up-regulated in several types of cancer. Studies suggest that specific pectin (arabinose and galactose ramified sugars) have the binding affinity to cancer cell receptors (Gal3), giving a promising application of pectin in the development of nanomaterials glycofunctionalized to direct drugs toward tumoral cells, increasing the efficiency of the anticancer therapy and reducing drug toxicity.

The multifaceted functions of native or modified pectins make them a good candidate for the nutrition of the future. In conclusion, pectin has evolved from a food ingredient to a potential healthy compound for highlighted applications.

Chapter 11 - Pectin represents a family of complex polysaccharides and commercially can be easily obtained by extraction from citrus peel, apple pomace, sugar beets, mango and other plants. Pectin is hydrophilic and has tendency to form gel at acidic conditions or in presence of divalent cations. Due to its high capacity to form gel, pectin is widely used as a gelling agent and stabilizer, but can also act as a water binder or thickener. Pectin has in recent years been getting a great attention for potential application in food packaging due to its nontoxicity, biodegradability, edibility, biocompatibility and selective gas permeability. However, the main disadvantages of pure pectin films are poor water vapor barrier and low mechanical properties. In order to overcome these problems, pectin has been investigated in combination with other polysaccharides such as chitosan, alginate or cellulose. This chapter will present an introduction of new concept of biodegradable food packaging materials, criteria for food packaging materials and overview of latest developments of pectin films intended for application in food packaging. Finally, the physical-mechanical and antibacterial properties of these films will also be discussed.

In: Pectin: Chemical Properties, Uses and Health Benefits
Editor: Phillip L. Bush
ISBN: 978-1-63321-438-5

Chapter 1

PECTIN GELS FOR BIOMEDICAL APPLICATION

R. Gentilini, F. Munarin, P. Petrini* and M. C. Tanzi
Laboratorio di Biomateriali, Dipartimento di Chimica, Materiali e Ingegneria Chimica 'G. Natta' and Unità di Ricerca Consorzio INSTM, Politecnico di Milano, Milan, Italy

ABSTRACT

Regenerative Medicine is an interdisciplinary field, which applies the principles of engineering and life sciences to the development of biological substitutes that restore, maintain, or improve tissue function, combining a scaffold/support material with appropriate cells and bioactive molecules. Briefly, a scaffold serves as temporary support for cell growth and to present stimuli directing the growth and formation of a new desired tissue. Depending on the specific application, the required scaffold material and its properties can vary. Within the possible materials to fabricate a scaffold, natural-based polymers are among the most attractive, mainly due to their similarities with the extracellular matrix, their degradability, hydrophilicity, biocompatibility and versatility as well as typically good biological performance. In particular, natural-based hydrogels, due to their capability of retaining water and other biomimetic properties, offer an attractive alternative for numerous biomedical applications, such as bioactive molecules delivery systems, 3D scaffolds and cell immobilization.

Recently, several studies have been focused on the production and characterization of different gels obtained with pectin, a polysaccharide non-conventional in regenerative medicine, which presents the peculiar characteristics for this application, such as tunable physical properties, high water content and ability to homogeneously immobilize cells, genes, proteins, drugs or growth factors.

Additionally, pectin was chemically modified by grafting an oligopeptide containing the RGD sequence, that is known to promote cell adhesion. A partial oxidation of pectin was aimed to obtain faster degradation rates of pectin gels. MC3T3 preosteoblasts and human mesenchymal stem cells were immobilized in RGD-modified pectin microspheres for bone tissue regeneration showing good cell viability, metabolic activity and osteogenic differentiation. In vivo experiments confirmed that RGD-modified pectin

* Corresponding author: Paola Petrini, Dip.di Chimica, Materiali e Ingegneria Chimica 'G. Natta' ed Unità di Ricerca Consorzio INSTM Politecnico di Milano Piazza L. da Vinci, 32. 20133 Milano (Italy), Tel +39 0223993386. E-mail: paola.petrini@polimi.it.

microspheres provided a complete adaptation to bone defect and induced bone regeneration avoiding the dispersion of cells after implantation.

1. Exploring the Multifold Potential of Pectin: Pectin Gels for Regenerative Medicine

Pectin, a natural polysaccharide present in the cell wall of most plants, is nowadays object of increasing interest for applications in the biomedical field. In particular, due to the peculiar gelling mechanism, low methoxy pectins have been proposed for the preparation of hydrogels for biomedical applications, namely drug delivery, gene delivery and regenerative medicine as implantable material for minimally invasive surgery [1, 2].

Regenerative medicine, including tissue engineering, is an interdisciplinary field which applies the principles of engineering and life sciences to the development of biological substitutes that repair, replace, or improve cells, tissues or organs functions, injured by different causes such as congenital defects, trauma or aging [3].

The traditional approach for engineering a tissue is shown in figure 1: appropriate scaffolds are seeded with cells and/or bioactive molecules, such as growth factors, oligopeptides and proteins, and then cultured *in vitro* and implanted *in vivo* to induce and direct the formation of a healthy tissue.

The scaffold provides an initial biomechanical profile for the replacement of the tissue until cells form, deposit and organize a new extracellular matrix. During the tissue formation, the scaffold is either degraded or metabolized, eventually leading to a vital organ or tissue that restores, maintains, or improves tissue function.

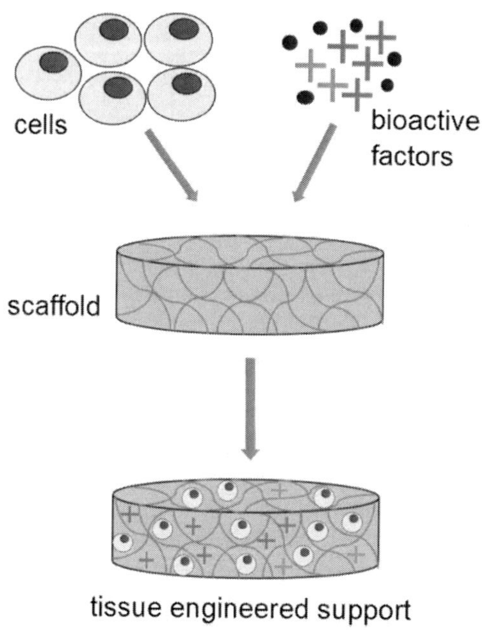

Figure 1. A scaffold, living cells, and/or biologically active molecules are used in different strategies to create a tissue-engineered support to promote tissue repair and regeneration.

The ability of an implanted biomaterial to induce appropriate host responses in a particular application defines its biocompatibility, that includes proper interactions between an implant and the surrounding cells (biomaterials and scaffold for tissue engineering). Among the different materials to be used for producing a scaffold, natural polymers, such as proteins and polysaccharides, are biologically active and typically promote cell adhesion and growth. Natural polymers can be easily engineered and modified to provide an optimal microenvironment to improve cell adhesion and tissue in-growth.

Among all possible choices, hydrogels represent promising materials for developing not conventional 3D scaffolds, owning a distinct efficacy as matrices for cell housing and due to their similar structure to the macromolecular-based components in the body [4-6]. Proper design of the hydrogel network structure and chemical composition allows to establish an environment suitable for the confined cells and to provide stability and structural support. The possibility to diffuse and transport oxygen and essential nutrients, as well as metabolic waste and excreted products, is also a peculiar characteristic of hydrogels.

Natural-based hydrogels, due to their capability of retaining water and other biomimetic properties, appear to be very attractive for numerous biomedical applications, such as bioactive molecules delivery systems, 3D scaffolds and cell immobilization [7, 8]. Natural hydrogels are typically based on proteins and ECM components such as collagen [9, 10], hyaluronic acid [11, 12], fibrin [13, 14], as well as materials derived from other biological sources such as alginate [15, 16], chitosan [17, 18], gelatin [19, 20] or silk fibroin [21, 22].

Due to their excellent properties, such as biocompatibility, flexibility of fabrication in different shapes, versatility, and physical characteristics similar to that of physiological environment, pectin gels find several biomedical applications. They can serve as scaffolds to be used as cell immobilization [5, 23] or drug/gene delivery [24], or acting as cell substrate for controlling of cell attachment [25].

2. Pectin Extraction for Regenerative Medicine

Pectin characteristics vary according to the plant species from which it is extracted, and pectins with different proprieties can derive from the same plant and even from the same cell wall [26]. Fruits and plants can be processed as sources for commercial pectins, depending on the yield, time and cost of the extraction process, the desired properties of the extracted pectins and the availability of the raw materials [27-30]. The extraction procedure can be performed with different methods: by use of chelating agents, such as ethylene-diamino-tetraacetic acid (EDTA) and sodium citrate; acids (like HCl) at high temperatures to increase the yield; bases (such as NaOH) that however cause extensive degradation; specific expensive enzymes (including arabinase, galactanase, polygalacturonase and rhamnogalatturonase) [27, 28, 31, 32]. The extraction process can modify the pectin properties, particularly the molecular weight. Previous studies have shown that extractions at room temperature lead to pectins with high molecular weight but low yield, while extraction at higher temperatures (over 80° C) results in lower molecular weight pectins, with an increase in yield [33]. We reported [34] that pectins for biomedical applications can be extracted from plants by use of non-toxic and biocompatible reagents and procedures. We developed a novel extraction procedure of pectin from Aloe vera (Aloe Barbadensis Miller), by modifying industrial protocols used for pectin extraction from different sources. In particular, the procedure was optimized by introducing a microwave pretreatment, inducing enzyme deactivation with the aim to preserve the quality of the extracted pectin in terms of molecular mass and intrinsic viscosity. Sodium citrate was employed as a chelating agent of the calcium ions found in the cell wall, thus inducing pectin dissolution from its insoluble, Ca-bound, form. This avoids to employ harsh conditions that depolymerize pectin during extraction. The use of other chelating agents, like EDTA, is known to increase the yield over 50% (w/w), giving pectin samples with a galacturonic acid content of 70% (w/w) [33]. However, EDTA has some limitations because of its cytotoxicity.

As a general rule, the process of extraction of pectin has to be tailored for the specific biomedical application purposes.

3. Bioactive Pectin for Regenerative Medicine

Pectin, as well as other natural polysaccharides (e.g. alginate or carboxymethylcellulose), is not adhesive for cells due to the presence of negatively charged carboxyl groups. As proliferation of many cell types is dependent on their adhesion to a suitable substrate, different approaches aimed to chemically modify the pectin structure are investigated. To tailor degradability of pectin gels, different methods for pectin modification are further described.

Controlled Enzymatic Degradation of Pectin for Surface Modification

Due to their structural complexity, pectins are modifiable by several enzymes, including hydrolases, lyases, and esterases. Pectin-degrading and -modifying enzymes may be used in a wide variety of applications to modulate pectin properties or produce pectin derivatives and oligosaccharides with functional interests. Pectin can undergo two different types of degradation: de-esterification as well as depolymerisation, corresponding to different enzymes. These enzymes are ubiquitous not only in plants but also in phytopathogenic organisms, helping to colonize the host plant.

Hairy regions of branched pectin can be separated by enzyme degradation to promote proliferation and differentiation of osteoblastic cells, in order to obtain an adequate osteointegration of implants. To achieve this, functionalisation of biomaterial surfaces by coating with pectin fragments is being investigated. For example, the hairy regions can be separated with various enzymes, such as rhamnogalacturonase, endopolygalacturonase, and pectin methylesterase, yielding to the so-called modified hairy regions (MHRs) [35-37].

Pectin coatings onto different biomaterials can be further tailored by enzymatically modifying the length of the hairy regions, which determine the wettability of the coated surfaces. Depending on their action site in the pectic polymer, pectin degrading enzymes can be shared between those degrading HG and those degrading RG-I.

In this context, the modification of branched regions of apple pectin was studied. Two different rhamnogalacturonan-rich modified hairy regions, MHR-A and MHR-B, which differ in the physical and chemical properties and biological characteristics, were obtained [38]. The two polysaccharide preparations, although showing the same structural elements, differ in relative amounts and lengths of the neutral side chains (long-haired MHR-A vs. short-haired MHR-B). Results of proliferation and differentiation of osteoblastic cells on titanium surfaces coated with oligosaccharides derived from pectins demonstrated that the use of pectin fragments implies different biological activities *in vitro*. In particular, osteoblastic cells prefer MHR-B coating than MHR-A modification. MHR-B is more hydrophobic than MHR-A. Side chain length probably affects protein adsorption profile so that a coating molecule carrying shorter side chains (MHR-B) allows protein adsorption more effectively because the hindering steric repulsion is lower [39].

The oligosaccharides derived from pectins can be similarly used to study the interaction of macrophages with the biomaterial surface. Macrophages influence medical device efficacy and also affect osteointegration of implants, as they are involved in the inflammatory response by producing cytokines that come in contact with the surface of the biomaterial [40]. Bussy [41] studied the capability of engineered rhamnogalacturonan- I (RG-I) fractions of apple pectin, differing in relative amounts and lengths of their neutral side chains and named MHR-α (long-haired, a new batch of MHR-A, differing from MHR-A in the relative content of residual monosaccharides) and MHR-B (short-haired), to control bone cell and macrophage behavior. On MHR- α, macrophages grew well, and they did not secrete either pro-inflammatory-cytokines or nitrites. Different results were gained from macrophages on MHR-B, except for nitrite secretion, suggesting that MHR-B, which allows osteoblast migration and differentiation, should also be able to stimulate osteoclast recruitment and maturation via macrophage activation.

Nagel [42] demonstrated how differently grafted and soluble pectin MHRs interact *in vitro* with fibroblasts, depending on the different structures. The dissimilarities were related to the diverse adsorption of serum-adhesive proteins in mediating cell responses. Five enzymatically-modified MHRs from apple (MHR-B, MHR-A and MHR-α), potato (MHR-P) and carrot (MHR-C), differing in relative amount and shape of neutral monosaccharidic residues in the side chains, were covalently linked to polystyrene (PS) Petri dishes. This work showed that MHR-B induces cell adhesion, proliferation and viability, in opposition to other modified pectins. Fibronectin was similarly adsorbed onto MHR-B and tissue culture polystyrene (TCPS) control, but when fibronectin is poorly adsorbed, and the integrin-binding site is therefore poorly available, cell aggregation, scarce proliferation and apoptosis are observed. The fibronectin cell binding site (i.e. RGD sequence) was more available on MHR-B than on TCPS control, but less on MHR-α. This study provides a basis for the design of intelligently-tailored biomaterial coatings able to induce specific cell functions.

Partial Oxidation of Pectin for Controlled Degradation

Biomaterials used as cell carrier, despite giving the mechanical support during the first days after implantation, need to be gradually degraded by the biological environment, so to release cells for the regeneration of the damaged tissues. It is well known that natural polymers, including pectin, are biocompatible, but, in some cases, their degradation cannot be easily tailored with tissue ingrowth [16, 43].

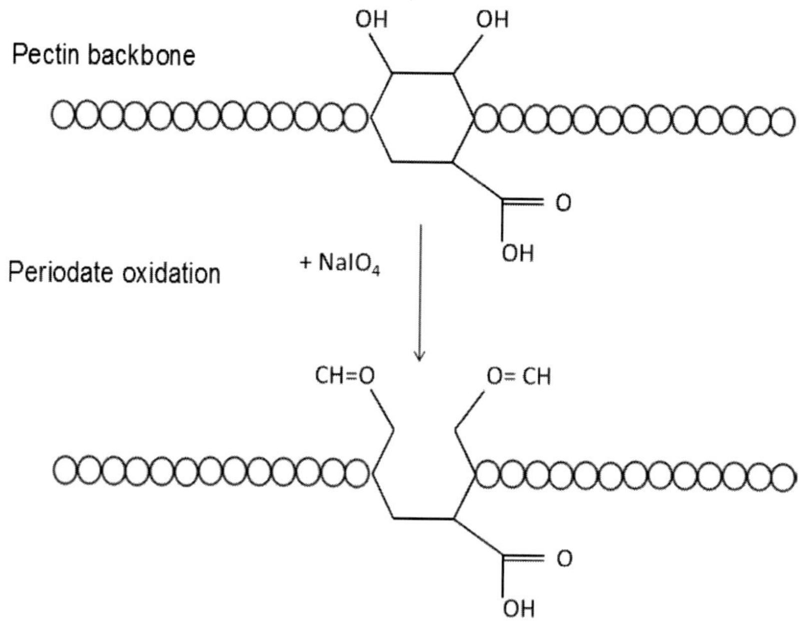

Figure 2. Periodate sodium cleaves the vicinal diols in the backbone of the pectin, forming to the di-aldehyde derivatives with an open chain.

Irradiation and oxidation are the most common techniques employed to accelerate the degradation of natural polyuronates. Despite oxidation, irradiation often lead to a significant decrease of polymer molecular weight [44] related to the rupture of the polysaccharidic chains, eventually leading to lowered mechanical properties of formed gels. Oxidation with sodium periodate was performed on different polysaccharides, such as alginates [45, 46] and other polysaccharides [47, 48] to induce their degradation. Thus, a study reported that alginate can be modified with a partial oxidation by sodium periodate, leading to the creation of acetal groups on alginate backbone, making the polymer more susceptible to hydrolysis [49].

Pectin was oxidized with sodium periodate to obtain faster degradable pectin microspheres [50]. Periodate oxidation cleaves the vicinal glycols in polysaccharides to form their di-aldehyde derivatives (Figure 2). Each cis-diol group consumes one molecule of periodate (mechanism of scission of alginate chains by periodate). Specifically, the mechanism for pectin oxidation involves the hydroxyl groups on carbons 2 and 3 of the repetitive galacturonic unit. The carbon–carbon bond of the cis-diol group in the uronic residue is broken by the partial oxidation of hydroxyl groups, forming two aldehyde groups in each oxidized monomeric unit. This study demonstrated that pectin microspheres could be formed from oxidized pectin, and their long term gelling, mechanical properties and biocompatibility were minimally affected. The mechanical properties of microspheres made with oxidized pectin resulted of the same order of magnitude of unmodified pectin microspheres, with no effects on cytocompatibility properties.

Pectin Modification to Promote Cell Adhesion

Hydrogels, including pectin based ones, are capable of suspending cells 3-dimensionally and supporting nutrient diffusion to cells, but may not provide an ideal environment for anchorage-dependent cells. The most commonly studied adhesive peptide, Arg-Gly-Asp (RGD), is found in cell-binding domains of extracellular matrix proteins. The RGD sequence is the minimal peptide sequence required for the adhesion of integrins to the ECM components [51-53]. Several studies described how alginate polymers can be modified by RDG-coulping, using NHS/EDC carbodiimide chemistry to covalently graft the oligopeptide to the carboxyl groups of alginate, promoting cell viability, metabolic activity, adhesion and differentiation [45, 54, 55].

In a previous study we reported pectin modification with RGD-containing oligopeptides for the preparation of microspheres, proposed as cell vehicles for bone tissue regeneration [50]. Aqueous carbodiimide chemistry was utilized to covalently couple the RGD sequence to pectin chains to prepare modified microspheres (Figure 3). Pectin microspheres represent an attractive model system to study highly specific cell-ligand interactions due to the low protein adsorption of the anionic polysaccharide. This study showed that this chemistry successfully initiates biological interactions between pectin polymers and mouse preosteoblasts, as demonstrated by culturing MC3T3 cells in contact with RGD-modified pectin. RGD-coupled pectin microspheres maintained cells alive, proliferating and differentiating up to 30 days of the experiment, and promoted the formation of a 3D extracellular matrix-like structure, bridging adjacent microspheres. In addition, we demonstrated that preosteoblastic and mesenchymal stem cells, immobilized inside RGD-pectin micropheres, are able to differentiate into the osteoblastic phenotype. Supported by histochemical analyses and gene expression, preosteoblasts and mesenchymal stem expressed specific markers of bone differentiation.

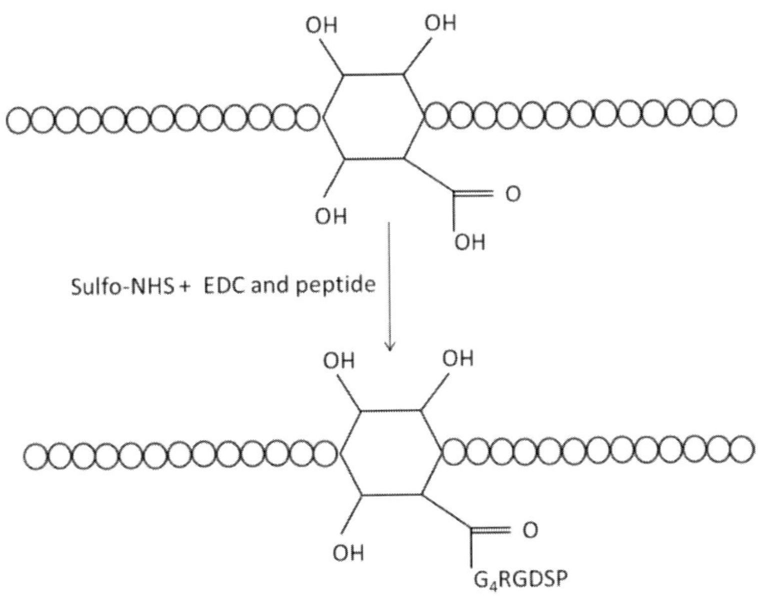

Figure 3. RGD grafting onto the pectin structure by use of aqueous NHS/EDC carbodiimide chemistry.

Functionalization of pectin can be used to develop effective and biodegradable wound dressings too. A complex set of properties is required for an efficient wound dressing, including exudate absorption capacity, good porosity for the permeation of water and antimicrobial properties. For this aim, Gupta [56] developed pectin and gelatin coatings on cotton fabric. The aldehyde groups introduced in pectin via periodate oxidation lead *in situ* reduction of silver nitrate to nanosilver, which is known as an excellent antimicrobial agent. This study examined the effect of various reaction parameters, such as the reaction time, temperature, pH and composition, on the efficiency of the *in situ* crosslinking reaction, in order to obtain a suitable product for wound healing application. The fabricated pectin/gelatin hydrogels were homogenous, without any phase separation. In addition, the Authors introduced plasticization by glycerol, which confered flexibility to the system, improving the handling ability.

4. Pectin-Based Microspheres

Relating to tissue regeneration applications, microspheres can immobilize drugs [57, 58], proteins [59, 60], growth factors [61, 62] and stem or precursor cells (possibly autologous), which may produce proteins or growth factors once injected and released in the pathological or defect site [50]. The use of microspheres allow to obtain homogeneous immobilization and the possibility of injecting microspheres suspended in a fluid carrier, without breaking the constructs trough the shear stress generated by the syringe and avoiding a premature release of cells or of the entrapped agents. Microspheres are particularly suitable for cell delivery: confining a high density of stem or precursor cells (of the same cell type or in co-cultures) in a microsphere can positively stimulate the formation of connections among cells and the subsequent cell proliferation, differentiation and formation of extracellular matrix, leading to the regeneration of the pathological or damaged tissue (Figure 4). Another advantage of the microspheres is the increased surface/volume ratio with respect of hydrogels with different morphologies, which maximize the diffusion of oxygen and nutrients within the microsphere and the outcome of catabolytes and active substances produced by cells. The interstices between the microspheres, appropriate in size, may also provide a space for both tissue and vascular ingrowth. Furthermore, the co-immobilization of cells and growth factors may encourage host cell migration, attachment, proliferation and differentiation in the pathological or defect site [41].

Munarin reports a comparative study of microspheres prepared with natural polymers to be used as cell carriers for the regeneration of different soft tissues [63]. The work proposes natural polymer microcapsules as vehicles for controlled cell delivery and release: in the encapsulation studies, cells survived to the immobilization process and remained alive after 7 days of incubation. By changing the composition of the gel forming material, the degradation of microspheres can be programmed, in order to achieve a better control on cell release *in vivo*.

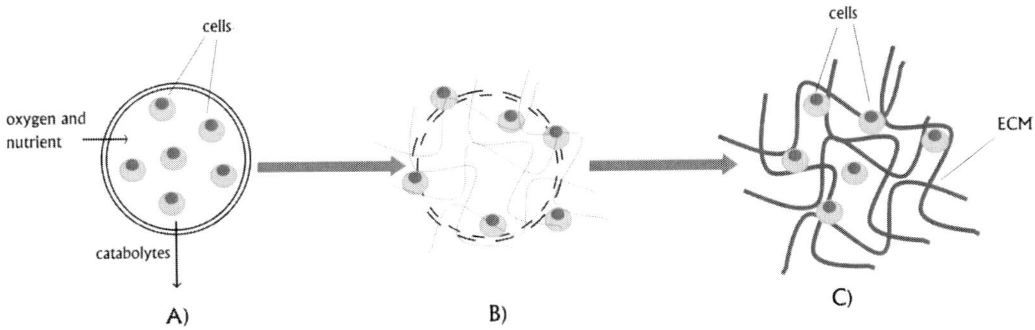

Figure 4. A) Cells are entrapped in microspheres by cross-linking the pectin. The microsphere morphology allows the diffusion of oxygen and nutrients, with the outcome of catabolytes. B) and C) While cells are producing their own extracellular matrix (ECM), the natural polymer is degrading, leaving only regenerated tissue in the defect site.

Calcium phosphates/pectin microspheres are of great interest for bone tissue regeneration, according to their ability to form hybrid systems composed by the internal polysaccharidic matrix and the external mineralized coating. In this context, Munarin developed a biomimetic method to obtain calcium phosphate/pectin microspheres to be used as injectable scaffolds, with the ability to promote the process of biomineralization [64]. As a result, the work highlights an internal matrix, useful for the immobilization of cells, genes, proteins or growth factors to be carried in the pathological situ, with a degradability of pectin allowing the modulation of the release. At the same time, the external mineral coating of pectin microspheres increased the structural stability of microspheres, by mimicking the structure of the bone and therefore increasing the integration with the host tissue.

5. Injectable Pectin Gels

Nowadays, one of the major challenge in hydrogel technology, is the development of injectable hydrogels, that, compared to the traditional scaffolds, allow to immobilize drugs, genes, cells or other active biomolecules needed for the tissue regeneration, distributing them within any defect size or shape, in some cases prior to gelation, and to obtain a controlled and sustained release. Immobilization of such active agents has been exploited for biochemical or functional tissue substitution, recombinant cell transplantation and tissue regeneration applications. The advantages of using injectable hydrogels rely on their high moldability (they can adapt themselves to the defect shape), on the possibility of in vivo delivery in a minimally invasive way (resulting in a faster recovery, smaller scar size and less pain for patients), and capacity of easy and effective encapsulation of cells and/or drugs (Figure 5). Thus, from a clinical point of view, the use of such injectable systems for regenerative purposes instead of traditional surgical procedures is very attractive as it reduces patient discomfort, risk of infection, scar formation and the cost of treatment [65].

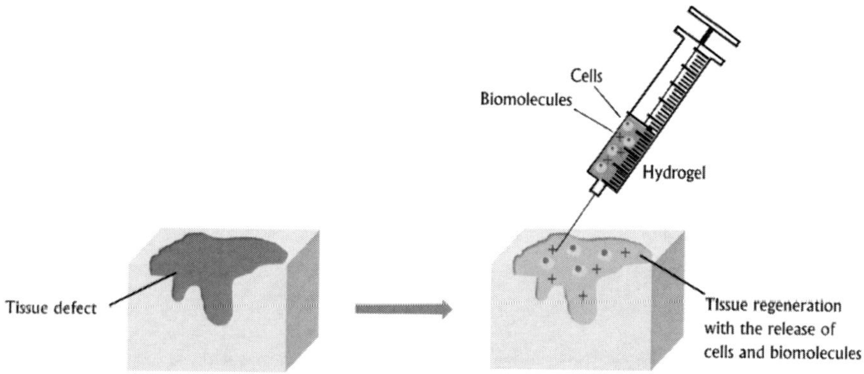

Figure 5. Tissue regeneration by using an injectable hydrogel loaded with cells and active biomolecules.

Despite these many advantageous properties, hydrogels also have several limitations too. In some cases, the homogeneity of distribution of cells or active molecules loaded into hydrogels, though achievable, may be limited. Ease of application can also be problematic, as many hydrogels are not sufficiently deformable to be injectable, needing surgical implantation.

Injectable hydrogels can be prepared with different methods, such as thermal gelation, photopolymerization, ionic interaction, physical self-assembly, and chemical crosslinking with agents such as glutaraldehyde or, as a non-toxic substitute, genipin [66-68].

Injectable hydrogels present a typical behavior of viscous liquids during injection and increase their viscosity in the absence of shear ("shear thinning"), allowing preformed hydrogels to be injected by application of shear stress (during injection) and quickly self-heal after removal of the shear [69]. For a suitable injectable hydrogel as cell carrier, the degradation rate and mechanical properties of the hydrogel must provide adequate support for cell adhesion for tissue growth. In particular, these properties can be fine-tuned through variations in the chemical structure and crosslinking density in hydrogels [4]. In the production of a hydrogel for cell immobilization, it is important to design tailor-made materials to prevent strong spatial constriction, a critical condition for cells, as it limits the movement and proliferation. Another key parameter is the spatial distribution of cells into the gel: a too slow gelling kinetic determines cell segregation due to the low viscosity of gel precursors that do not allow the suspension. On the other hand, a rapid gelling kinetic, leads to inhomogeneous distribution of the cells, due to the difficulty of mixing the cell suspension within the gel [70].

The gel formation can be carried out both prior to injection, both *in situ*, where the hydrogels are flowable aqueous solutions before administration, but once injected, rapidly gel under physiological conditions [4]. The two approaches present both advantages and disadvantages. The injectability of gels forming in situ is easier due to the lowest viscosity and they are more adaptable to the injection site. The injectable matrix can be implanted in the human body with minimal surgical wounds, and bioactive molecules or cells can be incorporated simply by mixing before injection. Following gelation, these matrices become drug delivery deposits in pharmaceutics or cell-growing depots for tissue regeneration.

The gelling kinetic is a fundamental parameter in designing homogeneous and composite gels, injectable systems, cell-loaded gels, in vivo gelling systems, and effective drug loading prior the formation of the gel. To this aim, Moreira [71] studied the rheological and biocompatible properties of pectin-based hydrogels. Briefly, pectin–calcium carbonate hydrogels were prepared by internal gelation, by fine-tuning of the $NaHCO_3$ and $CaCO_3$ content, to keep a tight control over the pH of the hydrogels thus controlling their gelling kinetics, in order to obtain homogeneous hydrogels, as well as to reach pH values compatible with cell viability. The results of this work show that the considered formulations are cytocompatible and can be obtained with inexpensive and easy preparation methods. Rheological analyses confirmed their injectability, as the gels exhibited a viscoelastic and shear-thinning behavior, thus allowing their injection through a needle.

The injectability of the same hydrogels was then studied by Munarin [72], when injectable pectin gels were produced by internal gelation with $CaCO_3$. The injectability through different needle size was evaluated for all of the tested samples by the analyses performed at the texture analyzer. The rheological parameters confirmed that pectin gels behave as soft-gels with mechanical properties similar to soft tissues. The 99% of the immobilized cells resulted viable after immobilization and homogeneously dispersed in the gel. Differences in cell viability were observed by extruding the gels from syringes with different needle sizes.

CONCLUSION

As an abundant raw material, pectin can be extracted from different sources and its characteristics vary according to the plant species from which it is extracted, and pectins with different proprieties can derive from the same plant and even from the same cell wall. There is the need of tailored methods for pectin extraction in view of the application as a biomaterial: the main characteristics of the appropriate process are the use of biocompatible chemicals and the possibility to preserve the peculiar structural characteristics such as the integrity of branched regions, which show an important role in cell interaction. A high molecular weight and a low degree of esterification need to be pursued to form stable, ionotropic gels, in conditions compatible with cell viability or biomolecules loading. As an example, we reviewed, in this chapter, the extraction of pectin from Aloe Vera for biomedical applications.

Pectin gels are proving wide applicability as biomaterials and recent advances in regenerative medicine application of pectin gels, such as injectable hydrogels and microspheres, are described in this chapter. Bioactive modifications, such as enzymatic degradation, partial oxidation and RGD functionalization of this polysaccharide are here mentioned to highlight the suitability and versatility of pectin gels for different biomedical applications.

Overall, pectin can be considered a novel and versatile biomaterial and the required tight control of a number of properties including swelling, degradation, cell attachment, and binding or release of bioactive molecules can be obtained from the deep knowledge of this versatile family of polysaccharides.

References

[1] Munarin F, Tanzi MC, Petrini P. 2012, Advances in biomedical applications of pectin gels. *Int. J. Biol. Macromol*; 51:681-9.

[2] Munarin F, Petrini P, Bozzini S, Tanzi MC. 2012, New perspectives in cell delivery systems for tissue regeneration: natural-derived injectable hydrogels. *J. Appl. Biomater. Func*; 10:67-81.

[3] Mason C, Dunnill P. 2008, A brief definition of regenerative medicine. *Regen. Med*; 3:1-5.

[4] Tan HP, Marra KG. 2010, Injectable, Biodegradable Hydrogels for Tissue Engineering Applications. *Materials*; 3:1746-67.

[5] Jen AC, Wake MC, Mikos AG. 1996, Review: Hydrogels for cell immobilization. *Biotechnol. Bioeng*; 50:357-64.

[6] Lee KY, Mooney DJ. 2001, Hydrogels for tissue engineering. *Chem. Rev*; 101:1869-79.

[7] Mano JF, Silva GA, Azevedo HS, Malafaya PB, Sousa RA, Silva SS, et al. 2007, Natural origin biodegradable systems in tissue engineering and regenerative medicine: present status and some moving trends. *J. R. Soc. Interface*; 4:999-1030.

[8] Dang JM, Leong KW. 2006, Natural polymers for gene delivery and tissue engineering. *Adv. Drug Deliv. Rev*; 58:487-99.

[9] Hesse E, Hefferan TE, Tarara JE, Haasper C, Meller R, Krettek C, et al. 2010, Collagen type I hydrogel allows migration, proliferation, and osteogenic differentiation of rat bone marrow stromal cells. *J. Biomed. Mater. Res. A*; 94:442-9.

[10] Yuan T, Zhang L, Li KF, Fan HS, Fan YJ, Liang J, et al. 2014, Collagen hydrogel as an immunomodulatory scaffold in cartilage tissue engineering. *J. Biomed. Mater. Res. B*; 102:337-44.

[11] Luo Y, Kirker KR, Prestwich GD. 2000, Cross-linked hyaluronic acid hydrogel films: new biomaterials for drug delivery. *J. Control Release*; 69:169-84.

[12] Gerecht S, Burdick JA, Ferreira LS, Townsend SA, Langer R, Vunjak-Novakovic G. 2007, Hyaluronic acid hydrogel for controlled self-renewal and differentiation of human embryonic stem cells. *Proc. Natl. Acad. Sci. U. S. A*; 104:11298-303.

[13] Hall H. 2007, Modified fibrin hydrogel matrices: Both, 3D-scaffolds and local and controlled release systems to stimulate angiogenesis. *Curr. Pharm. Design*; 13:3597-607.

[14] Janmey PA, Winer JP, Weisel JW. 2009, Fibrin gels and their clinical and bioengineering applications. *J. R. Soc. Interface*; 6:1-10.

[15] Augst AD, Kong HJ, Mooney DJ. 2006, Alginate hydrogels as biomaterials. *Macromolecular Bioscience*; 6:623-33.

[16] Rowley JA, Madlambayan G, Mooney DJ. 1999, Alginate hydrogels as synthetic extracellular matrix materials. *Biomaterials*; 20:45-53.

[17] Bhattarai N, Gunn J, Zhang M. 2010, Chitosan-based hydrogels for controlled, localized drug delivery. *Adv. Drug Deliv. Rev*; 62:83-99.

[18] Hong Y, Song HQ, Gong YH, Mao ZW, Gao CY, Shen JC. 2007, Covalently crosslinked chitosan hydrogel: Properties of in vitro degradation and chondrocyte encapsulation. *Acta Biomater*; 3:23-31.

[19] McCain ML, Agarwal A, Nesmith HW, Nesmith AP, Parker KK. 2014, Micromolded gelatin hydrogels for extended culture of engineered cardiac tissues. *Biomaterials*; 35:5462-71.

[20] Reno F, Rizzi M, Cannas M. 2012, Gelatin-based anionic hydrogel as biocompatible substrate for human keratinocyte growth. *J. Mater. Sci. Mater. Med*; 23:565-71.

[21] Yucel T, Cebe P, Kaplan DL. 2009, Vortex-induced injectable silk fibroin hydrogels. *Biophys. J*; 97:2044-50.

[22] Motta A, Migliaresi C, Faccioni F, Torricelli P, Fini M, Giardino R. 2004, Fibroin hydrogels for biomedical applications: preparation, characterization and in vitro cell culture studies. *J. Biomater. Sci. Polym. Ed*; 15:851-64.

[23] Wang C, Varshney RR, Wang DA. 2010, Therapeutic cell delivery and fate control in hydrogels and hydrogel hybrids. *Adv. Drug Deliv. Rev*; 62:699-710.

[24] Schmidt JJ, Rowley J, Kong HJ. 2008, Hydrogels used for cell-based drug delivery. *Journal of Biomedical Materials Research Part A*; 87A:1113-22.

[25] Hou Y, Matthews AR, Smitherman AM, Bulick AS, Hahn MS, Hou H, et al. 2008, Thermoresponsive nanocomposite hydrogels with cell-releasing behavior. *Biomaterials*; 29:3175-84.

[26] Coimbra P, Ferreira P, de Sousa HC, Batista P, Rodrigues MA, Corriea IJ, et al. 2011, Preparation and chemical and biological characterization of a pectin/chitosan polyelectrolyte complex scaffold for possible bone tissue engineering applications. *International Journal of Biological Macromolecules*; 48:112-8.

[27] Kurita O, Fujiwara T, Yamazaki E. 2008, Characterization of the pectin extracted from citrus peel in the presence of citric acid. *Carbohyd. Polym*; 74:725-30.

[28] Panouille M, Thibault JF, Bonnin E. 2006, Cellulase and protease preparations can extract Pectins from various plant byproducts. *J. Agr. Food Chem*; 54:8926-35.

[29] Zykwinska A, Boiffard MH, Kontkanen H, Buchert J, Thibault JF, Bonnin E. 2008, Extraction of green labeled pectins and pectic oligosaccharides from plant byproducts. *J. Agr. Food Chem*; 56:8926-35.

[30] Singthong J, Ningsanond S, Cui SW, Douglas Goff H. 2005, Extraction and physicochemical characterization of Krueo Ma Noy pectin. *Food Hydrocolloid*; 19:793-801.

[31] Liu Y, Shi J, Langrish TAG. 2006, Water-based extraction of pectin from flavedo and albedo of orange peels. *Chem. Eng. J*; 120:203-9.

[32] Iglesias MT, Lozano JE. 2004, Extraction and characterization of sunflower pectin. *J. Food Eng*; 62:215-23.

[33] Ni Y, Yates KM, Zarzycki R. 1999, *Aloe pectins*. US Patent: Carrington Laboratories, Inc., Irving, Texas. p.

[34] Gentilini R, Bozzini S, Munarin F, Petrini P, Visai L, Tanzi MC. 2014, Pectins from Aloe Vera: Extraction and Production of Gels for Regenerative Medicine. *Journal of applied polymer science*; 131.

[35] Schols HA, Vierhuis E, Bakx EJ, Voragen AG. 1995, Different populations of pectic hairy regions occur in apple cell walls. *Carbohydr. Res*; 275:343-60.

[36] Mutter M, Beldman G, Schols HA, Voragen AG. 1994, Rhamnogalacturonan alpha-L-rhamnopyranohydrolase. A novel enzyme specific for the terminal nonreducing rhamnosyl unit in rhamnogalacturonan regions of pectin. *Plant Physiol*; 106:241-50.

[37] van Alebeek GJ, van Scherpenzeel K, Beldman G, Schols HA, Voragen AG. 2003, Partially esterified oligogalacturonides are the preferred substrates for pectin methylesterase of Aspergillus niger. *Biochem. J.*; 372:211-8.
[38] Kokkonen H, Cassinelli C, Verhoef R, Morra M, Schols HA, Tuukkanen J. 2008, Differentiation of osteoblasts on pectin-coated titanium. *Biomacromolecules*; 9:2369-76.
[39] Morra M, Cassinelli C, Cascardo G, Nagel MD, Della Volpe C, Siboni S, et al. 2004, Effects on interfacial properties and cell adhesion of surface modification by pectic hairy regions. *Biomacromolecules*; 5:2094-104.
[40] Anderson JM, Rodriguez A, Chang DT. 2008, Foreign body reaction to biomaterials. *Semin. Immunol.*; 20:86-100.
[41] Bussy C, Verhoef R, Haeger A, Morra M, Duval JL, Vigneron P, et al. 2008, Modulating in vitro bone cell and macrophage behavior by immobilized enzymatically tailored pectins. *J. Biomed. Mater. Res. A*; 86A:597-606.
[42] Nagel MD, Verhoef R, Schols H, Morra M, Knox JP, Ceccone G, et al. 2008, Enzymatically-tailored pectins differentially influence the morphology, adhesion, cell cycle progression and survival of fibroblasts. *Bba-Gen Subjects*; 1780:995-1003.
[43] Lansdown AB, Payne MJ. 1994, An evaluation of the local reaction and biodegradation of calcium sodium alginate (Kaltostat) following subcutaneous implantation in the rat. *J. R. Coll. Surg. Edinb.*; 39:284-8.
[44] Sen M. 2011, Effects of molecular weight and ratio of guluronic acid to mannuronic acid on the antioxidant properties of sodium alginate fractions prepared by radiation-induced degradation. *Appl. Radiat. Isot.*; 69:126-9.
[45] Bidarra SJ, Barrias CC, Fonseca KB, Barbosa MA, Soares RA, Granja PL. 2011, Injectable in situ crosslinkable RGD-modified alginate matrix for endothelial cells delivery. *Biomaterials*; 32:7897-904.
[46] Grellier M, Granja PL, Fricain JC, Bidarra SJ, Renard M, Bareille R, et al. 2009, The effect of the co-immobilization of human osteoprogenitors and endothelial cells within alginate microspheres on mineralization in a bone defect. *Biomaterials*; 30:3271-8.
[47] Lu SY, Liu MZ, Ni BL. 2010, An injectable oxidized carboxymethylcellulose/N-succinyl-chitosan hydrogel system for protein delivery. *Chem. Eng. J*; 160:779-87.
[48] Maia J, Carvalho RA, Coelho JFJ, Simoes PN, Gil MH. 2011, Insight on the periodate oxidation of dextran and its structural vicissitudes. *Polymer*; 52:258-65.
[49] Boontheekul T, Kong HJ, Mooney DJ. 2005, Controlling alginate gel degradation utilizing partial oxidation and bimodal molecular weight distribution. *Biomaterials*; 26:2455-65.
[50] Munarin F, Petrini P, Tanzi MC, Barbosa MA, Granja PL. 2012, Biofunctional chemically modified pectin for cell delivery. *Soft Matter*; 8:4731-9.
[51] Ruoslahti E, Pierschbacher MD. 1987, New perspectives in cell adhesion: RGD and integrins. *Science*; 238:491-7.
[52] Hynes RO. 1992, Integrins: versatility, modulation, and signaling in cell adhesion. *Cell*; 69:11-25.
[53] Yamada KM. 1997, Integrin signaling. *Matrix Biol*; 16:137-41.
[54] Moshaverinia A, Chen C, Xu XT, Akiyama K, Ansari S, Zadeh HH, et al. 2014, Bone Regeneration Potential of Stem Cells Derived from Periodontal Ligament or Gingival

Tissue Sources Encapsulated in RGD-Modified Alginate Scaffold. *Tissue Engineering Part A*; 20:611-21.

[55] Nakaoka R, Hirano Y, Mooney DJ, Tsuchiya T, Matsuoka A. 2013, Study on the potential of RGD- and PHSRN-modified alginates as artificial extracellular matrices for engineering bone. *J. Artif. Organs*; 16:284-93.

[56] Gupta B, MythiliTummalapalli, B.L.Deopura, M.S.Alamb. 2014, Preparation andcharacterizationof in-situ crosslinked pectin–gelatin hydrogels. *Carbohydrate Polymers*; 106:312-8.

[57] Vodna L, Bubenikova S, Bakos D. 2007, Chitosan based hydrogel microspheres as drug carriers. *Macromolecular Bioscience*; 7:629-34.

[58] Esposito E, Menegatti E, Cortesi R. 2005, Hyaluronan-based microspheres as tools for drug delivery: a comparative study. *Int. J. Pharm*; 288:35-49.

[59] Godbey WT, Wu KK, Mikos AG. 1999, Poly(ethylenimine) and its role in gene delivery. *J. Control Release*; 60:149-60.

[60] Park TG, Yong Lee H, Sung Nam Y. 1998, A new preparation method for protein loaded poly(D, L-lactic-co-glycolic acid) microspheres and protein release mechanism study. *J. Control Release*; 55:181-91.

[61] Luginbuehl V, Meinel L, Merkle HP, Gander B. 2004, Localized delivery of growth factors for bone repair. *Eur. J. Pharm. Biopharm*; 58:197-208.

[62] Kim SE, Park JH, Cho YW, Chung H, Jeong SY, Lee EB, et al. 2003, Porous chitosan scaffold containing microspheres loaded with transforming growth factor-beta1: implications for cartilage tissue engineering. *J. Control Release*; 91:365-74.

[63] Munarin F, Petrini P, Fare S, Tanzi MC. 2010, Structural properties of polysaccharide-based microcapsules for soft tissue regeneration. *J. Mater. Sci.-Mater. M*; 21:365-75.

[64] Munarin F, Giuliano L, Bozzini S, Tanzi MC, Petrini P. 2010, Mineral phase deposition on pectin microspheres. *Mat. Sci. Eng. C-Mater*; 30:491-6.

[65] Hou QP, De Bank PA, Shakesheff KM. 2004, Injectable scaffolds for tissue regeneration. *J. Mater. Chem*; 14:1915-23.

[66] Drury JL, Mooney DJ. 2003, Hydrogels for tissue engineering: scaffold design variables and applications. *Biomaterials*; 24:4337-51.

[67] Nuttelman CR, Rice MA, Rydholm AE, Salinas CN, Shah DN, Anseth KS. 2008, Macromolecular monomers for the synthesis of hydrogel niches and their application in cell encapsulation and tissue engineering. *Prog. Polym. Sci*; 33:167-79.

[68] Brandl F, Sommer F, Goepferich A. 2007, Rational design of hydrogels for tissue engineering: impact of physical factors on cell behavior. *Biomaterials*; 28:134-46.

[69] Delair T. 2012, In situ forming polysaccharide-based 3D-hydrogels for cell delivery in regenerative medicine. *Carbohyd. Polym*; 87:1013-9.

[70] Balakrishnan B, Banerjee R. 2011, Biopolymer-based hydrogels for cartilage tissue engineering. *Chem. Rev*; 111:4453-74.

[71] Moreira HR, Munarin F, Gentilini R, Visai L, Granja P L, Tanzi MC, et al. 2104, Injectable pectin hydrogels produced by internal gelation: pH dependence of gelling and rheological properties. *Carbohydrate polymers*; 103:339-47.

[72] F. Munarin, R. Gentilini, Petrini P, Tanzi MC. 2013, *Injectable Pectin Hydrogels as Cell Carriers for Soft Tissue Regeneration*. 25th European Conference on Biomaterials. Madrid Spain.

In: Pectin: Chemical Properties, Uses and Health Benefits
Editor: Phillip L. Bush
ISBN: 978-1-63321-438-5

Chapter 2

PECTIN: DIETARY SOURCES, PROPERTIES AND HEALTH BENEFITS

Adriana Cuervo[1], Miguel Gueimonde[2], Abelardo Margolles[2] and Sonia González[1]

[1]Department of Functional Biology, University of Oviedo, Facultad de Medicina, C/Julián Clavería s/n, Oviedo, Asturias, Spain

[2]Department of Microbiology and Biochemistry of DairyProducts, Instituto de Productos Lácteos de Asturias, Consejo Superior de Investigaciones Científicas (IPLA-CSIC), Paseo Río Linares s/n, Villaviciosa, Asturias, Spain

ABSTRACT

The term dietary fiber includes a wide group of dietary compounds of plant origin and resistant to digestion by human gastrointestinal enzymes. Traditionally, this group of compounds has been classified according to their chemical structure and physiological activities, as its behavior in water, being insoluble ones the most frequently consumed in Westernized countries. The assessment of the different types of fiber provided from diet implies some difficulties since factors, such as fruit maturation or the intake of peeled fruit, may affect the fiber content of a food and they are impossible to quantify in a food composition table. These handicaps result in few nutritional studies providing a detailed intake of these dietary compounds.

From soluble fibers, pectins, provided by citric fruits and vegetables, have attracted a deal of attention in the last decades, giving the epidemiological evidences linking their intake with protection against cardiovascular disease, type II diabetes, colorectal cancer and gastrointestinal diseases. Also, pectins evade digestion by intestinal enzymes and passes directly into the colon, where they are metabolized by some intestinal bacteria such as *Bifidobacterium* and *Lactobacillus,* which constitute the traditional target of prebiotics, and contributing to the increase of short chain fatty acid production. Thus, pectin confers benefits upon host health by decreasing the risk of some diseases, such as the irritable bowel syndrome, inflammatory bowel disease, cardiovascular disease and cancer.

The contributors to this chapter will provide a brief description about the dietary sources of pectin in humans and review the most relevant literature discussing the prebiotic effect of this dietary compound together with its implications for health by means of the increase in the production of bacterial metabolites.

1. INTRODUCTION

1.1. Concept, Structure and Properties of Pectin

The term pectin (derived from Greek *pektikos*, dense, thick, curdled) refers to the most structurally and functionally complex family of polysaccharides in nature [1]. These compounds are present in the higher plant primary cell walls and middle lamella, representing one third of the cell dry weight, and frequently associated with other compounds, such as cellulose, hemicellulose and lignin, exerting a structural function.

Although the relationship between humans and pectic polysaccharides has a long history, because of its involvement in the preparation of jams and jellies as a way of food preservation, it was in 1825, when these compounds were firstly isolated and described by the French chemist and pharmacist, Henri Braconnot [2], finding that supposed the starting point for the subsequent industrial use of these compounds. Today, there is an industry dedicated to the extraction and processing of these fibers, mainly from orange peel and apple, for its use as a gelling agent in many foodstuffs and for the stabilization of acidified milk drinks and yogurts [3].

The structure of pectin is very difficult to determine, not only because it can change during its isolation, storage and processing [4], but also because it can be extremely heterogeneous between plants and tissues, and even within a single cell wall. The basic structure of pectin polysaccharides consists of linear chains of 300-1,000 residues of D-galacturonic acid (GalA) (comprising around 70% of total pectin), some of them methylated, and covalently α-1,4-linked [5] (Figure 1).

Pectin can be present in different polymeric forms:- *Homogalacturonan (HG)*: the most abundant, accounting for around 65% of pectins in plants, consists in a linear homopolymer of α-1,4-linked GalA, partially esterified with methyl and, in some cases, acetylated [6]; - *Rhamnogalacturonan-I (RG-I):* constitute 20-35% of pectin in plants. Is the only type of pectin not built by a linear chain of GalA but is branched with α-1,2-linked rhamnose (Rha) residues. These residues in the backbone can be substituted with β-1,4-galactan, branched arabinan, and/or arabinogalactan side chains [7]. The next two types of pectin are minor components, both of them accounting for less than 10%: - *Xylogalacturonan (XGA):* the backbone of homogalacturonan is substituted with xylose residues by β-1,4 linkages [8]; - *Rhamnogalacturonan-II (RG-II):* composed of 12 types of glycosyl residues, including glucuronic acid, aceric acid, apiose and fucose, linked together by, at least, 22 different glycosidic bonds [9].

The degree of esterification (DE) of HG pectin, which is defined by the ratio between the methylated galacturonic acid residues and the total of galacturonic acid units present in the molecule, is one of the most important characteristics of these fibers, because it determines their gelling properties and, therefore, their physiological properties.

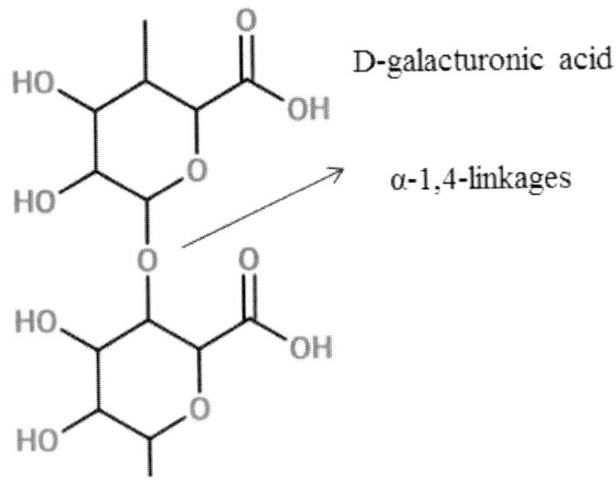

Figure 1. Basic structure of pectin, composed by residues of galacturonic acid kinked by glycosidic α-1,4 bonds.

According to DE, pectins can be classified in: high methoxylpectins (HM), with around 60-75% of esterification and low methoxylpectins (LM) with 20-40%. The non-esterified GalA residues can be both free or as salts with sodium, potassium, ammonium or calcium ions. Pectin gel is formed when chains of HG are cross-linked resulting in a three dimensional network in which water and other molecules are trapped. In HM pectins, junction zones are formed by hydrogen bridges and hydrophobic forces between methoxyl groups, favored by high sugar concentration and low pH, while in LM pectins junction zones are formed by calcium cross-linking between free carboxyl groups [3]. It is estimated that general DE in plants is around 60-90%, however, it can been modulated by some factors, such as plant variety or species, type of tissue, degree of maturation and factors related to pectin extraction.

1.2. Evaluating the Pectin Content in Foods

There are several methods for quantifying the content of pectin in plants. One of the most used is the procedure that includes the first hydrolysis of pectin in hot concentrated acid medium (H_2SO_4) [10], and the subsequent quantification of the resulting anhydrogalacturonic acid residues by a colorimetric method, using m-hydroxydiphenyl [11], carbazol [12] or 3,5dimethylphenol [13] as chromogenic agents. Other possibility, which avoid the use of H_2SO_4, is to hydrolyze pectin to galacturonic acid using pectin degrading enzymes, and then quantifying using a colorimetric method [12] by HPLC [14].

In any case, determining the pectin content of foods is not an easy task, not only because the amount varies depending on factors such as the plant variety, the type of fruit, its degree of ripeness and factors related to its processing and storage, but also because of the complex structure of pectin and the interferences caused by other carbohydrates present in the same food [15].

The difficulties involved in the evaluation of pectin content lead to the scarcity in food composition tables that include detailed information for this component.

Most of them only provide data for the total dietary fiber in foods or, at best, the soluble and insoluble portions. Consequently, studies evaluating the amount of pectin provided by the regular diet in humans are very sparse. Most data in this area of research comes from intervention studies that evaluate the impact of high doses of pectin extracts on health.

1.3. Food Sources of Pectin

It is estimated that the intake of pectin in western countries is around 4–5 g per day [16]. Fruits, such as avocado, citrus (orange, tangerine and lemon) or apple, have been identified as good sources of pectin; vegetables, as Brussels sprouts, artichokes, carrot, broccoli, pumpkin, eggplant, French beans or cabbage, etc., as well as legumes and nuts are also common components of the regular diet with a high content in pectin [17].

2. Physiological Effects of Pectin

Food sources of pectin are also rich in a number of bioactive compounds, such as vitamins and minerals, antioxidants, polyphenols, and other types of fibers. This explains why the research focused on evaluating the impact of pectin, and its food sources, on health has received much attention in the last few years. During this time, most epidemiological studies have reported the existence of an inverse association between pectin intake and the risk of some pathologies, including cardiovascular disease [18], type II diabetes [19], gastrointestinal disorders [20, 21] and colorectal cancer [22]. It has been considered that the role of pectin in these diseases depends on its physicochemical characteristics, especially its solubility in water, determining, in turn, the viscosity and fermentability of these compounds [23]. Nevertheless, the mechanisms underlying these associations are not well established at this moment. In the absence of more solid scientific evidence, it has been speculated that a part of the health benefits of pectin could be attributed to its impact on intestinal microbiota compositions and its metabolic activity.

3. Impact of Pectin Intake on Gastrointestinal Health

The human gastrointestinal tract harbors a very rich and complex microbial community which initial establishment begins at birth and is affected by several perinatal factors, such as mode of delivery and feeding habits [24]. This microbial colonization of the intestine is necessary for a proper maturation of the immune system [25] and host development [26]. In spite of the high inter-individual variability, recent studies have identified different specific human microbiota enterotypes [27] and microbiota alterations have been observed in different disease states [28]. Moreover, the intestinal microbiota displays an enormous metabolic versatility [29], allowing the use of different dietary and intestinal substrates, including mucins, non-digestible oligosaccharides or dietary fiber, and leading to the production of metabolic products, such as short chain fatty acids (SCFA), which may result beneficial to the host.

Therefore, the intestinal microbiota plays an important role in human health, not only by participating in the digestion but also by maintaining host immune and physiological homeostasis [28]. This role on health of the microbiota provides the rationale for developing dietary intervention strategies targeting to microbiota modulation. Among these strategies, the use of dietary supplementation with prebiotic fibers has been widely explored, but the impact of specific nutrients in the context of the general diet remains largely unknown.

Different dietary fibers seem to differ on their effects upon gut microbiota composition and activity. The insoluble fraction is only partially fermented, contributing mainly to increase fecal bulk and reducing transit time [30]. Soluble fiber, including pectin, is more easily fermented which leads to the production of bacterial metabolites such as SCFA, lactate, succinate, CO_2, methane and hydrogen, amongst others [31]. Some of these metabolites have been associated with important biological functions [32].

3.1. Microbial Metabolism of Pectin

The effect of some fibers upon gut microbiota composition has been studied in some detail and the microorganisms responsible for the fermentation have been identified. Fructo-oligosaccharides have been consistently shown to increase colonic bifidobacterial numbers [33] and the effect of resistant starch on some microbial groups, such as *Ruminococcus*, has been repeatedly reported [33]. Moreover, different authors have reported associations between a diet rich in resistant starch and high levels of butyrate production [34], which results interesting in the context of the reported beneficial effects of this SCFA [35]. However, the available information regarding other important dietary fibers, such as pectins, is still limited [36].

Some gut microorganisms metabolize complex glycans and polysaccharides, including those present in the plant cell wall. There is evidence that pectin-rich foods, such as apples, could modulate the gut microbiota composition in humans, and some researchers have attributed this effect to the presence of pectins. Shinohara and coworkers demonstrated that strains of *Bifidobacterium, Lactobacillus, Enterococcus*, and *Bacteroides* metabolize apple pectin; however, other intestinal isolates belonging to the species *Escherichia coli*, *Eubacterium limosum*, and *Clostridium perfringens*, do not. Also, when fecal cultures were incubated together with apple pectin, the numbers of *Bifidobacterium* and *Lactobacillus* significantly increased [37]. These and other works support the potential prebiotic and/or bifidogenic properties of pectins [38]. Increasing the populations of bifidobacteria and lactobacilli in the human gut could be a health benefit under some specific conditions.

During the last years some research, mainly using in vitro fermentation models to prove the impact of pectins in bacterial growth, have been focused on this aim. For instance, *Lactobacillus acidophilus NCFM* was able to grow and predominate over other bacterial populations in a model of human fresh fecal microbiota in which pectin-reach potato fiber was added [39]. Also, the evaluation of the prebiotic properties of pectin-oligosaccharides from apple, using fecal batch culture fermentations, showed that these sugars were able to increase the numbers of lactobacilli and bifidobacteria, and to reduce those of *Bacteroides* and clostridia [40]. Furthermore, in mixed fecal bacterial cultures oligosaccharides derived from orange peel pectin were found to increase bifidobacterial numbers [41], and the arabinan-rich fraction resulting from sugar beet pulp based pectin selectively stimulated bifidobacterial growth in human fecal fermentations [42]. Remarkably, it must be considered that the degree of polymerization could deeply influence the fermentation properties of pectins.

In fact, it has been suggested that pectic-oligosaccharides are a better prebiotic candidate than pectins [43]. Regarding the enzymes involved in the metabolism of pectins by probiotic bacteria, the available information is fragmentary. Currently, we know that extracellular pectinase, aldolase, galacturonase and esterase activities have been linked to pectin degradation in lactobacilli and bifidobacteria [44, 45].

These results suggest that pectins or pectin-derived oligosaccharides, independently of the vegetable source from which they have been obtained, have the potential to be metabolized and to increase the population of potential beneficial bacteria. However, although pectins are well tolerated as food ingredients in humans, evidence from intervention studies is scarce, and the impact on intestinal microbial profiles is still unclear [46].

3.2. Modulation of SCFAs Production by Pectins and Health Related Effects

The substrates available in the colon, such as non-digestible oligosaccharides or dietary fibers, are fermented by the intestinal microbiota to metabolic products such as organic acids, SCFA and other fermentation products, including gases and ethanol. Some of these metabolites such as unbranched SCFA (mainly butyric, propionic and acetic acids) have been reported to be important to host health [47-49]. In recent years it has been proven that dietary fiber exerts a large effect on gut microbiota composition and its metabolites [50-52]. Animal model studies demonstrate the effect of concentrated fiber from apple in increasing the concentrations of propionic, butyric and total SCFA in feces [53], and consumption of soluble fiber concentrates has been associated with higher acetic, propionic and butyric acid levels [54]. The balance among the different fermentation products is highly dependent on the available fibers, in general, those that are fermented quickly produce larger amounts of lactic and acetic whereas the substrates fermented more slowly lead to the production of more butyric acid as end fermentation product [55]. Moreover, other metabolites such as organic acids, mainly lactic, succinic and pyruvic, together with some partially degraded carbohydrate polymers, will also be metabolized to SCFA by cross-feeding mechanisms [56, 57]. In the colonic environment pectin seems to be extensively fermented [58]. Strains from different microbial species normally present in the human gut have been reported to be able to utilize pectin in pure culture [37, 59], although, in the colon is likely the fermentation is carried out with contributions from different microorganisms. In vitro studies have reported the ability of pectin to increase bifidobacteria and enterobacteria levels reducing those of Bacteroidaceae and Lachnospiraceae and resulting in an increase of acetate concentration [60], which is in agreement with animal studies reporting acetate as the main SCFA produced after consumption of a pectin-rich diet [49; 61]. Moreover, dietary pectin has been found to affect fecal SCFA levels in the elderly [36]. These results suggest an important role of dietary pectin in modulating gut microbiota composition and activity.

CONCLUSION

To summarize, any qualitative or quantitative change in the colon, either in the microorganisms or on the dietary carbohydrates available, could affect the production of microbial metabolism end-products and, therefore, human health. In this context the beneficial effects of pectin would result from the specific action in the human body of the SCFA produced by microbial fermentation of these substrates as well as from the ability of these compounds to modify the intestinal microbiota.

REFERENCES

[1] Mohnen D. Pectin structure and biosynthesis. *Curr. Opin. Plant. Biol.* 2008 Jun;11(3): 266-77.
[2] Braconnot H. Annales de chimie et de physique. *Annals of Chemistry and Physics* 1825; 28(2):173-8.
[3] Willats WGT, Knox JP, Mikkelse JD. Pectin: new insights into an old polymer are starting to gel. *Trends Food Sci Tech.* 2006; 17:97-104.
[4] Novosel'skaya IL, Voropaeva NL, Voropaeva LN, Rashidova SS. Trends in the science and applications of pectins. *Chem. Nat. Comp.* 2000; 36(1):1-10.
[5] Ridley BL, O'Neill MA, Mohnen D. Pectins: structure, biosynthesis, and oligogalacturonide-related signaling. *Phytochemistry* 2001 Jul;57(6): 929-67.
[6] Mukhiddinov ZK, Khalikov DK, Abdusamiev FT, Avloev CC. Isolation and structural characterization of a pectin homo and ramnogalacturonan. *Talanta* 2000 Oct 2; 53(1):171-6.
[7] Nakamura A, Furuta H, Maeda H, Takao T, Nagamatsu Y. Structural studies by stepwise enzymatic degradation of the main backbone of soybean soluble polysaccharides consisting of galacturonan and rhamnogalacturonan. *Biosci Biotechnol Biochem.* 2002 Jun; 66(6): 1301-13.
[8] Zandleven J, Beldman G, Bosveld M, Schols HA, Voragen AGJ. Enzymatic degradation studies of xylogalacturonans from apple and potato, using xylogalacturonan hydrolase. *Carbohydr. Polym.* 2006; 64: 495-503.
[9] O'Neill MA, Warrenfeltz D, Kates K, Pellerin P, Doco T, Darvill AG, et al. Rhamnogalacturonan-II, a pectic polysaccharide in the walls of growing plant cell, forms a dimer that is covalently cross-linked by a borate ester. In vitro conditions for the formation and hydrolysis of the dimer. *J. Biol. Chem.* 1996 Sep 13; 271(37): 22923-30.
[10] Anthon GE, Barrett DM. Combined enzymatic and colorimetric method for determining the uronic acid and methylester content of pectin: Application to tomato products. *Food Chemistry* 2008; 110:239-47.
[11] Blumenkrantz N, Asboe-Hansen G. New method for quantitative determination of uronic acids. *Anal. Biochem.* 1973 Aug; 54(2): 484-9.
[12] McComb E A, McCready R M. Colorimetric determination of pectic substances. *Anal. Chem.* 1952; 24:1630-3.

[13] Scott RW. Colorimetric determination of hexuronic acids in plant materials. *Anal. Chem.*1979;51(7): 936-41.

[14] Rumpunen K, Thomas M, Badiles N, Thibault JF. Validation of a combined enzymatic and HPLC method for screening of pectin in fruits of Japanese quince (Chaenomeles japonica). *Lebensmittle-Wissenschaft und Technologie* 2002; 55:490-6.

[15] Ibarz A, Pagán A, Tribaldo F, Pagán J. Improvement in the measurement of spectrophotometric data in the m-hydroxydiphenyl pectin determination methods. *Food Control* 2006; 17:890-3.

[16] Pilnik W. Pectin—a many splendoured thing. In: Phillips GO, Williams PA, Wedlock DJ, editors. *Gums and stabilizers for the food industry.* Oxford: Oxford University Press; 1990. p. 313-26.

[17] Marlett JA, Cheung TF. Database and quick methods of assessing typical dietary fiber intakes using data for 228 commonly consumed foods. *J. Am. Diet Assoc.* 1997 Oct; 97(10): 1139-48, 1151.

[18] Pereira MA, Pins JJ. Dietary fiber and cardiovascular disease: experimental and epidemiologic advances. *Curr. Atheroscler. Rep.* 2000 Nov; 2(6): 494-502.

[19] McCarty MF. Nutraceutical resources for diabetes prevention--an update. *Med. Hypotheses* 2005; 64(1):151-8.

[20] Bijkerk CJ, Muris JW, Knottnerus JA, Hoes AW, de Wit NJ. Systematic review: the role of different types of fibre in the treatment of irritable bowel syndrome. *Aliment Pharmacol. Ther.* 2004 Feb 1; 19(3):245-51.

[21] Nakao M, Ogura Y, Satake S, Ito I, Iguchi A, Takagi K, et al. Usefulness of soluble dietary fiber for the treatment of diarrhea during enteral nutrition in elderly patients. *Nutrition* 2002 Jan; 18(1):35-9.

[22] Tjonneland AM, Overvad K, Bingham SA, Day NE, Luben R, Ferrari P, et al. [Dietary fibers in food and protection against colorectal cancer in the European Prospective Investigation into Cancer and Nutrition (EPIC) study]. *Ugeskr. Laeger* 2004 Jun 14; 166(25):2458-60.

[23] Guillon F, Champ M. Structural and physical properties of dietary fibres, and consequences of processing on human physiology. *Food Res. Int.* 2000; 33(3):233-45.

[24] Johnson C, Versalovic J. The human microbiome and its potential importance to pediatrics. *Pediatrics* 2012;129:950-60.

[25] Hooper LV, Littman DR, Macpherson AJ. Interactions between the microbiota and the immune system. *Science* 2012 Jun 8;336(6086): 1268-73.

[26] Sommer F, Bäckhed F. The gut microbiota--master of host development and physiology. *Nat. Rev. Microbiol.* 2013;11:227-38.

[27] Arumugam M, Raes J, Pelletier E, Le PD, Yamada T, Mende DR, et al. Enterotypes of the human gut microbiome. *Nature* 2011 May 12; 473(7346): 174-80.

[28] Sekirov I, Russell SL, Antunes LC, Finlay BB. Gut microbiota in health and disease. *Physiol. Rev.* 2010 Jul; 90(3): 859-904.

[29] Qin J, Li R, Raes J, Arumugam M, Burgdorf KS, Manichanh C, et al. A human gut microbial gene catalogue established by metagenomic sequencing. *Nature* 2010 Mar 4; 464(7285): 59-65.

[30] Anderson JW, Baird P, Davis RH, Jr., Ferreri S, Knudtson M, Koraym A, et al. Health benefits of dietary fiber. *Nutr. Rev.* 2009 Apr;67(4): 188-205.

[31] Christian MT, Edwards CA, Preston T, Johnston L, Varley R, Weaver LT. Starch fermentation by faecal bacteria of infants, toddlers and adults: importance for energy salvage. *Eur. J. Clin. Nutr.* 2003 Nov; 57(11): 1486-91.

[32] Cuervo A, Arboleya S, Gueimonde M, González S. Microbiota modulation by diet in humans. Prebiotics, fibres and other compounds. *Agro Food Ind. Hi. Tech.* 2012; 23:23-6.

[33] Scott KP, Gratz SW, Sheridan PO, Flint HJ, Duncan SH. The influence of diet on the gut microbiota. *Pharmacol. Res.* 2013 Mar; 69(1):52-60.

[34] Jenkins DJ, Vuksan V, Kendall CW, Würsch P, Jeffcoat R, Waring S, et al. Physiological Effects of Resistan Starches on Fecal Bulk, Short Chain Fatty Acids, Blood Lipids and Glycemic Index. *J. Am. Coll. Nutr.* 1998; 17(6): 609-16.

[35] Hamer HM, Jonkers D, Venema K, Vanhoutvin S, Troost FJ, Brummer RJ. Review article: the role of butyrate on colonic function. *Aliment Pharmacol. Ther.* 2008 Jan 15; 27(2):104-19.

[36] Cuervo A, Salazar N, Ruas-Madiedo P, Gueimonde M, Gonzalez S. Fiber from a regular diet is directly associated with fecal short-chain fatty acid concentrations in the elderly. *Nutr. Res.* 2013 Oct; 33(10): 811-6.

[37] Shinohara K, Ohashi Y, Kawasumi K, Terada A, Fujisawa T. Effect of apple intake on fecal microbiota and metabolites in humans. *Anaerobe* 2010 Oct; 16(5):510-5.

[38] Koropatkin N, Cameron E, Martens E. How glycan metabolism shapes the human gut microbiota. *Nat. Rev. Microbiol.* 2012; 10:323-35.

[39] Knudsen A, van Zanten G, Jensen S, Forssten S, Saarinen M, Lahtinen S, et al. Comparative fermentation of insoluble carbohydrates in an in vitro human feces model spiked with Lactobacillus acidophilus NCFM. *Starch-Stärke* 2013; 65:346-53.

[40] Chen J, Liang RH, Liu W, Li T, Liu CM, Wu SS, et al. Pectic-oligosaccharides prepared by dynamic high-pressure microfluidization and their in vitro fermentation properties. *Carbohydr Polym.* 2013 Jan 2;91(1): 175-82.

[41] Manderson K, Pinart M, Tuohy KM, Grace WE, Hotchkiss AT, Widmer W, et al. In vitro determination of prebiotic properties of oligosaccharides derived from an orange juice manufacturing by-product stream. *Appl. Environ. Microbiol.* 2005 Dec; 71(12):8383-9.

[42] Holck J, Lorentzen A, Vigsnaes LK, Licht TR, Mikkelsen JD, Meyer AS. Feruloylated and nonferuloylated arabino-oligosaccharides from sugar beet pectin selectively stimulate the growth of Bifidobacterium spp. in human fecal in vitro fermentations. *J. Agric Food Chem.* 2011 Jun 22; 59(12):6511-9.

[43] Olano-Martin E, Gibson GR, Rastell RA. Comparison of the in vitro bifidogenic properties of pectins and pectic-oligosaccharides. *J. Appl. Microbiol.* 2002;93(3):505-11.

[44] Karam NE, Belarbi A. Detection of polygalacturonases and pectin esterases in lactic acid bacteria. *World J. Microbiol. Biotechnol.* 1995 Sep; 11(5):559-63.

[45] Slovakova L, Duskova D, Marounek M. Fermentation of pectin and glucose, and activity of pectin-degrading enzymes in the rabbit caecal bacterium Bifidobacterium pseudolongum. *Lett. Appl. Microbiol.* 2002; 35(2):126-30.

[46] Fanaro S, Jelinek J, Stahl B, Boehm G, Kock R, Vigi V. Acidic oligosaccharides from pectin hydrolysate as new component for infant formulae: effect on intestinal flora, stool characteristics, and pH. *J. Pediatr. Gastroenterol. Nutr.* 2005 Aug; 41(2):186-90.

[47] Fukuda S, Toh H, Hase K, Oshima K, Nakanishi Y, Yoshimura K, et al. Bifidobacteria can protect from enteropathogenic infection through production of acetate. *Nature* 2011 Jan 27; 469(7331):543-7.

[48] Gao Z, Yin J, Zhang J, Ward RE, Martin RJ, Lefevre M, et al. Butyrate improves insulin sensitivity and increases energy expenditure in mice. *Diabetes* 2009 Jul; 58(7):1509-17.

[49] Peng X, Li S, Luo J, Wu X, Liu L. Effects of dietary fibers and their mixtures on short chain fatty acids and microbiota in mice guts. *Food Funct.* 2013 Jun; 4(6): 932-8.

[50] Claesson MJ, Jeffery IB, Conde S, Power SE, O'Connor EM, Cusack S, et al. Gut microbiota composition correlates with diet and health in the elderly. *Nature* 2012 Aug 9; 488(7410): 178-84.

[51] De Fillipo C, Cavalieri D, Di PM, Ramazzotti M, Poullet JB, Massart S, et al. Impact of diet in shaping gut microbiota revealed by a comparative study in children from Europe and rural Africa. *Proc. Natl. Acad. Sci. U S A* 2010 Aug 17;107(33):14691-6.

[52] Wu GD, Chen J, Hoffmann C, Bittinger K, Chen YY, Keilbaugh SA, et al. Linking long-term dietary patterns with gut microbial enterotypes. *Science* 2011 Oct 7;334(6052):105-8.

[53] Sembries S, Dongowski G, Jacobasch G, Mehrlander K, Will F, Dietrich H. Effects of dietary fibre-rich juice colloids from apple pomace extraction juices on intestinal fermentation products and microbiota in rats. *Br. J. Nutr.* 2003 Sep;90(3):607-15.

[54] Guerin-Deremaux L, Li S, Pochat M, Wils D, Mubasher M, Reifer C, et al. Effects of NUTRIOSE(R) dietary fiber supplementation on body weight, body composition, energy intake, and hunger in overweight men. *Int. J. Food Sci. Nutr.* 2011 Sep;62(6):628-35.

[55] Van de Wiele T, Boon N, Possemiers S, Jacobs H, Verstraete W. Inulin-type fructans of longer degree of polymerization exert more pronounced in vitro prebiotic effects. *J. Appl. Microbiol.* 2007 Feb;102(2):452-60.

[56] De Vuyst L, Leroy F. Cross-feeding between bifidobacteria and butyrate-producing colon bacteria explains bifdobacterial competitiveness, butyrate production, and gas production. *Int. J. Food Microbiol.* 2011 Sep 1;149(1):73-80.

[57] Flint HJ, Bayer EA, Rincon MT, Lamed R, White BA. Polysaccharide utilization by gut bacteria: potential for new insights from genomic analysis. *Nat. Rev. Microbiol.* 2008 Feb;6(2):121-31.

[58] Cummings JH, Hill MJ, Bone ES, Branch WJ, Jenkins DJ. The effect of meat protein and dietary fiber on colonic function and metabolism. II. Bacterial metabolites in feces and urine. *Am. J. Clin. Nutr.* 1979 Oct; 32(10): 2094-101.

[59] Lopez-Siles M, Khan TM, Duncan SH, Harmsen HJ, Garcia-Gil LJ, Flint HJ. Cultured representatives of two major phylogroups of human colonic Faecalibacterium prausnitzii can utilize pectin, uronic acids, and host-derived substrates for growth. *Appl. Environ. Microbiol.* 2012 Jan; 78(2): 420-8.

[60] Yang J, Martinez I, Walter J, Keshavarzian A, Rose DJ. In vitro characterization of the impact of selected dietary fibers on fecal microbiota composition and short chain fatty acid production. *Anaerobe* 2013 Oct; 23:74-81.

[61] Berggren A, Björck I, Nyman E. Short-chain fatty acid content and pH in caecum of rats given various sources of carbohydrates. *J. Sci. Food Agr.* 1993; 67:397-407.

In: Pectin: Chemical Properties, Uses and Health Benefits
Editor: Phillip L. Bush
ISBN: 978-1-63321-438-5

Chapter 3

THE COMBINATION OF DIFFERENT SOURCES AND EXTRACTION METHODS AS A STRATEGY TO ENHANCE PECTIN PRODUCTION

***Elaine Berger Ceresino*[*], *Jéssika Gonçalves dos Santos*, *Paula de Paula Menezes Barbosa*, *Haroldo Yukio Kawaguti* and *Fabiano Jares Contesini*[†]**

Laboratory of Biochemistry. Department of Food Science.
State University of Campinas. Campinas, SP, Brazil

ABSTRACT

Pectins are biopolymers with multiple applications because of their structural diversity and complexity. Although pectins from different sources have some common structural characteristics, many aspects of the common structure change according to the species and the physiological stage of the plant. Moreover, the application of pectin is determined by its chemical features, including galacturonic acid content, methoxyl content, degree of esterification and acetyl value. The most traditional raw materials used for the extraction of pectins are either apple pomace or citrus peels that are supplied as by-products of juice production. Both materials contain significant amounts of pectic substances, but with different chemical characteristics that make them suitable for specific applications. Considering that pectin is widely used as a functional ingredient, many researchers have been testing the use of other materials and alternative methods of extraction for industrial exploitation. Among them, different waste materials have been tested, such as sugar beet and passion fruit pomace. The yield and quality of extracted pectins are essential for their commercialization and are highly affected by the method used. The usual extraction process is based on the combination of acidic solutions and high temperature. Moreover, it is a very time-consuming process - up to 12 h. Microwave-assisted extraction has been tested and presented good results in the

[*] Corresponding author: Elaine Berger Ceresino. Laboratório de Bioquímica. Departamento de Ciência de Alimentos - FEA, Universidade Estadual de Campinas. Rua Monteiro Lobato, 80. Cx. Postal 6121. 13083-862. Campinas-SP, Brasil. Tel./fax: +55 19 3521 2175, E-mail: elaineceresino@gmail.com.

[†] Fabiano Jares Contesini e-mail: fabiano.contesini@gmail.com.

extraction of pectins from passion fruit peel, berries and from watermelon waste fruit rinds. The extraction of pectin from navel orange peel assisted by ultra-high pressure showed that the intrinsic viscosity and viscosity-average molecular weight were much higher than those extracted by traditional heating, microwave and commercial pectin. In order to obtain time-saving, and eco-friendly extraction methods, the use of microbial enzymes has attractive properties that justify more extensive researches. Some studies confirm that the implementation of a biotechnological method would greatly increase the pectin production and contribute to free the processing plants from expensive works to neutralize the acidic components of the traditional technology. Given the importance of this biopolymer regarding the wide application in medicinal and food products, this chapter reviews current issues regarding the prospect for new sources of pectin and the advances in their extraction methods.

1. INTRODUCTION

Pectin or pectic substances are a group of closely associated polysacchari-des present in plant cell walls. They are presented in primary cell walls and in the middle lamellae of plants, where it helps bind cells together by the regulation of intercellular adhesion (Willats et al., 2001). Chemically, pectins are a group of polysaccharides that are rich in galacturonic acid (GalA) that often display different degrees of methyl esterification involving the C-6 carboxyl group (Domozych, 2007). Pectin is widely used in the food industry due to its gelling properties being applied in jams and jellies, fruit preparations for yogurts, drinks and fruit juice concentrates, fruit desserts and milk, jellied milk products, confectionery, as well as stabilizers or fermented acidified milk products and yogurt (Willats et al., 2006). The natural pectic polysaccharides act as thickeners, promoting increased viscosity solutions, gelling agents, stabilizers as well as in foods and beverages. The pectic substances act in preventing flotation in fruit preparations, stability of bakery products in protein stabilization, texture improving the softness of the product, increasing the volume and the prevention and control of syneresis (Voragen et al., 2009). The ability of pectins to form gel depends on the molecular size and degree of esterification (DE), that is the ratio of esterified carboxylic acid units to total carboxylic acid units in pectin (Walter, 1991). The occurrence and proportions of pectin in different sources is variable among plant species. Due to variations in these parameters, pectins from different sources do not have the same characteristics and gelling ability (Voragen et al., 2009). The main sources for commercial pectin production are apple pomace and citrus peels, both by-products from juice or cider manufacturing. The production of pectin from food industry by-products are considered beneficial from both the economic and ecological scope encouraging the study of many other agricultural by-products or wastes (Schieber et al., 2003). In addition, some studies are being performed to obtain it from other raw materials such as seeds and oil-cakes (Liang et al., 2012). The process used for the extraction of this biopolymer also has great influence in its structural and technological characteristics (MacDougall and Ring, 2004). Severe extraction processes are used at the industrial level of pectin production, which are frequently detrimental to pectin structure resulting in de-esterification and depolymerization. The extraction in acid medium at high temperatures is the most common method used for the extraction of pectins from agro-industrial waste of fruit juices; however, emerging eco-friendly technologies have been studied about the extraction of pectin in order to cause less damage to the pectin structure and avoid environ-mental contamination (Kratchanova et al., 2004).

Among them, we highlight the use of microwave, high pressure processing, enzymes and subcritical water.

2. INDUSTRIAL AND UNCONVENTIONAL SOURCES OF PECTIN

The most important raw materials for the extraction of commercial pectin are citrus peel and apple pomace, both by-products from juice or cider manufacturing (Visser and Voragen, 1996; May, 1990; May, 2000; Mesbahi et al., 2005; Kurita et al., 2008). When dried, they contain about 15-20% and 30-35% of this polymer (Silva and Rao, 2006).

Currently, many works have been done in order to discover, improve and optimize the extraction of pectins from traditional and new sources. Taking into consideration the impact of the use of industrial by-products to obtain commercial pectin, the characteristics of pectins from different sources have been studied extensively. The residues that would be discarded, creating problems of disposal in the environment, are being studied as a source of pectin. The differences in size of the polygalacturonic acid chain and the esterification degree of its carboxyl groups vary considerably, depending on the source that is extracted, affecting its ability to form gels in the pectin structure. The extraction procedure, portion of plant tissue, and neutral sugars content also determine considerable variability in the final features of the pectin (Barrera et al., 2002). Table 1 shows some sources of pectin and their characteristics.

Table 1. Pectin yield, degree of methoxylation and molecular weight from different sources of pectin

Source	Pectin yield (%)	Degree of methoxylation (DM, %)	Molecular weight (kDa)	Reference
Citrus peel (commercial)	25	65.00	224.48	Torralbo et al., 2012
Fraction of oxalate-soluble pectin in Dou fu chai	20.61	14.90	980.67	Liang et al., 2014
Murta	27	48	334	Taboada et al., 2010
Fraction of high-methoxyl pectin in passion fruit	11.3	26.3	51.29	Yapo; Koffi, 2006
Fraction of chelator-soluble pectin in papaya *Solo*	49.48	82.22	411-430	Koubala et al., 2014
Wolf apple	33.7	77.15	177.76	Torralbo et al., 2012

Different types of natural sources of pectin have been studied, among them, traditional and other different and exotic such as: apple, citrus, banana, beetroot, cocoa, fig, mango, murta, mulberry, papaya, passion fruit, pomelo, pumpkin, sisal, palm, watermelon and others.

Apple

According to literature (Hang and Woodans, 1984; Wang et al., 2014), the primary by-product of the apple (*Malus domestica*) juice industry, apple pomace, is recognized as one of the main sources of commercial pectin. Wang et al. (2014) analyzed pectin properties after subcritical water treatment, and the maximum yield of apple pomace pectin was 16.68% at 150 °C. The neutral sugar content was significantly affected by the temperature at 170 ºC, and the percentages of them were: 4.45% of xylose, 1.90% of mannose and 40.94 % of glucose. At the same temperature, the highest value to the degree of methoxylation obtained was 89.69%. Regarding gel characteristics, the pectin extracted at 130 ºC presented the highest molecular weight and galacturonic acid content, which affected the pectin's attributes such as gelling and rheological properties (Lim et al., 2012).

Rha et al. (2011) obtained a yield of 9.5%, with a DM of 70.5% to pectin extracted from apple pomace. These values were lower than those observed by Wang et al. (2014). However, the neutral sugar content was quite different: arabinose (66.9%), glucose (32.6%) and galactose (30.8%). Physicochemical, functional properties and the yield of pectin extraction are dependent on the source and affected by the nature of the extraction process used (Shin and Hwang, 2002; Kumar and Chauhan, 2010).

The extraction of pectic substances from apple pomace has been extensively discussed in literature, including the works performed by Renard et al. (1990); Kratchanova et al. (1994); Renard et al. (1995); Joye and Luzio (2000); Sato et al. (2011); Zhang et al. (2013).

Citrus

In the same manner, citrus fruits, especially their albedo tissue, are largely studied because they are rich sources of pectic substances (Ralet and Thibault, 1994; Ros et al., 1996; Schröder et al., 2004; Liu et al., 2006; Yapo et al., 2007; Prabasari et al., 2011).

Schröder et al. (2004) studied the pectins from albedo of unripe Lemon (*Citrus limon* (L.) Burm. cv. Yen Ben) in fractioned form. The authors verified the high water-binding capacity of lemon pectin by microscopic techniques. Tightly packed cells with little intercellular space and thick cell walls were observed, and the authors concluded that the pectin of the cell wall was responsible for the high water-binding capacity. This property is dependent on factors such as: the sum of free negative charges in galacturonic acid, molecular size, presence and length of side chains (Pagán et al., 1999).

According to Bain (1958), the exceptionally high level of pectin in the lemon albedo throughout the cell walls from early development of fruit is a notable feature.

Prabasari et al. (2011) determined the composition of orange albedo alcohol insoluble solid. They reported that 85% of orange albedo was constituted of carbohydrate polymers, a higher or similar value when compared to literature (Brillouet et al., 1988; Ros et al., 1996;

Ralet and Thibault, 1994; Yapo et al., 2007). The orange albedo also presented galacturonic acid and glucose as main sugars as well as 73% of DM.

On the other hand, Wang et al. (2014) found 68.88% of galacturonic acid when the citrus peel extracted at 120 °C was evaluated. When the temperature was increased from 100 °C to 140 °C, the DM also increased, followed by a decrease. The highest DM of 74.74% was obtained at 120 °C. This DM value enables the classification of this pectin into high methoxyl pectin. These characteristics imply in the formation of gels that require higher sugar content and low pH (Liu et al., 2010). Regarding the gel formation, the extraction at 120 °C resulted in gel with increased hardness, strength, viscous force and stickiness. High DM, molecular weight and galacturonic acid content lead to more hydrogen bond formation and hydrophobic interactions to form a tight network in gel.

Dou Fu Chai

Dou fu chai is the common name of *Premna microphylla turcz*, a deciduous shrub that is widely distributed in the mountainous regions in the East, Central and South of China (Zhan et al., 2009). Its juice has been used to prepare a "green tofu" by local people (Wang et al., 2008).

Chen et al. (2014) studied the *Premna microphylla turcz* as a source of low methoxy pectin (LMP). The extraction was made according to the modified method of Taboada et al. (2010), using high temperature and acidic medium. The preparation of alcohol insoluble solids and various pectin fractions were studied. The pectin was fractioned in water-soluble pectin, oxalate-soluble pectin, acid-soluble pectin and alkali-soluble pectin.

Ammonium oxalate was found to be the most effective extracting agent, reflecting a high yield (20.61%), whereas the results from oxalate-soluble pectin presented high galacturonic acid content (76.15%), average molecular weight (980.67 kDa), low neutral sugar content (6.41%) and a low degree of methoxylation (14.90%).

Passion Fruit

Passion fruit is present in tropical regions throughout the world. One of the main edible species cultivated for a commercial purposes is yellow passion fruit (*P. edulis* f. *flavicarpa* Degener), commonly used for juice production, generating a great amount of discarded peels. Yapo and Koffi (2006) studied the preparation of alcohol insoluble solids and different pectin fractions from passion fruit: water-soluble pectin, chelating agent-extracted pectin and high-methoxyl pectin.

The yield of alcohol insoluble solid preparation represented 82.3% of the starting dried peel, and the principal sugars were glucose (30.8%) and xylose (12.3%). The high-methoxyl pectin yield was about 70%.

It was observed that the predominance of polysaccharides in the cell wall was related to its gelling capacity. The fraction chelating agent-extracted pectin presented greater gel strength, probably due to the high calcium content that is involved in the gelling process.

Passion fruit peel as a source of pectin was also studied by Seixas et al., (2014) which extracted pectin in acidic medium (using three different acids: nitric, acetic and tartaric) with the aid of microwave heating. The use of tartaric acid presented the highest pectin yield (15.32 - 30.29%).

Although the tartaric acid has provided the highest yields, pectin extracted with this agent presented undesirable features, such as low molar mass, in spite of the obtainment of a high degree of esterification.

Pomelo

Pomelo is a citrus fruit native to Southeast Asia that has been considered a potential pectin source according to Methacanon et al. (2013), who studied pomelo peels pectin extraction through experimental design. The DM value of pomelo peels pectins was higher than 50%, being considered high methoxyl pectin. The molecular weight was in the range of 440-645 kDa, depending on the acid used in the extraction. The samples were mainly composed of galacturonic acid and neutral sugars in small content. The authors concluded that pH was the most significant factor in the extraction, and the optimum conditions, when nitric acid was employed, were pH 2 at 90 °C for 90 minutes.

The results obtained suggested that the pomelo peels could be a potential source of pectin, decreasing the generation of waste.

Sisal

Sisal, *Agave sisalana Perrine*, is known worldwide as a source of hard fibers. The process of removing fibers from the sisal leaf generates 95% waste, which has been studied as a potential pectin source since some studies have indicated the presence of pectin in sisal leaves (Aspinall et al., 1958; Silva and Beltrão, 1999). Santos et al. (2013) investigated the aqueous extraction of pec-tin from sisal waste using the response surface methodology.

The author analyzed the influence of the liquid/solid ratio, temperature and extraction time in the pectin yield. The highest pectin yield (19.21%) was obtained when sisal waste was extracted at 85 °C for 60 minutes and using a liquid/solid ratio of 2%.

According to the authors, the yield from sisal waste was higher when compared with other noncommercial sources of pectins.

Palm

Asian Palmyra palm is found in Southeast Asia countries. The sugar of this plant can be obtained from young inflorescence (Yujaoren et al., 2008). Sugar palm is one of the most versatile palm species because almost all parts of the tree can be used (Siregar, 2005).

Yujaoren et al. (2008) evaluated the extraction of young, ripened sugar palm meat pectin by microwave and compared it to the conventional heating process.

By traditional method, the ripened sugar palm resulted in 20% yield (pH 2 and 80 °C), therefore, the optimum conditions used for microwave extraction was 800 W, pH 2 and 3 minutes, resulting in 19.6-23.5% of pectin.

Watermelon

Watermelon (*Citrullus lanatus*) peel is usually discarded (Al-Sayed and Ahmed, 2013) and contributes to environmental problems. The conversion of waste *C. lanatus* fruit peel into a product such as pectin, along with the optimization of the process, was the aim of Maran et al. (2013) who used the microwave to extract pectin. The use of experimental design improved the conditions of extraction, and four factors were evaluated: microwave power, irradiation time, pH and solid/liquid ratio. The highest yield (25.79%) was obtained using 477 W, 128 s; 1.52; 1:20.3 g/ml, respectively.

Zhang et al. (2008) explain that the effects are based on the fact that microwave irradiation accelerates cell rupture by increasing the temperature and pressure inside the plants' cells, favoring the exudation of pectin to the surrounding solvent.

Many authors that have studied the residues such as peel and pomace, as sources of pectin, concluded that these new sources have great potential of exploitation. This is a positive aspect in the environmental point of view, since by-products can be reused, minimizing their disposal in the environment besides adding value.

3. Traditional Physicochemical Method for Pectin Extraction

The production of pectin was initiated and developed during the twentieth century in Europe and the United States, mainly using the raw material of apple pomace (Kertesz, 1951). The extraction of pectin can be accomplished by aqueous acidic basic, with chelating agents or by the action of enzymes.

The basic process for extracting pectin results in a low degree of esterification, as a result of the saponification of ester groups, while the acid extraction process usually results in a high degree of pectin esterification (Joye and Luzio, 2000). The industrial production of commercial pectin may be considered a chemical hot extraction by hydrolysis with dilute acid, and the conditions vary depending on the feedstock and the desired characteristics from the extracted pectin (Sakai et al., 1993; Thakur 1997).

Currently, the industrial process for obtaining pectin is based on the extraction and solubilization of protopectin from apple pomace and peel of citrus fruits such as: lime, lemon, orange and grapefruit, and performed in slightly acidic conditions under heating (Oliveira et al., 2002; Thibault and Ralet, 2003). During the extraction, protopectin is transformed into soluble pectin in the initial phase, and these chains are broken down into smaller units. Commercial pectin powder can be classified as high methoxyl with a percentage of the groups esterified greater than 50%, being common between 50 and 75%, or low methoxy, with lower degree of esterification of 50%, usually between 20 and 45% (Sundar Raj et al., 2012; Willats et al., 2006).

Industrially, the method used to obtain pectins from agro-industrial waste of fruit juices is performed in acidic aqueous medium under heating (Thibault and Ralet, 2003), and the conditions are dependent upon the raw material. However, the process can be summarized in the following steps, starting with the extraction of vegetable pectin in acidified aqueous media, purification of the extracted liquid containing pectin and separation of the extract from pectin using precipitation (Christensen, 1984). Most of the water soluble pectin in the juice remains and is mainly composed of the remaining insoluble fraction.

The solubilization of this fraction involves physical and chemical processing, accompanied by the removal of neutral sugar side chain, as well as the hydrolysis of ester bonds (Voragen et al., 1995).

The hot extraction in acid medium is the method used industrially for the extraction of pectins from agro-industrial waste of fruit juices. Mineral acids are usually added to hot water, but organic acids can be used as alternatives such as tartaric, citric or lactic.

The parameters of the extraction process and varying factors like temperature, pH, time, and the type of acid can influence not only the yield of pectin, but also the chemical structure of the final product. They usually have pH in the range of 1.0-3.0 for 30 minutes to 6 hours with a range of temperatures of 80 - 90 ºC (Pagan et al., 2001; Levigne et al., 2002). The ratio of liquid and solid fraction dehydrated extraction is usually 1:18, with about 1:15 to 1:35 apple pomace and bagasse for citrus fruits. The viscosity increases with the concentration of pectin and the molecular weight.

Industrially, extracted pectin is typically separated from the pomace using hydraulic presses or a centrifugation process; afterwards the extract is filtered and concentrated (Sakai et al., 1993; Sundar Raj et al., 2012). Pectins with fast gelation, a greater degree of methoxylation of 70%, are normally extracted at pH 2.5 at 100 °C for 45 minutes. Pectins with medium or slow speed of gelation, methoxylated groups containing 60% to 70%, are extracted at lower temperatures for longer periods, because at low temperatures the de-esterification procedure is faster than depolymerization. The extract normally contains between 0.3 and 0.5 % pectin (Voragen et al., 1995).

When preparing the pectin powder, concentrated liquid extract is treated with organic solvents or certain metallic salts to precipitate the polymers (Sakai et al., 1993). The pectin is precipitated in alcohol concentrations higher than 45% (w/v). To minimize the volume of alcohol, the clarified extract can be concentrated up to 3-4% of pectin content. The precipitate obtained by the addition of alcohol is subsequently washed to remove contaminants in the form of sugars, phenolic compounds, pigments, heavy metals, residues of pesticides, acids, and other materials insoluble in alcohol (Voragen et al., 1995). The precipitated pectin is collected, dried and milled. In general, the stored pectins may have some depolimerizations and demethylations, a self-hydrolysis process, especially if the acid form is in pectin and the moisture content is above 5%. The pH stability is between 3.5 and 4.5 (Sundar Raj et al., 2012).

Yapo and Koffi (2014) studied the extraction of pectin from cashew apple pomace (*Anacardium occidentale L.*). Pectin was extracted from the insoluble material obtained by treating the pulp with boiling alcohol. Pectin extracted under different extraction with nitric acid, pH 1.0, 1.5 and 2.0 using a 1:25 ratio (w / v) fraction of dry and acid temperature of 75 °C for 90 minutes. It was found that pectin amount ranging from 10% - 25% could be extracted depending on the extraction force. The extracted pectin contained large amounts of galacturonic acid (69.9% - 84.5%) with some neutral sugars rham-nose (1.3% -2.5%), arabinose (2.6% - 5.4%) and galactose (4.7% - 8.6%). The degree of methoxylation ranged from 28% to 46% and was only slightly affected by the extraction force; thus, indicating the isolation of naturally low methoxyl pectins.

In terms of gelation ability, the extracted pectin yielded firmer gels with added calcium ions compared to the commercial citrus pectin containing low-methoxyl pectins. Therefore, cashew apple pomace appears to be potentially viable for the possible production of "non-chemical or enzymatically-tailored" pectins with a low-methoxyl source. Scabio et al. (2007) studied the influence of different conditions of time (3-37 minutes), temperature (63-97 °C) and a nitric acid concentration (8 - 92 mM) to extract pectin from dried apple pomace. Extraction was performed with a solid-liquid ratio of 1:20 (w/v) in 200 mL of nitric acid. The precipitation fraction of acid-soluble pectin was made with ethanol. The effects of these factors on the gravimetric yield and the degree of esterification were studied with a central composite experimental design. The selected coordinates of the center assay (20 min, 80 °C, 50 mM) conditions resulted in less damaged polysaccharides with a gravimetric yield of 9.05 g/100 g pectin from pulp, and a 74.39% degree of esterification.

All studied factors caused an increase in yield and a decrease in the degree of esterification. The reaction temperature may be used as an operating parameter to control the gravimetric yield of pectin with an expected degree of esterification. The set of extracted samples was analyzed by titration, and pectin classified as high methoxyl content with a larger fraction of acid that is usually found in commercial preparations, suggesting that the industrial conditions used are less severe.

Muñoz et al. (2008) used guava (*Psidium guajava L.*) for the extraction of pectin. The authors used a composite rotational design methodology to determine the extraction yield of pectin flour pulp and pulp with peel from the guava plant. The extraction was performed in 4 g of flour to 200 mL of a solution of citric acid at different concentrations and different times of extraction at the temperature of 97 °C. The best extraction conditions were: citric acid concentration of 5 g/100 g and extraction time of 60 minutes. The extraction of pectin with citric acid and alcoholic precipitation provided yield above 11% for flour pulp and guava pulp with peel. The pectins obtained had a degree of esterification below 50% were considered low esterification. However, the galacturonic acid content was close to commercial pectin. The authors stated that pectins extracted could be used in the gelation of foods with low sugar content, such as soluble dietary fiber, thickener and stabilizer in food.

Kliemann et al. (2009) extracted pectin from passion fruit peel (*Passiflora edulis flavicarpa*) using three different acids 0.5 M (citric, hydrochloric or nitric) at different temperatures (40–90 °C), pH (1.2–2.6) and extraction times (10–90 min), with and without skins using a 2^4 factorial design. Citric acid was the best acid for the extraction of pectin. Temperature, pH and extraction time had highly significant effects on the pectin yield. A central composite design with face centring was used to optimize the extraction process conditions for citric acid without skins. The best pectin yield (70%) was obtained from citric acid. The optimal conditions for maximization of pectin yield included the use of citric acid at 80 °C at pH 1 with an extraction time of 10. The extracted pectins with citric acid were rich in anhydrogalacturonic acid and had a low degree of methoxylation.

Sotanaphun et al. (2012) investigated the effects of temperature and pH on the extraction of pectin from the fruit peel of *Citrus maxima*. The most suitable condition was extraction at 80 °C without pH adjustment (pH was about 4.5) in 20 times by volume of water. Amberlite XAD-16 polystyrene was used to remove phenolic compounds before concentration and precipitation of pectin. This suggested that the method was simple and inexpensive.

The yield of the obtained pectin was 7.23±0.19% and the viscosity of 1% solution was 4.52±1.36 centipoise. Its galacturonic acid content and degree of esterification were 74.12±2.07% and 76.30±3.38%, respectively, indicating that the extracted pectin was high-methoxyl pectin.

Salam et al. (2012) studied lemon peel to extract pectin. Grounded lemon peel was digested in a solution of 0.01 N HCl at a temperature of 80-90°C for 1.5 hours. The solid mass was filtered out and the filtrate was treated with different low molecular weight alcohol such as methanol, ethanol and isopro-panol to precipitate the pectin.

The precipitate was dried at 40 °C under vacuum. The yield of pectin from fresh and dried lemon peel differed according to the precipitating agent used. The yield of the pectin was found in the range of 1.08 (ethanol and dried lemon peel) - 2.218% (isopropanol and fresh lemon peel).

Among the precipitating agents (95% ethanol, methanol, isopropanol), the yield was highest for ethanol. The pectin obtained by precipitation with 95% ethanol was used for characterization, and the degree of esterification for the isolated pectin was 88.624%.

In spite of the industrial production, commercial pectin can be considered a chemical extraction by hydrolysis with hot dilute mineral acid; the extraction of pectic substances is possible by simple dissolution in aqueous medium (Yapo and Koffi, 2006).

Canteri et al. (2010) studied two methods to extract pectin from pericarp of passion fruit (*Passiflora edulis*). The first acid extraction method was used with 50 mM nitric acid at 80 °C, or 20 minutes in the bleached raw material, and the second method was with aqueous extraction of fresh raw material. The study investigated the chemical composition and physical properties of pectin extracted by the aqueous process cold compared to the hot acid process.

The gravimetric yield of pectin extracted with hot nitric acid was about 200 g/kg. The yield of cold aqueous extraction was considerably lower than 29 g/kg, respectively. The sugar profile of pectin cold without added acid and extracted with hot acid yellow passion fruit proved to be very similar, with a predominance of glucose and galactose as major sugars.

Both pectins showed more than 65% of galacturonic acid. The analysis results indicate that bleaching had an important, positive role in the composition and physical properties of extracted pectin.

4. Alternative Methods for Pectin Extraction: Microwave and High Pressure Processing

The traditional extraction of this polysaccharide, thermal/acid extraction, can cause damage to the molecule structure, changing functional and physic-chemical properties of pectin (Koubala et al., 2008). Besides that, high energy consumption and a large amount of acidic residue formation are also dis-advantages faced by industries that apply this technique for pectin extraction (Panouillé et al., 2006). Therefore, alternative methods of pectin obtainment are studied in order to increase yield, reduce the degradation of this poly-saccharide, and improve its technological and biological properties.

Microwave assisted extraction (MAE) is an alternative pectin method of extraction with more benefits than a traditional method of extraction: shorter time, less solvent and lower costs (Prakash Maran et al., 2013). This technique has also shown favorable yield results, even better than traditional pectin extraction (Gong and Yang, 2004; Bélafi-Bakó et al., 2011; Jiang et al., 2012). The explanation is based on the technological features. The electromagnetic microwave energy is mainly converted into high water molecules vibration, heating the whole material (Kratchanova et al., 2004).

Plant cells' temperature increase, and consequently, the internal pressure increases as well, leading to the collapse of the cell and the release of pectin and other constituents (Zhongdong et al., 2006). As a result, the MAE process ensures greater and faster permeability of the extracting agent in the matrix, increasing the recovery of this compound.

Microwave power and exposure time are studied regarding their influence in quantitative (yield of extraction) and qualitative (degree of esterification, galacturonic acid content, molar weight) characteristic of pectin extraction (Wu et al., 2009). Bagherian et al. (2011), showed that for the pectin extraction of grapefruit, higher microwave power and heating-time produce higher yield, degree of esterification (DE) and galacturonic acid content (GalA). However, long heating-time started pectin degradation and decreased yield extraction. In another study, microwave power, extraction time and type of organic acid, significantly affected the characteristics of pectin from passion fruit and the yield of extraction. Higher yields were obtained in higher microwave power (628W) and heating-time (9 min). Passion fruit pectin extracted by the micro-wave process exhibits medium to high DE (50.00% to 64.56%) (Seixas et al., 2014).

High pressure processing (HPP), also known as high hydrostatic pressure (HHP), is a novel technology (Considine et al., 2008), which has been used for food preservation, reducing spoilage microorganism load, and undesirable enzyme activities (Bang and Chung, 2010). This treatment usually ranges from 100 MPa to 800 MPa (Xi, 2006), and the main advantage is the maintenance of sensory and nutritional characteristics, such as flavor, color and vitamins that could be modified or completely lost during thermal treatments, being minimally altered by high-pressure processing (Hendrickx et al., 1998).

The high pressure process can cause deprotonation of charged groups and disruption of salt bridges and hydrophobic bonds, leading to conformational changes like cellular wall, membrane and organelle collapse (Xi, 2013), improving the internal mass transfer in cells, and proving to be a fast and efficient method to extract bioactive and other natural compounds (Prasad et al., 2009a; Guo et al., 2012; Xi, 2013). Many studies demonstrated that the extraction of bioactive compounds by HPP provides better yields in less time, when compared to other methods (Zhang et al., 2007a; Zhang et al., 2007b; Adil et al., 2008; Corrales et al., 2009; Qadir et al., 2009; Prasad et al., 2009b; Prasad et al., 2009c; He et al., 2010; Lee et al., 2010). There are few studies about this new method for pectin extraction.

As in the HPP bioactive compounds extractions, the extraction of pectin by HPP shows good results, improving pectin technological application and extraction yields (Guo et al., 2012; Guo et al., 2014).

Figure 1 shows the procedure employed by the cited authors to extraction by HPP and its purification.

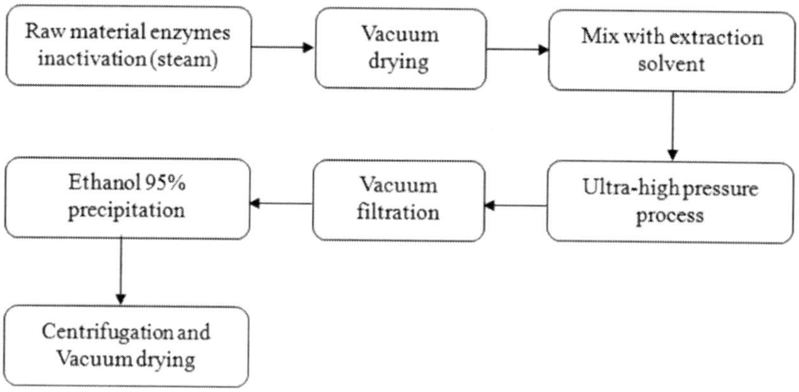

Guo et al., 2012; Guo et al., 2014.

Figure 1. Pectin HPP extraction and purification procedure.

Guo et al. (2012) performed the first study of HHP pectin extraction and its optimization. Pectin was extracted from fresh navel orange peel (*C. sinensis* Osbeck), and the extraction efficiency was evaluated by extraction yield and viscosity. The optimum conditions for pectin extraction were defined as pressure 500 MPa, temperature 55 ºC and pressure-holding time of 10 min. A comparison was then carried out for the same raw material, traditional thermal extraction (THE) and MAE. HHP showed better results for extraction yield and time of extraction (20.44% ± 0.64, 10 min) than THE and MAE extraction (15.47% ± 0.26, 60 min and 18.13% ± 0.23, 21min, respectively).

Guo et al. (2014) conducted a comparative study on emulsion stabilizing properties of pectins of honey pomelo (*Citrus grandis* Osbeck), by three methods: HHP, high-speed shearing homogenization (HSHE) and THE (traditional thermal extraction). The results showed that pectin extracted by the HHP process had the largest value of molecular weight and apparent viscosity of the solutions and emulsions formed (oil in water), minor emulsion particle size, and subsequently, emulsion with increased stability.

These studies show that HHP pectin extraction is a promissory technique, affording better technological application and yield extraction, when compared to pectin extracted by conventional heating techniques.

Therefore, more studies are required about this novel method of pectin extraction, with different raw materials.

As shown, it can be concluded that contrary from traditional pectin techniques of extraction, MAE and HPP are favorable alternative techniques for pectin obtainment, saving time and solvent consumption. Moreover, higher yield extraction and pectin with remarkable technological quality are achieved.

5. Biotechnological and Subcritical Water Extraction of Pectin

As reported in detail in this chapter, pectin is a very important poly-saccharide intensively used in food industries, which is commonly produced from citrus peel or fruit (Hoshino et al., 2009). Despite the fact that the use of strong acids results in high extraction yields and time-saving advantages, environmental problems including the disposal of acidic wastewater can be caused. Hence, thermal and/or mechanical treatments have been studied and applied to extract pectin.

They include ultrasound (Panchev et al., 1988) and autoclaving (Ooster-veld et al., 2000). In this context, the use of enzymes for pectin extraction has been studied as an environmentally friendly alternative. An enzyme-hydrolytic technology seems environmentally safe and more effective in terms of pectin yield (Ptichkina et al., 2008).

Among the enzymes, endo-polygalacturonase (Contreras-Esquivel et al., 2006), (hemi) cellulase (Shkodina et al., 1998; Zykwinska et al., 2008), and protease (Zykwinska et al., 2008) have been studied.

Lim et al. (2012) reported that the extraction of pectin from Yuza (*Citrus junos*) pomace combining physical and enzymatic treatment (Viscozyme® L) that is a multienzyme complex prepared from *Aspergillus aculeatus.* The enzymatic extraction resulted in a 7.3% of yield, producing pectin with 55% of galacturonic acid. Furthermore, the pectin obtained by this method showed a higher degree of esterification (46%) compared to chemically-extracted pectin (41%). The authors also observed less change in the pasting properties in the wheat flour–water system containing pectin prepared by enzymatic treatment.

In the work of Jeong et al. (2014), the extraction of pectin from rapeseed cake was carried out by a combination process consisting of a fat removal process and enzymatic hydrolysis using the commercial enzymes Celluclast and Alcalase, a cellulase and protease, respectively.

Different parameters such as enzymatic hydrolysis time, enzyme-rapeseed cake ratio, and Celluclast-Alcalase ratio were studied to evaluate degradation of rapeseed cake and pectin yield. When the hydrolysis condition reached 270 min of hydrolysis time or an enzyme-RSC ratio of 1:50, defatted rapeseed was suitably decomposed and the loss of liberated reducing sugars was minimized.

The authors observed that Alcalase led to the destruction of protein-carbohydrate complex, while Celluclasts lightly cleaved some linkages of carbohydrate. Therefore, when using the Celluclast-Alcalase ratio of 1:4, the highest pectin yield was obtained (6.85%).

Complex enzyme preparations have been studied for pectin extraction including the one obtained from *Aspergillus awamori* with cellulase, xylanase, β-glucosidase, endopolygalacturonase and pectinesterase activity (Ptichkina et al., 2008). This enzyme complex degraded cellulose and other insoluble constituents of the plant tissue, also showing some pectinesterase activity. This is very interesting since it allows the degree of esterification to be obtained, depending on the digestion time. When the 3h of hydrolysis was applied, a 53% degree of esterification was observed. Additionally, reduction in degree of esterification at longer times should yield pectin with a higher content of unesterified galacturonate residues.

Another important alternative for extracting pectins is the use of subcritical water, when the water is under subcritical temperatures and pressures with dielectric constant and the ion product is greatly changed (Teo et al., 2010), which according to studies has proven to be effective for pectin extraction from citrus peel.

The popularity of this solvent to extract a variety of organic compounds has grown over the last ten years, and hence, several works have reviewed the use subcritical water as an effective solvent, catalyst and reactant for hydro-lytic conversions and extractions (Carr et al., 2011).

This is an effective solvent for both polar and nonpolar compounds, and its versatility as a solvent is related to the tunable polarity of water, which is directly dependent upon the temperature. The polarity of water decreases when the temperature of water is increased.

Therefore, the solubility of nonpolar organics increases, and the solubility of polar organics decreases (Fernández-Prini et al., 1991). In this case, some studies show interesting results of pectin extraction using subcritical water as described below.

Pectin from citrus peel and apple pomace was extracted using subcritical water in the work of Wang et al. (2014). The best results observed were 21.95% and 16.68% of yield of extraction from citrus peel pectin and apple pomace pectin, respectively.

After the extraction, the endothermic properties of pectins were affected by extraction temperature, while exothermic properties were only affected by its constituents and raw materials.

The extracted pectins of both sources showed interesting bioactive potential including scavenging more than 60% DPPH radical and showing the highest proliferation inhibition rates of colon cancer cells.

Wang and Lü (2014) reported the optimization of extraction of pectic polysaccharides from apple pomace by subcritical water using response surface methodology. In optimal conditions, the levels of the parameters were obtained as follows: extraction temperature 140 °C, extraction time of 5 min, substrate: water ratio 1:14.

The results indicated that the pectic polysaccharides extracted from apple pomace were lower while ash content, endothermic transition temperature and fusion heat of the extracted when compared with commercial pectin, while the content of neutral sugars were higher in comparison with the same sample.

In addition, the extracted pectic polysaccharides showed higher in vitro antioxidant capability and inhibitory effect on HT-29 colon adenocarcinoma cells than commercial pectin.

Conclusion

Pectin may be found in many different sources, among them fruits and their wastes. As shown, the residues from the juice industry and the processing of other raw materials supply pectins that are suitable for specific applications. The type of polymer obtained depends on the source and extraction process.

Alternative processes have been studied to achieve good yields, since the traditional treatment presents great results considering the global quantity, but the heating combined to acidic conditions may result in hydrolysis of pectic substances.

In many cases, the combination of unconventional sources and processes improved the qualitative and quantitative characteristics of the extracted pectin including yield, degree of esterification, galacturonic acid content and viscosity.

References

Adil, I. H., Yener, M. E., Bayndrl, A. Extraction of total phenolics of sour cherry pomace by high pressure solvent and subcritical fluid and determination of the antioxidant activities of the extracts. *Separation Science and Technology,* 2008, 43, 1091–1110.

Al-Sayed, H. M., Ahmed, A. R. Utilization of watermelon rinds and sharlyn melon peels as a natural source of dietary fiber and antioxidants in cake. *Annals of Agricultural Sciences,* 2013, 58, 83-95.

Aspinall, G. O., Cañas-Rodriguez, A. Sisal pectic acid. *Journal of the Chemical Society,* 1958, 810, 4020-4027.

Bain, J. M. Morphological, anatomical, and physiological changes in the developing fruit of the Valencia orange, *Citrus sinensis* (L.) Osbeck. *Australian Journal of Botany*, 1958, 6, 1-23.

Bang, W. S., Chung, H. J. Effect of high hydrostatic pressure on the enzyme activities in *Saccharomyces cerevisiae* and *Escherichia coli. New Bio-technology,* 2010, 27, 440–444.

Bagherian, H., Zokaee Ashtiani, F., Fouladitajar, A., Mohtashamy, M., Comparisons between conventional, microwave- and ultrasound-assisted methods for extraction of pectin from grapefruit. *Chemical Engineering and Processing: Process Intensification,* 2011, 50, 1237–1243.

Bélafi-Bakó, K., Cserjési, P., Beszédes, S., Csanádi, Z., Hodúr, C. Berry Pectins: Microwave-Assisted Extraction and Rheological Properties. *Food and Bioprocess Technology,* 2011, 5, 1100–1105.

Barrera, A. M., Ramírez, J. A., González-Cabriales, J. J., Vázquez, M. Effect of pectins on the gelling properties of surimi from silver carp. *Food Hydrocolloids,* 2002, 16, 441-447.

Brillouet, J. M., Rouau, X., Hoebler, C., Barry, J. L., Carre, B., Lorta, E. A new method for determination of insoluble cell walls and soluble non-starchy polysaccharides from plant materials. *Journal of Agricultural and Food Chemistry*, 1988, 36, 969-979.

Canteri, M. H. G., Scheer, A. P., Ginies, C., Renard, C. M. G. C., Wosiacki, G. Importância do tratamento térmico na casca de maracujá para extração de pectina. *Revista Brasileira de Tecnologia Agroindustrial*, 2010, 4, 109-121.

Carr, A. G., Mammucari, R., Foster, N. R. A review of subcritical water as a solvent and its utilization for the processing of hydrophobic organic compounds. *Chemical Engineering Journal*, 2011, 172, 1–17.

Chen, J., Liang, R. H., Liu, W., Luo, S. J., Liu, C. M., Wu, S. S., Wang, Z. J. Extraction of pectin from *Premna microphylla turcz* leaves and its physicochemical properties. *Carbohydrate polymers*, 2014, 102, 376-384.

Christensen, S. H. Pectins. In: Glicksman, M., editor. *Food Hydrocolloids*. Boca Ratorn, FL: CRC Press; 1984.

Considine, K. M., Kelly, A. L., Fitzgerald, G. F., Hill, C., Sleator, R. D. High-pressure processing–effects on microbial food safety and food quality. *FEMS Microbiology Letters*, 2008, 289, 1–9.

Contreras-Esquivel, J. C., Voget, C. E., Vita, C. E., Espinoza-Perez, J. D., Renard, C. M. G. C. Enzymatic extraction of lemon pectin by endo-poly-galacturonase from *Aspergillus niger. Food Science and Biotechnology*, 2006, 15, 163–167.

Corrales, M., Avelina, F. G., Peter, B., Bernhard, T. Extraction of anthocyanins from grape skins assisted by high hydrostatic pressure. *Journal of Food Engineering*, 2009, 90, 415–421.

Domozych, D. S., Serfis, A., Kiemle, S. N., Gretz, M. R. The structure and biochemistry of charophycean cell walls: I. Pectins of *Penium margaritaceum. Protoplasma,* 2007, 230, 99-115.

Fernández-Prini, R. J., Corti, H. R., Japas, M. L. *High-Temperature Aqueous Solutions: Thermodynamic Properties*. Boca Raton, FL: CRC Press, 1991.

Gong, S. Z., Yang, Z. R. Optimization on micro wave-assisted extraction of pectin from shaddock peel. *Linchan Huaxue Yu Gongye/Chemistry and Industry of Forest Products*, 2004, 24, 87–90.

Guo, X., Han, D., Xi, H., Rao, L., Liao, X., Hu, X., Wu, J. Extraction of pectin from navel orange peel assisted by ultra-high pressure, microwave or traditional heating: A comparison. *Carbohydrate Polymers*, 2012, 88, 441–448.

Guo, X., Zhao, W., Pang, X., Liao, X., Hu, X., Wu, J. Emulsion stabilizing properties of pectins extracted by high hydrostatic pressure, high-speed shearing homogenization and traditional thermal methods: A comparative study. *Food Hydrocolloids*, 2014, 35, 217–225.

Hang,Y. D., Woodams, E. E. Apple pomace: a potential substrate for citric acid production by *Aspergillus niger. Biotechnology Letters*, 1984, 6, 763-764.

He, X., Kim, S. S., Park, S. J., Seong, D. H., Yoon, W. B., Lee, H. Y., Park, D. S., Ahn, J. Combined effects of probiotic fermentation and high-pressure extraction on the antioxidant, antimicrobial, and antimutagenic activities of Deodeok (*Codonopsis lanceolata*). *Jounal of Agricultural and Food Chemistry,* 2010, 58, 1719–1725.

Hendrickx, M., Ludikhuyze, L., Van Den Broeck, I., Weemaes, C. Effects of high pressure on enzymes related to food quality. *Trends In Food Science and Technology*, 1998, 9, 197–203.

Hoshino, M., Tanaka, M., Terada, A., Sasaki, M., Goto, M. Characteristics of pectin extracted from citrus peel using subcritical water. *Journal of Bio-science and Bioengineering,* 2009, 108, Supplement 1, S144–S145.

Jeong, H.-S., Kim, H.-Y., Ahn, S.-H., Oh, S.-C., Yang, I., Choi, I.-G. Optimization of enzymatic hydrolysis conditions for extraction of pectin from rapeseed cake (*Brassica napus* L.) using commercial enzymes. *Food Chemistry,* 2014, 157, 332-338.

Jiang, L. N., Shang, J. J., He, L. B., Dan, J. M. Comparisons of microwave-assisted and conventional heating extraction of pectin from seed water-melon peel. *Advanced Materials Research*, 2012, 550-553, 1801–1806.

Joye, D. D., Luzio, G. A. Process for selective extraction of pectins from plant material by differential pH. *Carbohydrate Polymers*, 2000, 43, 337-342.

Kratchanova, M., Panchev, I., Pavlova, E., Shtereva, L. Extraction of pectin from fruit materials pretreated in an electromagnetic field of super-high frequency. *Carbohydrate Polymers*, 1994, 25, 3, 141-144.

Kratchanova, M., Pavlova, E., Panchev, I. The effect of microwave heating of fresh orange peels on the fruit tissue and quality of extracted pectin. *Carbohydrate Polymers*, 2004, 56, 181-185.

Kertesz, Z. I. *The pectic substances*. New York: Interscience; 1951.

Kliemann, E., Simas, K. N., Amante, E. R., Prudêncio, E. S., Teófilo, R. F., Ferreira, M. M., C., Amboni, R. D. M. C. Optimization of pectin acid extraction from passion fruit peel (*Passiflora edulis* flavicarpa) using response surface methodology. *International Journal of Food Science and Technology*, 2009, 44, 476-483.

Koubala, B. B., Mbome, L. I., Kansci, G., Tchouanguep Mbiapo, F., Crepeau, M. J., Thibault, J. F., Ralet, M. C. Physicochemical properties of pectins from ambarella peels (*Spondias cytherea*) obtained using different extraction conditions. *Food Chemistry*, 2008, 106, 1202-1207.

Koubala, B. B., Christiaens, S., Kansci, G., Van Loey, A. M., Hendrickx, M. E. Isolation and structural characterization of papaya peel pectin. *Food Research International*, 2014, 55, 215-221.

Kumar, A., Chauhan, G. S. Extraction and characterization of pectin from apple pomace and its evaluation as lipase (steapsin) inhibitor. *Carbo-hydrate Polymers*, 2010, 82, 454-459.

Kurita, O., Fujiwara, T., Yamazaki, E. Characterization of the pectins extracted from citrus peel in the presence of citric acid. *Carbohydrate Polymers,* 2008, 74, 725-730.

Lee, H. Y., He, X., Ahn, J. Enhancement of antimicrobial and antimutagenic activities of Korean barberry (*Berberis koreana* Palib.) by the combined process of high-pressure extraction with probiotic fermentation. *Journal of the Science of Food and Agriculture,* 2010, 90, 2399–2404.

Lévigne, S., Ralet, M. C., Thibault, J. F. Characterization of pectins extracted from fresh sugar beet under different conditions using an experimental design. *Carbohydrate Polymers*, 2002, 19, 145-153.

Liang, R.-H., Chen, J., Liu, W., Liu, C.-M., Yu, W., Yuan, M., et al. Extraction, characterization and spontaneous gel-forming property of pectin from creeping fig (*Ficus pumila* Linn.) seeds. *Carbohydrate Polymers,* 2012, 87, 76–83.

Lim, J., Yoo, J., Ko, S., Lee, S. Extraction and characterization of pectin from Yuza (*Citrus junos*) pomace: A comparison of conventional-chemical and combined physical–enzymatic extractions. *Food Hydrocolloids*, 2012, 29, 160-165.

Liu, L., Cao, J., Huang, J., Cai, Y., Yao, J. Extraction of pectins with different degrees of esterification from mulberry branch bark. *Bioresource Techno-logy,* 2010, 101, 3268-3273.

Liu, Y., Shi, J., Langrish, T. A. G. Water-based extraction of pectin from flavedo and albedo of orange peels. *Chemical Engineering Journal*, 2006, 120, 203-209.

MacDougall, A. J., Ring, S. G. Pectic polysaccharides. In: Tomasik, P., editor. *Chemical and Functional Properties of Food Saccharides*. Boca Raton, FL: CRC Press; 2004; 181-195.

Maran, P. J., Sivakumar, V., Thirugnanasambandham, K., Sridhar, R. Micro-wave assisted extraction of pectin from waste *Citrullus lanatus* fruit rinds. *Carbohydrate polymers*, 2014, 101, 786-91.

May, C. D. Industrial pectins: sources, production, and applications. *Carbohydrate Polymers,* 1990, 12, 79-99.

May, C. D. Pectins. In: Phillips, G. O., Williams, P. A., editor. *Handbook of Food Hydrocolloids.* Cambridge, England: Woodhead Publishing Limited; 2000; 1-472.

Mesbahi, G., Jamalian, J., Farahnaky, A. A comparative study on functional properties of beet and citrus pectins in food systems. *Food Hydrocolloids*, 2005, 19, 4, 731-738.

Methacanon, P., Krongsin, J., Gamonpilas, C. Pomelo (*Citrus maxima*) pectin: Effects of extraction parameters and its properties. *Food Hydrocolloids*, 2014, 35, 383-391.

Oosterveld, A., Beldman, G., Schols, H. A., Voragen, A. G. J. Characterization of arabinose and ferulic acid rich pectic polysaccharides and hemi-celluloses from sugar beet pulp. *Carbohydrate Research*, 2000, 328, 185–197.

Oliveira, L. F., Nascimento, M. R. F., Borges, S. V., Ribeiro, P. C. N., Ruback, V. R. Aproveitamento alternativo da casca do maracujá-amarelo (*Passiflora edulis* f. flavicarpa) para produção de doce em calda. *Ciência e Tecnologia de Alimentos*, 2002, 22, 259-262.

Pagán, J., Ibarz, A., Llorca, M., Coll, L. Quality of industrial pectin extracted from peach pomace at different pH and temperatures. *Journal of the Science of Food and Agriculture,* 1999, 79, 1038-1042.

Pagán, J., Ibarz, A., Llorca, M., Pagán, A., Barbosa-Cánovas, G. V. Extraction and characterization of pectin from stored peach pomace. *Food Research International,* 2001, 34, 605-612.

Panchev, I., Kirchev, N., Kratchanov, C. Improving pectin technology. *Inter-national Journal of Food Science and Technology*, 1988, 23, 337–341.

Panouillé, M., Thibault, J. F., Bonnin, E. Cellulase and protease preparations can extract pectins from various plant byproducts. *Journal of Agricultural and Food Chemistry,* 2006, 54, 8926–8935.

Prabasari, I., Pettolino, F., Liao, M. L., Bacic, A. Pectic polysaccharides from mature orange (*Citrus sinensis*) fruit albedo cell walls: Sequential extraction and chemical characterization. *Carbohydrate Polymers*, 2011, 84, 484-494.

Prakash Maran, J., Sivakumar, V., Thirugnanasambandham, K., Sridhar, R. Optimization of microwave assisted extraction of pectin from orange peel. *Carbohydrate Polymers*, 2013, 97, 703–709.

Prasad, K. N., Yang, B., Zhao, M., Wei, X., Jiang, Y., Chen, F. High pressure extraction of corilagin from longan (*Dimocarpus longan* Lour.) fruit pericarp. *Separation and Purification Technology*, 2009a, 70, 41–45.

Prasad, K. N., Yang, B., Zhao, M., Ruenroengklin, N., Jiang, Y. M. Applica-tion of ultrasonication or high-pressure extraction of flavonoids from Litchi fruit pericarp. *Journal of Food Process Engineering*, 2009b, 32, 828–843.

Ptichkina, N. M., Markina, O. A., Rumyantseva, G. N. Pectin extraction from pumpkin with the aid of microbial enzymes. *Food Hydrocolloids*, 2008, 22, 192–195.

Qadir, S. A., Kwon, M. C., Han, J. G., Ha, J. H., Chung, H. S., Ahn, J., Lee, H. Y. Effect of different extraction protocols on anticancer and antioxidant activities of berberis koreana bark extracts. *Journal of Bioscience and Bio-engineering*, 2009, 107, 331–338.

Ralet, M. C., Thibault, J. F. Extraction and characterization of very highly methylated pectins from lemon cell walls. *Carbohydrate Research*, 1994, 260, 283-296.

Renard, C. M. G. C., Voragen, A. G. J., Thibault, J. F., Pilnik, W. Studies on apple protopectin. I. Extraction of insoluble pectin by chemical means. *Carbohydrate Polymers*, 1990, 12, 9-25.

Renard, C. M. G. C., Lemeunier, C., Thibault, J. F. Alkaline extraction of xyloglucan from depectinised apple pomace: Optimisation and characterization. *Carbohydrate Polymers*, 1995, 28, 209-216.

Rha, H. J., Bae, I. Y., Lee, S., Yoo, S. H., Chang, P. S., Lee, H. G. Enhancement of anti-radical activity of pectin from apple pomace by hydroxamation. *Food Hydrocolloids*, 2011, 25, 545-548.

Ros, J. M., Schols, H. A., Voragen, A. G. J. Extraction, characterization, and enzymatic degradation of lemon peel pectins. *Carbohydrate Research,* 1996, 282, 271–284.

Sakai, T., Sakamoto, T., Hallaert, J., Vandamme, E. Pectin, pectinase and protopectinase: production, properties and applications. *Advances in Applied Microbiology,* 1993, 39, 213-294.

Salam, M. A., Jahan, N., Islam, M. A., Hoque, M. M. Extraction of pectin from lemon peel: technology development. *Journal of Chemical Engineering*, 2012, 27, 25-30.

Santos, J. D. G., Espeleta, A. F., Branco, A., Assis, S. Aqueous extraction of pectin from sisal waste. *Carbohydrate polymers*, 2013, 92, 1997-2001.

Sato, M. F., Rigoni, D. C., Canteri, M. H. G., Petkowicz, C. L. O., Nogueira, A., Wosiacki, G. Chemical and instrumental characterization of pectin from dried pomace of eleven apple cultivars. *Acta Scientiarum Agronomy*, 2011, 33, 383-389.

Scabio, A., Fertonani, H. C. R., Schemin, M. H. C., Petkowicz, C. L. O., Carneiro, E. B. B., Nogueira, A., Wosiacki, G. A model for pectin extraction from apple pomace. *Brazilian Journal of Food Technology*, 2007, 10, 259-265.

Schieber, A., Hilt, P., Steker, P., Endre, B., H.-U.; Rentschler, C. A new process for the combined recovery of pectin and phenolic compounds from apple pomace. *Innovative Food Science and Emerging Technologies*, 2003, 4, 99-107.

Schröder, R., Clark, C. J., Sharrock, K., Hallett, I. C., MacRae, E. Pectins from the albedo of immature lemon fruitlets have high water binding capacity. *Journal of Plant Physiology*, 2004, 161, 371-379.

Seixas, F. L., Fukuda, D. L., Turbiani, F. R. B., Garcia, O. S., Petkowicz, C. L. D. O., Jagadevan, S., Gimenes, M. L. Extraction of pectin from passion fruit peel (*Passiflora edulis* f. *flavicarpa*) by microwave-induced heating. *Food Hydrocolloids*, 2014, 38, 186–192.

Shin, H. H., Hwang, J. K. Modeling of rheological properties of pectins by side branches. *Korean Journal of Food Science and Technology*, 2002, 34, 583-589.

Shkodina, O. G., Zeltser, O. A., Selivanov, N. Y., Ignatov, V. V. Enzymic extraction of pectin preparations from pumpkin. *Food Hydrocolloids*, 1998, 12, 313–316.

Silva, J. A. L., Rao, M. A. Pectin: structure, functionality, and uses. In: Stephen, A. M., Phillips, G. O., Williams, P. A., editor. *Food polysaccha-rides and their applications*, New York: Taylor and Francis Group; 2006; 353-411.

Silva, O. R. R., Beltrão, N. E. M. O. *Agronegócio do sisal no Brasil.1*. Brazil: Brasília; Embrapa-CNPA, 1999.

Siregar, J. P. Tensile and flexural properties of Arenga Pinnata filament (*Ijuk Filament*) reinforced epoxy composites. *MS thesis*. Universiti Putra Malaysia, 2005.

Sotanaphun, U., Chaidedgumjorn, A., Kitcharoen, N., Satiraphan, M., Asavapichayont, P., Sriamornsak, P. Preparation of pectin from fruit peel of citrus maxima. *Silpakorn University Science and Technology Journal*, 2012, 6, 1.

Sundar Raj, A. A., Rubila, S., Jayabalan, R., Ranganathan, T. V. A review on pectin: chemistry due to general properties of pectin and its pharmaceuti-cal uses. *Open Access Scientific Reports*, 2012, 1, 12.

Taboada, E., Fisher, P., Jara, R., Zúñiga, E., Gidekel, M., Cabrera, J. C., Cabrera, G. Isolation and characterization of pectic substances from murta (*Ugni molinae turcz*) fruits. *Food Chemistry*, 2010, 123, 669-678.

Teo, C. C., Tan, S. N., Yong, J. W., Hew, C. S., Ong, E. S. Pressurized hot water extraction (PHWE). *Journal of Chromatography A*, 2010, 1217, 2484–2494.

Thakur, B. R., Singh, R. K., Handa, A. K. Chemistry and uses of pectin - a review. *Critical Reviews in Food Science and Nutrition*, 1997, 37, 47-73.

Thibault, J. F., Ralet, M. C. Physico-chemical properties of pectins in the cell walls after extraction. In: Voragen, F., Schols, H., Visser, R., editors. *Advances in Pectin and Pectinase Research*, Dordrecht: Kluwer Academic Publishers; 2003; 91-105.

Torralbo, D. F., Batista, K. A., Di-Medeiros, M. C. B. Fernandes, K. F. Extraction and partial characterization of *Solanum lycocarpum* pectin. *Food Hydrocolloids*, 2012, 27, 378-383.

Visser, J., Voragen, A. G. J. *Progress in biotechnology: Pectin and pectinases*. 14. The Netherlands: Elsevier; 1996.

Voragen, G. J., Pilnik, W., Thibault, J. F., Axelos, M. A. V., Renard, C. M. G. C. Food polysaccharides and their applications. In: Stephen, A. M., editor, *Pectins*. New York: Marcel Dekker Inc.; 1995; 287-339.

Voragen, A. G., Coenen, G. J., Verhoef, R. P., Schols, H. A. Pectin, a versatile polysaccharide present in plant cell walls. *Structural Chemistry*, 2009, 20, 263-275.

Walter, R. H. Analytical and graphical methods. In: Walter, R. H., editor. *The chemistry and technology of pectin*, New York: Academic Press Inc; 1991; 190-218.

Wang, C. W., Xu, W., Wang, W. G., Zhang, J. F., Wu, Y. X., Chi, R. A. Kinetics of pectin extraction from *Premna microphylla* turcz leaves. *Chemistry and Industry of Forest Products*, 2008, 28, 16-20.

Wang, X., Chen, Q., Lü, X. Pectin extracted from apple pomace and citrus peel by subcritical water. *Food Hydrocolloids*, 2014a, 38, 129-137.

Wang, X., Lü, X. Characterization of pectic polysaccharides extracted from apple pomace by hot-compressed water. *Carbohydrate Polymers*, 2014b, 102, 174–184.

Willats, W. G., McCartney, L., Mackie, W., Knox, J. P. Pectin: cell biology and prospects for functional analysis. *Plant Molecular Biology*, 2001, 47, 9–27.

Willats, W. G., Knox, J. P., Mikkelsen, J. D. Pectin: new insights into an old polymer are starting to gel. *Trends in Food Science and Technology*, 2006, 17, 97-104.

Yapo, B. M., Koffi, K. L. Extraction and characterization of highly gelling low methoxy pectin from cashew apple pomace. *Foods*, 2014, 3, 1-12.

Yapo, B. M., Koffi, K. L. Yellow passion fruit rind - a potencial source of low-methoxyl pectin. *Journal of Agricultural and Food Chemistry*, 2006, 54, 2738-2744.

Yapo, B. M., Lerouge, P., Thibault, J. F., Ralet, M. C. Pectins from citrus peel cell walls contain homogalacturonans homogenous with respect to molar mass, rhamnogalacturonan I and rhamnogalacturonan II. *Carbohydrate Polymers*, 2007, 69, 426-435.

Yujaroen, P., Subjaroenkul, U., Rungrodnimitchai, S. Extraction of Pectin from Sugar Palm Meat. *Thammasat International of Science and Techno-logy*, 2008, 13, 44-47.

Xi, J. Application of high hydrostatic pressure processing of food to extracting lycopene from tomato paste waste. *High Pressure Research*, 2006, 26, 33–41.

Xi, J. High-pressure processing as emergent technology for the extraction of bioactive ingredients from plant materials. *Critical Reviews in Food Science and Nutrition*, 2013, 53, 837–852.

Zhan, Z. J., Tang, L., Shan, W. G. A new triterpene glycoside from *Premna microphylla*. *Chemistry of Natural Compounds*, 2009, 45, 197-199.

Zhang, B., Yang, R., Liu, C. Z. Microwave-assisted extraction of chlorogenic acid from flower buds of *Lonicera japonica* Thunb. *Separation and Puri-fication Technology*, 2008, 62, 480-483.

Zhang, L., Ye, X., Ding, T., Sun, X., Xu, Y., Liu, D. Ultrasound effects on the degradation kinetics, structure and rheological properties of apple pectin. *Ultrasonics Sonochemistry*, 2013, 20, 222-231.

Zykwinska, A., Boiffard, M.-H., Kontkanen, H., Buchert, J., Thibault, J. F., Bonnin, E. Extraction of green labeled pectins and pectic oligosaccharides from plant byproducts. *Journal of Agricultural and Food Chemistry*, 2008, 56, 8926–8935.

In: Pectin: Chemical Properties, Uses and Health Benefits
Editor: Phillip L. Bush
ISBN: 978-1-63321-438-5

Chapter 4

PECTIN: AN EFFICIENT MATRIX FOR CELL AND ENZYME IMMOBILIZATION

Fabiano Jares Contesini**, Ricardo Rodrigues de Melo, Danielle Branta Lopes, Jose Valdo Madeira Junior, Haroldo Yukio Kawaguti and Elaine Berger Ceresino

Laboratory of Biochemistry. Department of Food Science,
State University of Campinas, Campinas, SP, Brazil

ABSTRACT

Pectins are polysaccharides containing D-galacturonic acid and galacturonic acid with methyl ester residues that can be acetylated to some degree. This biopolymer has been used as a gelling agent for the last two centuries and is extensively applied in food and pharmaceutical industries. In this case, pectins with a methylation degree lower than 50%, called low-methoxyl pectin (LMP), form gel in the presence of calcium ions, and hence, may be used as a gelling agent in numerous types of products such as: low-calorie jams and jellies, confectionery jelly products, and other food applications. However, one highlighted use of LMP is for the entrapment, encapsulation or immobilization of enzymes and cells for biotechnological applications. The encapsulation of a lipase in pectin gels cross-linked with calcium ions brought three to four times more enzymatic activity in water miscible organic co-solvents compared with aqueous systems. In another study, α-amylase and glucoamylase enzymes were immobilized to pectin by covalent binding showing greater thermal and pH stability over the free enzyme system with the complete retention of original activities. The immobilized enzymes showed the highest release of glucose compared with free enzymes when applied in starch hydrolysis. Another important use of LMP is in the entrapment of microbial cells for biocatalytic/ bio-transformation and fermentation uses. When the cells of the *Nocardia tartaricans* bacterial strain were immobilized in pectate gel to obtain L-tartrate, higher *cis*-epoxysuccinate hydrolase activity was observed compared with the free cells. An

* Corresponding author: Fabiano Jares Contesini. Address: Laboratório de Bioquímica. Departamento de Ciência de Alimentos - FEA, Universidade Estadual de Campinas. Rua Monteiro Lobato, 80. Cx. Postal 6121. 13083-862. Campinas-SP, Brasil. Tel./fax: +55 19 3521 2175, e-mail: fabiano.contesini@gmail.com.

additional study reports the immobilization of *Saccharomyces cerevisiae* cells in pectin gel for ethanol produc-tion, indicating that no significant changes occurred. Cells maintained their growth capacity, and the beads could be reutilized several times in successive batch fermentations, which is one of the major advantages of cell immobilization. The uses of pectin will be reviewed in this chapter since different high-added-value-compounds can be obtained showing the remarkable relevance of this matrix for biocatalysts immobilization.

Keywords: Pectin, enzyme immobilization, cell immobilization, gelification

1. INTRODUCTION

Pectins or pectic substances are collectively known as one of the main plant cell wall components, contributing to tissue integrity and rigidity, and are probably the most complex macromolecules in nature. Pectins are composed of heteropolysaccharides, predominantly containing galacturonic acid residues which may present methyl esterified. In general, these polysaccharides can be defined as a chain structure of axial-axial α-(1,4)-linked D-galacturonic acid units, containing rich regions of L-rhamnose, mainly with arabinose, galactose and xylose as side chains (Voragen et al., 2009; Jolie et al., 2010; DiCosimo et al., 2013).

The degree of esterification (DE) of pectins, which corresponds to the ratio of esterified galacturonic acid units to total galacturonic acid units, has an expressive influence on their properties. Depending on the DE, pectins can be divided into two major groups: high methoxyl pectins (HMP) and low metho-xyl pectins (LMP). Most pectins have degrees of esterification of about 50–80% (high metho-xyl pectins, DE > 50%). If the degree of esterification is lower than 50% (low-methoxyl pectins, DE < 50%), these compounds behave like a completely new family of polymers. In this way, pectins can form two types of gels depending on their degree of esterification. HMP will form gels in acid pH and in the presence of high concentrations of sugar (e.g. sucrose or glucose); while LMP require a divalent cation such as calcium (Fraeye et al., 2010; Videcoq et al., 2011; Mishra et al., 2012).

Moreover, pectin molecules are regarded as safe products for human consumption and have been used successfully for many years in food and pharmaceutical industries (Sriamornsak et al., 2010). In the food industry, pectins are presented as a high-value functional food ingredient, widely used as a gelling agent and stabilizer. Commercially, this bio-polymer is known primarily as a gelling agent and is widely used in the production of jams and jellies, low-calorie jams, fruit juice, confectionary products and bakery fillings. In addition, it is used as a stabilizer in acidified milk drinks and yogurts, or as thickener to improve the texture of sauces (Willats et al., 2006). The other major use of pectins is in the pharmaceutical industries as an effective agent for drug delivery. Among natural polymers, pectin has interesting properties for drug delivery applications, such as the mucoadhesiveness, the ease of dissolution in basic environments and the ability to form gels in acid environments (Sriamornsak et al., 2010; Munarin et al., 2012).

Immobilization techniques are methods with great potential both scientifically and industrially because of their broad technological and economic importance. For industrial purposes, immobilization of biocatalysts (enzyme and/or cells) offers several advantages, including reusability, easy product separation and enhancement of enzyme stability (Dolejš et al., 2014). Immobilization procedures are relatively simple; the materials used are bio-compatible, widely available, and acceptable for use in food and pharmaceutical applications. Various natural and synthetic polymers can be used for this purpose, but natural materials, mainly polysaccharides, are the materials of choice, due to the advantage of being nontoxic, biocompatible, relatively stable and bio-degradable (Yunyu et al., 2004). In this context, pectins, which are polysaccharides commonly found in nature, have been described by several works in literature as an important and effective matrix for the entrapment of cells and other biocatalysts (Giordano et al., 2008; Voo et al., 2011; Contesini et al., 2012). Table 1 reports the use of pectin for encapsulation of enzymes, cells and other compounds.

Table 1. Uses of pectin for encapsulation/immobilization of different compounds

Support	Material encapsulated	Use	Industrial field	References
Chitosan microgels coated with pectin layers	5-Fluorouracil	Oral and topical chemotherapy	Pharmaceutical industry	Puga et al. (2013)
Pectin beads	Virulent factor Cwp84	Oral vaccine against *Clostridium difficile* infection	Pharmaceutical industry	Sandolo et al. (2011)
Pectin–whey protein microparticles	*Lactobacillus acidophilus* La5	Prebiotic	Food industry	Gebara et al. (2013)
PVA-pectin cryogels	Keratinase	Treatment of wounds and eschars	Pharmaceutical industry	Martínez et al. (2013)
Pectin microcapsules	Glucosyltransferase	Conversion of sucrose into isomaltulose	Food industry	Contesini et al. (2012)
Pectin gel	Glucoamylase and *Saccharomyces cerevisiae*	Ethanol production	Fuel industry	Giordano et al. (2008)

Thus, this chapter has the purpose of summarizing the research conducted in recent years on the application of pectins as efficient supports for cell and enzyme immobilization. The use of these polysaccharides as a matrix will be reviewed with the aim of studying the immobilization of different biocatalysts, which are applied to produce compounds with high-added-value.

2. Extraction of Pectins

Pectin is a major polysaccharide in cell walls with great applications in industries of different segments, mainly the pharmaceutical and food sectors. Its importance in the food sector lies in its ability to form gel in the presence of Ca^{2+} ions or a solute at low pH (Thakur et al., 1997). It is presented in primary cell walls and in the middle lamellae of plants, where it helps to bind cells together by the regulation of intercellular adhesion. Pectic substances are usually associated with other cell wall components such as cellulose, hemi-cellulose and lignin (Willats et al., 2001). They can be found especially in fruits and young tissues, from where they are usually extracted for commercial purposes. The general biochemical definition of pectin is that it is a group of polyssacharides that are rich in galacturonic acid (GalA) and often displays different degrees of methyl esterification involving the C-6 carboxyl groups (Domozych et al., 2007).

The main sources for commercial pectin production are apple pomace and citrus peels, both by-products from juice or cider manufacturing. Apple pomace contains 10-15% of pectin on a dry matter basis, whilst citrus peel contains 20-30%. The chemical characteristics of both pectins are similar and equivalent from the application point of view (Sriamornsak, 2003). The production of pectin from food industry by-products are considered beneficial from both an economic and ecological perspective (Schieber et al., 2003).

Sugar beet pulp is considered a promising source for pectin extraction considering its low value after sugar refining. It contains 15–30% pectin in dry weight (Lv et al., 2013) and satisfying properties when applied as a thickener or as an agent to increase viscosity in fluid products. However, when compared with commercial citrus pectin, it does not have the ability to form firm gels in food (Mesbahi et al., 2005).

Many others agricultural by-products or wastes have been studied, such as cacao pod husks (Chan and Choo, 2013), peach pomace (Faravash and Ashtiani, 2008) and passion fruit peel (Seixas et al., 2014). Nevertheless, these sources present discrepancies in satisfying the complete industrial requirements in terms of yield and functional properties.

Severe extraction processes are used at the industrial level of pectin production, which are frequently detrimental to pectin structure. The connection between pectin and other polymer components in the cell wall inhibits their release from the cell matrix, and preprocessing of the plant material is used to facilitate the extraction (Kratchanova et al., 2004). Pre-treatments such as blanching, washing and drying are applied before extraction of the pectin from the raw material in order to inactivate enzymes that would increase the rate of degradation of pectin molecules. The pre-treatments of the raw material also increase stability during transportation and storage (Stephen and Phillips, 2010). Kratchanova et al. (1994) pre-treated orange, lemon and apple wastes in an electromagnetic field of super-high frequency and concluded that fresh pectinous raw materials subjected to this pre-treatment before drying presented higher pectin yield, in addition to higher values for degree of esterification and gel strength. The explanation for these phenomena lies first in the partial disintegration of the plant tissue and hydrolysis of protopectin, and second in the rapid inactivation of pectolytic enzymes.

Acid treatment using hydrochloric acid, nitric acid or, less common, sulfuric acid (pH 1.5 to 3) at high temperatures (70 - 80 °C) is the most common extraction method used in industries. The precise conditions for extraction vary according to the raw material and the type of pectin desired, as well as the manufacturer's facilities, in order to obtain an efficient process (Joye and Luzio, 2000; Faravash and Ashtiani, 2008). The separation of the hot pectin extract is a critical step because the solids in the liquid phase form a viscous solution. The viscosity is affected by the pectin concentration and its molecular weight. Moreover, the pectin extract may be further clarified by filtration through a filter aid and concentrated under a vacuum. The commercial pectin is usually sold in a powder form that can be produced by mixing the concentrated liquid from either apple or citrus with an alcohol, usually ethanol in a concentration higher than 45%. The pectin is separated as a stringy gelatinous mass, which is pressed and washed to remove the mother liquor, dried and ground. This process yields pectin of around 70% esterification (or methoxylation) (Canteri et al., 2012).

3. Immobilization of Enzymes Using Pectin and Derivatives

Enzymes are natural catalysts that are extremely important not only for their environmentally friendly behavior (physiological pH and temperature) but also for their great uses in several industries such as food, dairy, pharmaceutics, detergent, textile, pulp and paper, animal feed, leather, and cosmetics (Houde et al., 2004; Jamal et al., 2013). They are capable of specifically reducing hazardous wastes and, consequently, are key to new processes (Jamal et al., 2013). The industrial application of these enzymes is sometimes hampered by the nonexistence of long-term operational stability and the difficulty of their recovery and re-use (Sheldon, 2007). Research in the area of enzyme technology has provided significant evidence and strategies that facilitate the optimal use of enzymes at large scale by entrapping and immobilizing them (Gómez et al., 2006; Husain and Husain, 2008). Enzymes entrapped in porous polymeric matrices present inherent limitations of enzyme leaching; however, by controlling the pore dimensions, such leaching can be minimized. Alternatively, entrapping cross-linked or pre-immobilized enzyme preparations could be a better and more practical option (Betancor et al., 2005). There are many reasons to immobilize enzymes, such as greater ease of handling and separation from the product, reduction or removal of protein contamination, and facility of recovery and reuse of cost-efficient enzymes (Sheldon, 2007).

There are several methods for enzyme immobilization, which can be divided into three traditional methods: binding to a support (carrier), entrapment (encapsulation) and cross-linking. Support binding can be physical (such as hydrophobic and Van der Waals interactions), ionic, or covalent, and the support can be a biopolymer, a synthetic resin, or an inorganic polymer such as silica or a zeolite (Sheldon, 2007). Entrapment is conducted via inclusion of an enzyme in a polymer network, such as a natural one like alginate and low-methoxyl pectin, or synthetic polymers, such as polyvinyl alcohol and poly (ethylene oxide). Nevertheless, this technique has been more commonly used for the immobilization of cells rather than for enzymes (Contesini et al., 2012). Entrapment requires the synthesis of the polymeric system in the presence of the enzyme and has the benefit of higher protection of the protein structure and biological activity compared to the adsorption method. Finally, cross-linking of enzyme aggregates or crystals is carried out using a bifunctional reagent to prepare carrierless macroparticles. The use of a carrier inevitably leads to a dilution of activity and loss of more than 50% of native activity (Sheldon, 2007).

The interaction between enzymes and polysaccharides is a great immobilization method where the polysaccharides provide rigidity and hydration to the enzymes and increase their stability (Jadhav and Singhal, 2013). Different polysaccharides have been used for numerous enzymes to increase their pH and thermal stability (Gómez et al., 2000; Darias and Villalonga, 2001; Altikatoglu et al., 2009). Pectins can offer significant advantages to encapsulate active proteins instead of chemically-modified matrices, including low cost of equipment, less expensive waste treatments, use of soft techniques for cross-linking, and tailorability of molecular structure (Costas et al., 2008).

Many researches present the use of pectin for immobilization of a great number of enzymes that can be employed for different purposes. An example of these varied applications is in the treatment of burn wounds, as shown by Martínez et al. (2013). Enzymatic debridement of dead wound tissue is a great alternative for improving the topical penetration of antibiotics administered, thus preserving the spontaneous epithelialization potential (Krieger et al., 2012). Some enzymes, like proteases, which catalyze the degradation of tissue proteins, favor wound cleaning; moreover, they exhibit anti-inflammatory, fibrinolytic, and antiedemic effects (Vernikovskii and Stepanova, 2012). The use of immobilized enzymes in these cases allows for restriction of protease action located in a specific region of the body, in addition to enhancing their stability under the conditions of the wound healing process (Martínez et al., 2013). Considering this panorama, Martínez's group of researchers immobilized the enzyme keratinase from *Paecilomyces lilacinus* and enrofloxacin (EF) loaded on pectin PolyVinyl Alcohol cryogel patches for antimicrobial treatment (Martínez et al., 2013). Pectins with different esterification and biopolymer concentrations were tested for optimum enzymatic and antibiotic release. The release of keratinase was 63.8% from the PolyVinyl Alcohol (PVA) cryogel presenting 55.0% degree of esterification (DE) of pectin in 180 minutes, while the amount of enzyme released in PVA cryogels having 33.0%, 62.0% and 71.7% DE of pectin was 29.1%, 37.3% and 26.0%, respectively, in 3 h. In addition, no interference between keratinase and enrofloxacin was shown, allowing the dual immobilization of the antibiotic and the enzyme in the film for the controlled release purposes. It was observed that the release of enrofloxacin without the enzyme was faster (15.4% after 5 h of incubation) than the hydrogel containing both EF and the enzyme at the same time (6.9%).

Another research related to the use of pectin in enzyme immobilization was conducted on the decolorization of synthetic dyes. Dye wastewater from textile and dyestuff industries is very difficult to treat. The synthetic ones, classified by their chromophore as azo, anthraquinone, triphenylmethane, heterocyclic or phthalocyanine, are quite stable and resistant to microbial attack, making it difficult to remove them from effluents by conventional biological processes (Pala and Tokat, 2002). Enzymes are able to act on specific recalcitrant pollutants, removing them by precipitation or transformation into other products, changing the characteristics of a particular waste to make it more amenable for treatment (Karam and Nicell, 1997). Jamal et al. (2013) established a simple, inexpensive and high yield technique for glycosylated *Trichosanthes dioica* peroxidase immobilization with lectin Concanavalin A and entrapment with calcium alginate-pectin beads for use in effective color removal of industrial effluent contaminated with dyes. This immobilized biocatalyst complex retained only 56% of the original activity. Optimum concentration (418 U/mL) was sufficient for maximum expression of peroxidase activity by entrapped preparation. The stability exhibited by the complex was significantly higher when compared to soluble peroxidase, and, therefore, immobilized enzyme preparations could be explored for developing bioreactors for the treatment of phenolic and other aromatic pollutants, including synthetic dyes present in industrial effluents.

Satar et al. (2008) also immobilized peroxidase, employing calcium alginate pectin. The Entrapped enzyme complex retained 51% of the original activity. The soluble and immobilized peroxidase showed maximum activity at 40 °C and pH 5.5, although the immobilized enzyme retained a better fraction of catalytic activity at higher temperatures and revealed significant enhancement in pH activity profiles, representing a marked intensification in its stability. Because of these characteristics exhibited for the immobilized preparations, they can be applied in the treatment of pollutants (phenolic and aromatic) present in agro-industrial wastewaters.

Jadhav and Singhal (2013) screened nine polysaccharides (agar, carrageenan, carboxymethyl cellulose, dextran, gellan, guar gum, gum Arabic, pectin and xanthan) to test their capacity to conjugate with α-amilase from *Bacillus licheniformis* (350 U/mg) by covalent binding, under previously optimized conditions, in order to improve their thermal and pH stability. α-Amylase is a starch-degrading enzyme capable of catalyzing the hydrolysis of internal α-1,4 and α-1,6-glycosidic bindings in starch in low molecular weight products, such as glucose, maltose and maltotriose units, which can be obtained from several sources, such as plants, animals and microbes. This enzyme is among the most important ones and is of great value for biotechnology, constituting an industrial class of enzymes (Kathiresan and Manivannan, 2006). α-Amylase bound to carboxymethyl cellulose and gellan showed 100% retention of original activity; whereas that conjugated to pectin and xanthan showed a marginal increase in specific activity. All the poly-saccharides conjugated with α-amylase preparations were more stable than free α-amylase at 60 °C, 70 °C and 80 °C for 15 min. In the same way, the study of pH stability showed that all the conjugated α-amylases were more stable towards extreme acidic (pH 4,0) and alkaline (pH 10,0) conditions.

To immobilize glucosyltransferase from *Erwinia* sp. D12, Contesini et al. (2012) used two different supports, employing adsorption onto Celite 545 and entrapment in microcapsules of low-methoxyl pectin. This enzyme is applied for the conversion of sucrose into isomaltulose, an interesting substitute for sucrose in the food industry, as it is considered non-cariogenic. Glucosyltransferase immobilized in microcapsules of low-methoxyl pectin with the addition of fat material (butter and oleic acid) was able to convert 30% of sucrose into isomaltulose using a batch at 20 °C and 130 rpm.

Another group of enzymes that present relevant industrial application is lipases. This hydrolase is responsible for catalyzing the hydrolysis of long-chain triacylglycerols at lipid-water interfaces. In the organic chemistry field, lipases are well-known and attractive amongst the most widely used bio-catalysts, because they can catalyze several unnatural and remarkable reactions in non-aqueous media, such as esterification (Lopes et al., 2011; Stergiou et al., 2013) and transesterification (Speranza and Macedo, 2012; Garlapati et al., 2013). In the field of biotechnology, they are acquiring more attention due to their enantioselectivity, substrate specificity, and physicochemical properties, having been used in different fields ranging from detergents to food, pharmaceutical and chemical industries, with more than a billion dollar market (Grbavčić et al., 2007; Franken et al., 2010; Singh and Mukhopadhyay, 2012). Costas et al. (2008) studied the immobilization of the lipase from *Brevibacillus agri* 52, which was used as a model to explore the enzyme stability in binary and ternary water-miscible and -immiscible organic solvent systems and encapsulated in pectin gels in the presence of organic solvents. They observed that the enzyme encapsulation in pectin gels cross-linked with calcium ions brought three to four times more enzymatic activity in 70% water-miscible organic solvents compared to aqueous systems, which can be easily scaled up with the benefit of recycling the biocatalyst.

4. Immobilization of Microbial Cells Using Pectin and Derivatives

There are many examples in literature that demonstrate that cell physiology and morphology are affected by immobilization. The factors that may contribute to these changes include: different microenvironments (such as ionic strength, ionic charges, pH, water activity) created by the gel matrix, compared to those that cells encounter in suspension cultures; physical stress exerted by closely packed cells growing on one another; and mass transfer limitation (oxygen, substrate, product) imposed by the gel matrix (Kurillová et al., 2000).

Usually, cell immobilization is carried out in preformed carriers that involve passive immobilization, usually *in situ* in a bioreactor or culture environment. Most of the carriers are porous with a wide range of pore sizes to suit immobilization of various organisms. For passive immobilization, cells are inoculated into the sterilized medium containing empty preformed carriers. Depending on the cell and the carrier type, immobilization then takes place in a combination of filtration, adsorption, growth, and colonization processes (Kurillová et al., 2000).

Various porous matrices have been described for living cell immobilization. The choice usually depends on the cell type used and the kind of application. For example, immobilized cells support high-pressure drop in a reactor or provide excellent scaffold for cell attachment (Kurillová et al., 2000).

Some works are related and show what kind of immobilization is employed, using pectin and its derivatives, microorganisms, substrate and product of interest.

Rosenberg et al. (1999) showed the biotransformation of cis-epoxysuccinate to L-tartaric acid using immobilized *Nocardia tartaricans* in pectate gel. The group of compounds of the tartaric class is commonly used in the food and pharmaceutical industries. The L-tartaric acid could be obtained by enzymatic reaction of cis-epoxysuccinate hydrolase; therefore the increase in L-tartaric acid production could be directly related to the increase in enzymatic activity. However, the cis-epoxysuccinate hydrolase is intra-cellular, hence this immobilized microbial process represents an optimal model to be studied. In addition, this process is independent from aeration and neutralization during the conversion. The work tested the immobilized microbial conversion by enzymatic activity, number of conversion cycles, detergent compounds during immobilization treatment and L-tartaric acid production. The results showed that immobilization possessed cis-epoxysuccinate hydrolase activity after 450 days and concluded that cross-linked calcium pectate gel has an advantage in preparation of spheric particles. Also, the addition of detergent gradually permeabilized in repeated bioconversions, which led to relatively good stability of the cells and increase of L-tartaric acid production. However, the detergent-treated cells have apparently shorter lifespan compared to the cells without the addition of detergent. In addition, organic acids produced by *N. tartaricans* during the process increased the permeabilization of the cells and simultaneously slowly decreased enzymatic activity. The results showed that immobilized microbial biotransformation of L-tartaric acid is a process with interesting advantages in some industrial fields.

Wu and Yu (2007) reported on the immobilization of fungus *Phanerochaete chrysosporium* for biotransformation. The paper studied pectin matrices as support for the removal of 2,4-dichlorophenol from wastewater. Biosorption of heavy metals in aqueous solution has received increasing attention as a potential alternative to eliminate pollutants in environment. Several studies report on *Phanerochaete chrysosporium* regarding elimination of xenobiotics and detoxification of effluents. The immobilization process represents an important way to increase biotrans-formation efficiency. Several characteristics may be studied, such as the mechanical strength of immobilization, rigidity, size, porosity and resistance to the environment. The results of the biotransformation process showed that the efficiency rate was about 45% with 3% pectin as adsorbent. Despite the process not representing the greatest efficiency, the pectin showed good matrix support, and it was an inexpensive process. In conclusion, the pectin immobilization was a moderate, efficient and promising method for the repetitive use of fungal biomass, as the reduction in adsorption efficiency and weight loss of biosorbent were negligible in the repetitive adsorption/ desorption of 2,4-dichlorophenol.

Voo et al. (2011) studied a comparative experiment between alginate and pectin as supports for production of poultry probiotic cells. Usually, lactic acid bacteria are used as poultry probiotic cells for fermentation processes; however, repeated batch fermentation could decrease the process yield, due to metabolic inhibitory end-products and damage to recovery cells during centrifugation. Immobilization is an interesting alternative to resolve these issues. The authors studied three types of materials for immobilization: alginate, pectinate and alginate/pectinate. According to the results, the pectin based beads were found to be more stable than other matrices. The cell concentration in pectin was similar to that in the alginate, however, the pectin gave significantly lower cell concentration in the growth medium for the initial fermentation cycles. In conclusion, pectin presents great potential as encapsulation material for probiotic cell production, due to its stability and favorable microenvironment for cell growth.

Berger and Rühlemann (1988) immobilized cells of *Saccharomyces cere-visiae* and *Streptococcus thermophilus* SC5 using different polymers, such as alginate, citrus pectin and pectic acid. The experiments showed that calcium pectate beads or aluminum pectate prepared from pectic acid of high molecular weight having a very low content of methoxy groups were suitable for immobilizing cells. The potential application of calcium pectate was found as a matrix of immobilized yeast cells for use in the production of ethanol by continuous fermentation, which was compared with cells immobilized in calcium alginate under the same conditions. The fermentation of ethanol in the horizontal reactor column using alginate beads containing yeast was stopped after 30 hours due to the expansion of the alginate beads, which caused blockage of the fixed bed, meaning the continuous flow of liquid medium was not guaranteed. In contrast, there was only slight swelling of the pectate granules and, therefore, no blockage of the fixed bed. Liquid medium steadily flowed through the column from the beginning until the seventh day of fermentation, where after the fermentation was stopped due to clogging of the bed caused by free cells.

Sriamornsak et al. (1997) found that the type of pectin was important in the formation of granules. In the partially esterified pectin with a lower degree of esterification, spherical granules formed in the presence of calcium ions and were used for the immobilization of *Sacchararomyces cerevisiae* cells. The effect of storage conditions on the viability of immobilized cell beads was also investigated, and it was found that after storage at 4 °C or -40 °C for 1 month, the beads retained sufficiently stable cell suspension when compared to non-immobilized yeast cells.

Dias et al. (2000) studied the immobilization of *Candida guilliermondii* UFMG - Y65 yeast cells on different support materials, such as alginate, *k*- carrageenan and low methoxy citrus pectin for the degradation of acetonitrile. The suspension of citrus pectin was dissolved in 18 mL of distilled water. After the sterilization process, the polymer suspension was added to a 6 mL aliquot containing 108 cells / mL at 30 °C to gel the pectin. The suspension was dripped into a sterile cross-linking solution containing $BaCl_2$. The beads formed were measured at approximately 2 mm diameter and were maintained in the cross-linking solution for 10 minutes to 24 hours at 10 °C. The degradation tests were performed in acetonitrile Erlenmeyer flasks containing selective medium and acetonitrile under stirring at 120 rpm and 25 °C for 120 hours. The rate of degradation of acetonitrile was monitored by the growth of yeast and generation of ammonia.

Panesar et al. (2007) studied the use of immobilized cell pectate gel to produce L (+)-lactic acid from whey. The authors found that the application of pectate gel for the immobilization of cells for lactic acid fermentation is promising because of its good level of stability at low pH levels and acceptability for applications in food products. The immobilization of *Lactobacillus casei* NBIMCC 1013 cells in spheres of calcium pectate gel was performed using commercial citrus pectin with low ester content. Bacterial biomass was carefully mixed with a pectate solution, and the resulting solution was dropped in calcium chloride, 0.2 M. The resulting granules with a diameter of approximately 4 mm were washed with distilled sterile water to remove excess calcium ions and the cells that did not immobilize. The obtained granules were kept overnight at 4 °C and then washed with a solution of 0.1 M aluminum nitrate and sterile water. Parameters for immobilization and fermentation were studied using a univariate sequential methodology for the optimization of the lactic acid production process. The maximum conversion of lactose (84.37%) lactic acid was 28.34 g / L, immobilized cells with 3% (w / v) pectate gel showing high stability. It was found that cells immobilized on beads of diameter 2.41 to 2.79 mm showed production of lactose at 88.12%. Increasing the concentration of cells in the immobilization process and stirring the reaction medium did not result in increased production. The best pH to obtain the lactic acid was 6.5. The optimal temperature range was between 37 - 40 °C, yielding 94.37% (w / v) conversion and a lactic acid production of 32.91 g / L. The immobilized system showed no decrease in the conversion of lactose into lactic acid for up to 16 batches, which proved its high stability and potential application for commercial use.

CONCLUSION

Pectins and pectic materials are complex macromolecules present in the middle lamellae of plants, helping to bind cells together. It is possible to be extracted from different types of sources, but citrus and pomaces are the most relevant, which results in interesting yields from an industrial point of view. These compounds find their main target in food and pharmaceutical industries, as they present an immense number of applications, including use as a gelling agent and stabilizer and for the production of jams. However, one highlighted focus of study for these polysaccharides is their use in the encapsulation and immobilization of enzymes and cells for biocatalytic and fermentation purposes. This is truly relevant from the economical point of view, taking into consideration the fact that immobilized enzymes and cells can be reused for several batches or in continuous form and the stability or catalytic properties of the immobilized biocatalysts can be greatly improved, which results in better yields. In addition to this, pectin is nontoxic and biodegradable. Therefore, pectins and pectic materials must be intensively studied for developing more elaborated, cost-effective and feasible techniques for their industrial applications.

REFERENCES

Altikatoglu, M., Arioz, C., Basaran, Y., Kuzu, H. Stabilization of horseradish peroxidase by covalent conjugation with dextran aldehyde against tempe-rature and pH changes. *Central European Journal of Chemistry*, 2009 7 (3), 423-428.

Berger, R., Rühlemann, I. Stable ionotropic gel for cell immobilization using high molecular weight pectic acid. *Acta Biotechnologica*, 1988 8(5), 401-405.

Betancor, L., López-Gallego, F., Hidalgo, A., Fuentes, M., Podrasky, O., Kuncova, G., Guisán, J. M., Fernández-Lafuente, R. Advantages of the pre-immobilization of enzymes on porous supports for their entrapment in sol-gels. *Biomacromolecules*, 2005 6(2), 1027-1030.

Canteri, M. H. G., Moreno, L., Wosiacki, G., De Scheer, A. P. Pectin: From raw material to the final product. *Polimeros*, 2012 22(2), 149-157.

Chan, S. Y., Choo, W. S. Effect of extraction conditions on the yield and chemical properties of pectin from cocoa husks. *Food Chemistry*, 2013 141(4), 3752-3758.

Contesini, F. J., Ibarguren, C., Grosso, C. R. F., Carvalho, P. D. O., Sato, H. H. Immobilization of glucosyltransferase from *Erwinia sp.* using two different techniques. *Journal of Biotechnology*, 2012 158(3), 137-143.

Costas, L., Bosio, V. E., Pandey, A., Castro, G. R. Effects of organic solvents on immobilized lipase in pectin microspheres. *Applied Biochemistry and Biotechnology,* 2008 151(2-3), 578-586.

Darias, R., Villalonga, R. Functional stabilization of cellulase by covalent modification with chitosan. *Journal of Chemical Technology and Biotechnology*, 2001 76(5), 489-493.

Dias, J. C. T., Rezende, R. P., Linardi, V. R. Biodegradation of acetonitrile by cells of *Candida guilliermondii* UFMG-Y65 immobilized in alginate, k-carrageenan and citric pectin. *Brazilian Journal of Microbiology*, 2000 3161-66.

DiCosimo, R., McAuliffe, J., Poulose, A. J., Bohlmann, G. Industrial use of immobilized enzymes. *Chemical Society Reviews*, 2013 42(15), 6437-6474.

Dolejš, I., Rebroš, M., Rosenberg, M. Immobilisation of *Clostridium spp.* for production of solvents and organic acids. *Chemical Papers*, 2014 68(1), 1-14.

Domozych, D. S., Serfis, A., Kiemle, S. N., Gretz, M. R. The structure and biochemistry of charophycean cell walls: I. Pectins of Penium margaritaceum. *Protoplasma*, 2007 230(1-2), 99-115.

Faravash, R. S., Ashtiani, F. Z. The influence of acid volume, ethanol-to-extract ratio and acid-washing time on the yield of pectic substances extraction from peach pomace. *Food Hydrocolloids*, 2008 22(1), 196-202.

Fraeye, I., Duvetter, T., Doungla, E., Van Loey, A., Hendrickx, M. Fine-tuning the properties of pectin–calcium gels by control of pectin fine structure, gel composition and environmental conditions. *Trends in Food Science and Technology,* 2010 21(5), 219-228.

Franken, L. P. G., Marcon, N. S., Treichel, H., Oliveira, D., Freire, D. M. G., Dariva, C., Destain, J., Oliveira, J. V. Effect of treatment with compressed propane on lipases hydrolytic activity. *Food and Bioprocess Technology*, 2010 3(4), 511-520.

Garlapati, V., Kant, R., Kumari, A., Mahapatra, P., Das, P., Banerjee, R. Lipase mediated transesterification of *Simarouba glauca* oil: A new feed-stock for biodiesel production. *Sustainable Chemical Processes Journal,* 2013 1(11), 1-6.

Gebara, C., Chaves, K. S., Ribeiro, M. C. E., Souza, F. N., Grosso, C. R. F., Gigante, M. L. Viability of *Lactobacillus acidophilus* La5 in pectin–whey protein microparticles during exposure to simulated gastrointestinal conditions. *Food Research International*, 2013 51(2), 872-878.

Giordano, R. L. C., Trovati, J., Schmidell, W. Continuous production of ethanol from starch using glucoamylase and yeast co-immobilized in pectin gel. *Applied Biochemistry and Biotechnology*, 2008 147(1-3), 47-61.

Gómez, J. L., Bódalo, A., Gómez, E., Bastida, J., Hidalgo, A. M., Gómez, M. Immobilization of peroxidases on glass beads: An improved alternative for phenol removal. *Enzyme and Microbial Technology*, 2006 39(5), 1016-1022.

Gómez, L., Ramírez, H. L., Villalonga, R. Stabilization of invertase by modification of sugar chains with chitosan. *Biotechnology Letters*, 2000 22(5), 347-350.

Grbavčić, S. Ž., Dimitrijević-Branković, S. I., Bezbradica, D. I., Šiler-Marinković, S. S., Knežević, Z. D. Effect of fermentation conditions on lipase production by *Candida utilis*. *Journal of the Serbian Chemical Society,* 2007 72(8-9), 757-765.

Houde, A., Kademi, A., Leblanc, D. Lipases and their industrial applications: An overview. *Applied Biochemistry and Biotechnology* 2004 118(1-3), 155-170.

Husain, M., Husain, Q. Applications of redox mediators in the treatment of organic pollutants by using oxidoreductive enzymes: A review. *Critical Reviews in Environmental Science and Technology,* 2008 38(1), 1-42.

Jadhav, S. B., Singhal, R. S. Screening of polysaccharides for preparation of α-amylase conjugate to enhance stability and storage life. *Carbohydrate Polymers*, 2013 92(2), 1724-1729.

Jamal, F., Singh, S., Khatoon, S., Mehrotra, S. Application of immobilized pointed gourd (*Trichosanthes dioica*) peroxidase-concanavalin a complex on calcium alginate pectin gel in decolorization of synthetic dyes using batch processes and continuous two reactor system. *Journal of Bioprocessing and Biotechniques*, 2013 3(2), 131-136.

Jolie, R. P., Duvetter, T., Van Loey, A. M., Hendrickx, M. E. Pectin methylesterase and its proteinaceous inhibitor: A review. *Carbohydrate Research*, 2010 345(18), 2583-2595.

Joye, D. D., Luzio, G. A. Process for selective extraction of pectins from plant material by differential pH. *Carbohydrate Polymers*, 2000 43(4), 337-342.

Karam, J., Nicell, J. A. Potential applications of enzymes in waste treatment. *Journal of Chemical Technology and Biotechnology*, 1997 69(2), 141-153.

Kathiresan, K., Manivannan, S. α-Amylase production by *Penicillium fellutanum* isolated from mangrove rhizosphere soil. *African Journal of Biotechnology*, 2006 5(10), 829-832.

Kratchanova, M., Pavlova, E., Panchev, I. The effect of microwave heating of fresh orange peels on the fruit tissue and quality of extracted pectin. *Carbohydrate Polymers*, 2004 56(2), 181-185.

Kratchanova, M., Panchev, I., Pavlova, E., Shtereva, L. Extraction of pectin from fruit materials pretreated in an electromagnetic field of super-high frequency. *Carbohydrate Polymers*, 1994 25(3), 141-144.

Krieger, Y., Bogdanov-Berezovsky, A., Gurfinkel, R., Silberstein, E., Sagi, A., Rosenberg, L. Efficacy of enzymatic debridement of deeply burned hands. *Burns*, 2012 38(1), 108-112.

Kurillová, L., Gemeiner, P., Vikartovská, A., Miková, H., Rosenberg, M., Ilavský, M. Calcium pectate gel beads for cell entrapment. 6. Morphology of stabilized and hardened calcium pectate gel beads with cells for immobilized biotechnology. *Journal of Microencapsulation*, 2000 17(3), 279-296.

Lopes, D. B., Duarte, M. C. T., Macedo, G. A. Biosynthesis of oleyl oleate wax ester by non-commercial lipase. *Food Science and Biotechnology*, 2011 20(5), 1203-1209.

Lv, C., Wang, Y., Wang, L. J., Li, D., Adhikari, B. Optimization of production yield and functional properties of pectin extracted from sugar beet pulp. *Carbohydrate Polymers*, 2013 95(1), 233-240.

Martínez, Y. N., Cavello, I., Hours, R., Cavalitto, S., Castro, G. R. Immobilized keratinase and enrofloxacin loaded on pectin PVA cryogel patches for antimicrobial treatment. *Bioresource Technology*, 2013 1452 80-284.

Mesbahi, G., Jamalian, J., Farahnaky, A. A comparative study on functional properties of beet and citrus pectins in food systems. *Food Hydrocolloids*, 2005 19(4), 731-738.

Mishra, R. K., Banthia, A. K., Majeed, A. B. A. Pectin based formulations for biomedical applications: A review. *Asian Journal of Pharmaceutical and Clinical Research*, 2012 5(4), 1-7.

Munarin, F., Tanzi, M. C., Petrini, P. Advances in biomedical applications of pectin gels. *International Journal of Biological Macromolecules*, 2012 51 (4), 681-689.

Pala, A., Tokat, E. Color removal from cotton textile industry wastewater in an activated sludge system with various additives. *Water Research*, 2002 36 (11), 2920-2925.

Panesar, P. S., Kennedy, J. F., Knill, C. J., Kosseva, M. R. Applicability of pectate-entrapped Lactobacillus casei cells for L(+) lactic acid production from whey. *Applied Microbiology and Biotechnology Journal*, 2007 74 (1), 35-42.

Puga, A. M., Lima, A. C., Mano, J. F., Concheiro, A., Alvarez-Lorenzo, C. Pectin-coated chitosan microgels crosslinked on superhydrophobic surfaces for 5-fluorouracil encapsulation. *Carbohydrate Polymers*, 2013 98(1), 331-340.

Rosenberg, M., Miková, H., Krištofíková, L. Production of L-tartaric acid by immobilized bacterial cells *Nocardia tartaricans*. *Biotechnology Letters*, 1999 21(6), 491-495.

Sandolo, C., Péchiné, S., Le Monnier, A., Hoys, S., Janoir, C., Coviello, T., Alhaique, F., Collignon, A., Fattal, E., Tsapis, N. Encapsulation of Cwp84 into pectin beads for oral vaccination against Clostridium difficile. *European Journal of Pharmaceutics and Biopharmaceutics*, 2011 79(3), 566-573.

Satar, R., Matto, M., Husain, Q. Studies on calcium alginate-pectin gel entrapped concanavalin A-bitter gourd (*Momordica charantia*) peroxidase complex. *Journal of Scientific and Industrial Research*, 2008 67(8), 609-615.

Schieber, A., Hilt, P., Endreß, H U., Rentschler, C., Carle, R. A new process for the combined recovery of pectin and phenolic compounds from apple pomace. *Innovative Food Science and Emerging Technologies*, 2003 4(1), 99-107.

Seixas, F. L., Fukuda, D. L., Turbiani, F. R. B., Garcia, P. S., Petkowicz, C. L. D. O., Jagadevan, S., Gimenes, M. L. Extraction of pectin from passion fruit peel (*Passiflora edulis f. flavicarpa*) by microwave-induced heating. *Food Hydrocolloids*, 2014 38189-192.

Sheldon, R. A. Enzyme immobilization: The quest for optimum performance. *Advanced Synthesis and Catalysis*, 2007 349(8-9), 1289-1307.

Singh, A. K., Mukhopadhyay, M. Overview of fungal lipase: A review. *Applied Biochemistry and Biotechnology*, 2012 166(2), 486-520.

Speranza, P., Macedo, G. A. Lipase-mediated production of specific lipids with improved biological and physicochemical properties. *Process Biochemistry,* 2012 47(12), 1699-1706.

Sriamornsak, P. Chemistry of pectin and its pharmaceutical uses: A review. *Silpakorn University International Journal*, 2003 3(1-2), 206-228.

Sriamornsak, P., Puttipipatkhachorn, S., Prakongpan, S. Calcium pectinate gel coated pellets as an alternative carrier to calcium pectinate beads. *International Journal of Pharmaceutics*, 1997 156(2), 189-194.

Sriamornsak, P., Wattanakorn, N., Takeuchi, H. Study on the mucoadhesion mechanism of pectin by atomic force microscopy and mucin-particle method. *Carbohydrate Polymers*, 2010 79(1), 54-59.

Stephen, A. M., Phillips, G. O. *Food polysaccharides and their applications.* CRC Press; 2010.

Stergiou, P. Y., Foukis, A., Filippou, M., Koukouritaki, M., Parapouli, M., Theodorou, L. G., Hatziloukas, E., Afendra, A., Pandey, A., Papamichael, E. M. Advances in lipase-catalyzed esterification reactions. *Biotechnology Advances,* 2013 31(8), 1846-1859.

Thakur, B. R., Singh, R. K., Handa, A. K. Chemistry and uses of pectin - A review. *Critical Reviews in Food Science and Nutrition*, 1997 37(1), 47-73.

Vernikovskii, B. V., Stepanova, E. F. Immobilized proteases for wound cleaning. *Russian Journal of General Chemistry*, 2012 82(3), 572-578.

Videcoq, P., Garnier, C., Robert, P., Bonnin, E. Influence of calcium on pectin methylesterase behaviour in the presence of medium methylated pectins. *Carbohydrate Polymers*, 2011 86(4), 1657-1664.

Voo, W.-P., Ravindra, P., Tey, B.-T., Chan, E.-S. Comparison of alginate and pectin based beads for production of poultry probiotic cells. *Journal of Bioscience and Bioengineering,* 2011 111(3), 294-299.

Voragen, A. G. J., Coenen, G. J., Verhoef, R. P., Schols, H. A. Pectin, a versatile polysaccharide present in plant cell walls. *Structural Chemistry*, 2009 20(2), 263-275.

Willats, W. G. T., Knox, J. P., Mikkelsen, J. D. Pectin: New insights into an old polymer are starting to gel. *Trends in Food Science and Technology*, 2006 17(3), 97-104.

Willats, W. G. T., McCartney, L., Mackie, W., Knox, J. P. Pectin: Cell biology and prospects for functional analysis. *Plant Molecular Biology*, 2001 47 (1-2), 9-27.

Wu, J., Yu, H. Q. Biosorption of 2,4-dichlorophenol by immobilized white-rot fungus *Phanerochaete chrysosporium* from aqueous solutions. *Bioresource Technology*, 2007 98(2), 253-259.

Yunyu, Y., Ronald, J. N., Denis, P. Immobilization of Cells in Polysaccharide Gels. In: *Polysaccharides*. CRC Press; 2004.

In: Pectin: Chemical Properties, Uses and Health Benefits
Editor: Phillip L. Bush
ISBN: 978-1-63321-438-5

Chapter 5

ORAL DRUG RELEASE SYSTEMS BASED ON PECTIN

Beatriz Stringhetti Ferreira Cury**, *Andréia Bagliotti Meneguin, Valéria Maria de Oliveira Cardoso and Fabíola Garavello Prezotti
Graduate Program in Pharmaceutical Sciences, Department of Drugs and Pharmaceuticals, School of Pharmaceutical Sciences, São Paulo State University – UNESP, Araraquara, SP, Brazil

"We dedicate this chapter to the memory of our dear Professor Raul Cesar Evangelista"

ABSTRACT

Pectin is a natural polysaccharide and its specific enzymatic degradability by colonic microbiota makes it a promising material for designing drug release systems, mainly those intended for targeting drugs to the colon. However, in despite of pectin resistance against proteases and amylases, remaining as aggregates of macromolecules in acid medium, a great challenge to optimize the performance of pectin in such systems lies in its high hydrophilicity that, in several times, results in an undesirable premature release of drugs. Blends of pectin with other polysaccharides and cross-linking reactions are valuable tools to modulate such properties of pectin, particularly reducing its solubility. These approaches have been focus of important researches of our research group and our findings have been published in important scientific journals. Blends of pectin and retrograded starch (RS) allowed the preparation of free films with suitable mechanical properties and reduced dissolution of films in acid media, while their high resistance against enzymatic digestion by pancreatin was demonstrated. The same polymer association was exploited for preparing tablets containing sodium diclofenac (SD), and the presence of pectin reduced significantly the drug dissolution in acid medium. In another study with free films, the blends of pectin-high amylose starch (HAS) cross-linked with sodium trimetaphosphate (STMP) contributed to the reduction of their hydrophilicity. This polymer association was also exploited for preparing hydrophilic matrices from which the drug release rates in acid medium were lowered. In addition, this

* Corresponding author: E-mail address: curybsf@fcfar.unesp.br. Phone: +55 16 3301 6961; Fax: +55 16 3322-0073.

same cross-linked HAS/pectin blend was employed for preparing microparticles loaded with SD by immersion and the mixtures containing the same proportion of polymers allowed a more effective control of drug release rates. Furthermore, microparticles obtained by physical mixture of polymers showed the lower percentage of drug released in acid medium and this behavior was attributed to the pectin that provides a diffusion layer of high viscosity that reduces the drug release rate. The association of pectin with gellan gum for preparing mucoadhesive beads by ionotropic gelation provided a pH dependent dissolution behavior, allowing reduced drug release rates in acid media. The purpose of this review is to evidence the importance of pectin as a carrier in the design of different drug release systems, aiming the targeting of drugs. Besides, the association of pectin with other polysaccharides and the cross-linking reaction are demonstrated to be reliable strategies to modulate the properties of the systems according to specific therapeutic needs.

INTRODUCTION

The oral administration of drugs is safer and more comfortable with a higher patient compliance to the treatment. Despite of inherent advantages of oral route, it offers a large number of limitations as those related to the instability or low permeability of drug in acid conditions of the stomach, which result in reduced treatment efficiency [1-3].

Site-specific drug release systems show significant biopharmaceutical and pharmacokinetic advantages compared to conventional systems, as reduction of required dose, drug protection against degradation, improvement of the bioavailability, reduction of side effects and optimization of pharmacological effects [4, 5].

Among the site-specific drug release systems, colonic systems have been developed for the local treatment of bowel diseases, oral administration of proteins and for improving systemic absorption of drugs, since the colon offers a favorable environment than the upper portions of GIT, with a pH more near to neutrality, reduced proteolytic activity and transit longer transit time [5-9].

To design a system that target a drug to the colon successfully it is necessary a triggering element sensible to physiological changes in order to protect the drug from premature release and/or degradation in upper portion of the GIT, releasing it in the proximal colon. For this purpose, various approaches have been attempted and the exploitation of enzymatic activity of colonic microbiota as a triggering element to promote the drug release represent a more reliable strategy because colonic microorganisms show a minor interindividual variability in relation to pH values and transit time [5, 9, 10].

The large availability, biodegradability, biocompatibility, security and low cost of natural polymers as the polysaccharides states the great interest in these materials as carriers for the designing of more efficient drug release systems that promote the targeting of drugs to specific organs or tissues, and/or the control of release rates, aiming different therapeutic needs. These materials aggregate an important feature of resisting to the drastic upper GIT enviroment, being later digested by colonic microbiota [11-14].

Pectins are natural anionic heteropolysaccharides composed mainly by homogalacturonans (HG) and rhamnogalacturonans (RG) [15-17]. HG represents the linear fraction or the smooth region of the structure, constituted by residues of galacturonic acid (GalA) linked by glycosidic bonds of type α-(1$\rightarrow$4). Residues of GalA can be esterified in C6 position and/or acetylated on O2 and O3 [18, 19]. According to their esterification degree (DE), pectin can be labeled as high methoxyl pectins (HMP) (DE $\geq$ 50%) or low methoxyl pectins (LMP) (DE < 50%) [8, 16, 20-22].

Rhamnose residues are randomly inserted to the main backbone by (1→2) α-L linkages, promoting a torsion of the structure in another linear chain, in which the arabinans and galactans residues can be joined [16, 18, 23-25].

The structural conformation of pectin allows to it some flexibility, providing important functional properties in cell plants and influencing on its applications in food and biomedical fields [23, 26, 27].

The structural complexity of pectins, as well as their modifications by chemical or physical approaches, results in a wide range of physicochemical and structural properties that allow this polysaccharide to be suitable for different uses. Besides of the properties variation according to structural conformation of pectins and their modifications, parameters as pH, temperature, dissolved solids, ionic strength and metal ions can also affect hardly the functional properties of pectins [28].

Pectins are mainly used as gelling and thickening agent, and the viscosity reached by dispersions will be dependent on its structure, molecular weight, DE, concentration and temperature [16, 27, 29, 30]. Thus, pectins with low DE form gels in the presence of bi or multivalent ions that crosslink the galacturonic acid chains whereas those with high DE form gels in acidic media, in the presence of sugars, as sucrose and glucose [12, 31-34].

Pectin, modified or not, has also been widely exploited as carrier in pharmaceutical field, particularly in the designing of drug release systems, mainly those intended for targeting drugs to the colon because it is specifically degraded by colonic microbiota while is resistant to amylases and proteases digestion. Moreover, it allows the preparation of both matrix and reservoir systems, showing also important mucoadhesive and swelling properties, that make it a promising material to release drugs in a controlled manner [34-45].

Despite of these favorable properties of pectin that fit well with the features required for suitable controlled drug release systems, the great challenge for reaching an effective control of drug release is the high solubility of this polysaccharide in aqueous acid media, that generally results in premature and undesirable release of drug on upper portions of the GIT [5, 9, 11, 12].

Chemical modifications as cross-linking reactions are key strategies to modulate physicochemical and mechanical properties of polysaccharides according to specific uses and these approaches do not affect the biodegradability of these materials [5, 12, 30, 46, 47].

On the other hand, the blend of polymers with well-known properties represents a rational way to reach new materials in which important properties as swelling, erosion, solubility and viscosity can be adjusted, allowing the design of innovative and effective drug release systems that attend to specific therapeutic needs [35, 48]. Moreover, changes in polymers ratio and cross-linking degree can provide different drug release profiles to achieve specific goals [35].

Our research group has exploited blends of pectin with other polysaccharides and the cross-linking reaction as approaches for designing of novel drug release systems, which were systematically characterized according to their physicochemical properties and performance as controlled drug release systems. The main findings about such systems are globally described throughout this chapter.

Pectin-Retrograded Starch Blends

Free Films

Films based on polymers can be employed as coating for solid dosage forms with the aim of controlling the drug release or even to protect the drug from external factors or some drastic condition in the organism, such as the acid environment of the stomach. Moreover, polymeric films can be used as dosage forms that can be administered by different routes, such as the buccal, releasing the drug for both local and systemic effects [49, 50].

Pectin and high amylose starch (HAS) were blended at different ratios and dispersed in aqueous media for preparing free films, by solvent casting method [51]. This study was innovative because the influence of pectin on the retrogradation of HAS had not been reported up to date of the research. Moreover, retrograded starch (RS) have been scarcely exploited in pharmaceutical field, mainly in the designing of drug release systems.

The polysaccharide blends were submitted to the retrogradation process under hydrothermal treatment in alternating cycles in order to obtain a material with high content of RS, which represents a starch fraction that resists against the digestion in stomach and duodenum, but is specifically degraded by colonic microbiota. In the retrogradation process, the pregelatinized starch (amorphous) changes to a more organized crystalline form [52-55].

The presence of pectin favored the retrogradation process of HAS, increasing the RS content (about 65.80 % to 96.68 %), which was maximum when pectin and HAS were in equal proportion (1:1) (Figure 1).

The high resistance of free films against the enzymatic digestion by α-amylase pancreatic (Figure 1) was evidenced by the *in vitro* test and those films prepared with equal polymer proportion presented the lowest digestibility, which can be attributed to enzymatic resistance inherent of the pectin associated to the high contents of RS in the films.

Films prepared with higher pectin proportion (4:1) presented the best mechanical properties (Figure 1), which were evaluated according to their resistance to perforation (puncture strength), indicating that pectin was responsible for building more flexible structures, which allow an extensive structural rearrangement until the break point was reached [56]. Films prepared with other pectin-HAS ratios also showed suitable mechanical properties, which are essential to the films perform effectively their protective barrier function.

Films prepared with lower pectin-HAS concentration showed the lowest values of water vapor permeability (WVP), because they were thinner and this decreased polymeric mass was not able to absorb many water molecules from the environment. Additionally, scanning electronic microscopy (SEM) showed that these films have a continuous structure without porous and fissures, which are facilitators of the diffusion process.

Indeed, the presence of pectin was essential to filmogenic properties of the polymer dispersions because when this polysaccharide was absent, a discontinuous and brittle structure was built, which was related to the hard nature of starch crystals [57]. Therefore, it was concluded that pectin aids create a more flexible network with low polymeric entanglement which allow higher inter chains mobility [58].

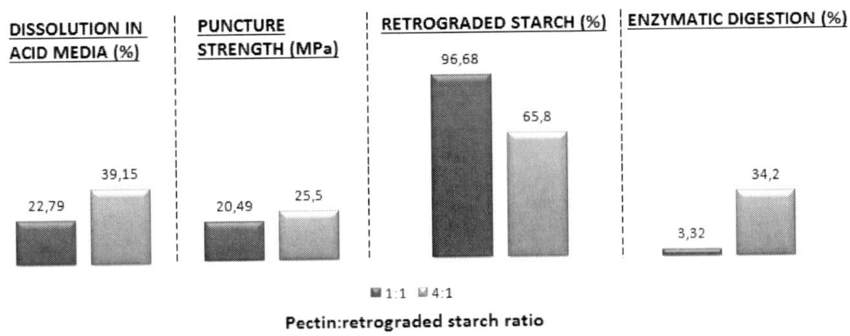

Figure 1. Properties of pectin/retrograded starch films.

This finding was corroborated by the rheological study of pectin-RS film forming dispersions, in which all of them showed the loss modulus (G") higher than the storage modulus (G') along the whole frequency range (0.6–623 rad s^{-1}), indicating the predominance of viscous behavior [59, 60]. Furthermore, the filmogenic dispersions with high pectin proportion (4:1) presented the lowest G' values (about 1000 x) in relation to 1:1 samples, indicating that the increased amount of pectin resulted in weaker structures [61, 62], and this feature was determinant for the films formation.

Despite of protective function that polymeric films can offer when applied to solid dosage forms such as tablets, capsules, pellets, microparticles, another fundamental function of the coating films is to act as a physical barrier that can control the diffusion rates of drug throughout them, leading to the designing of advanced systems able to control the drug release rates and/or target the drug to a specific organ or tissue, according to specific therapeutic needs [49].

In order to predict the controlling release role of the pectin-RS films, the dissolution of free films was analyzed in media with different pH values, simulating the ranging along the GIT.

Films obtained with equal pectin-HAS ratio showed the lowest dissolution values in both acid medium (Figure 1) and phosphate buffer pH 7.4. However, the increase of pectin proportion in the films promoted the raising of dissolution values, probably due to the high water solubility of this polysaccharide. The same trend was verified in relation to liquid uptake (LU) ability of the films.

This feature evidences the pH-responsive dissolution behavior of these films, which make them promising material for the designing of new drug release systems that aim the control of release rates along the GIT.

Matrix Compacted Systems

Blends of pectin and RS (1:1) were also investigated as excipient for preparing tablets containing sodium diclofenac (SD) and the drug release rates were significantly reduced in acid medium (about 50%) in comparison with tablets prepared only with retrograded starch, demonstrating the ability of pectin to control the release rates due to the building of a thick gel layer that restricts the drug diffusion to dissolution media [63].

PECTIN-HAS BLENDS CROSS-LINKED WITH SODIUM TRIMETAPHOSPHATE

Free Films

Pectin and HAS blends (1:1) were cross-linked with sodium trimetaphosphate (STMP) in alkaline aqueous medium as a strategy to reduce pectin water solubility, a major drawback that can lead to premature drug release in the upper GIT when developing drug release systems intended to controlled release at the colon [36].

Free films were prepared from aqueous dispersions (3, 4 and 5%, w/v) of cross-linked polymer blends by solvent casting technique. Free films with uncross-linked polymer blends were also prepared as control.

All films were very homogeneous, translucent, colorless and flexible with continuous and smooth surfaces. Films thickness increased linearly with the rising in polymer concentration and the cross-linking process led to tighter structures, due to the introduction of covalent bonds inter and intra polymer chains.

Films presented high resistance to enzymatic digestion by pancreatin, during the *in vitro* test. At low polymer concentrations (3 and 4%), cross-linked films were the most resistant (Figure 2) because the introduction of covalent bonds created denser regions in the polymer network, making the enzyme access more difficult, reducing the films digestibility. Moreover, pectin plays an important role in this feature due to its inherent resistance to enzymes present in the upper GIT [9, 11].

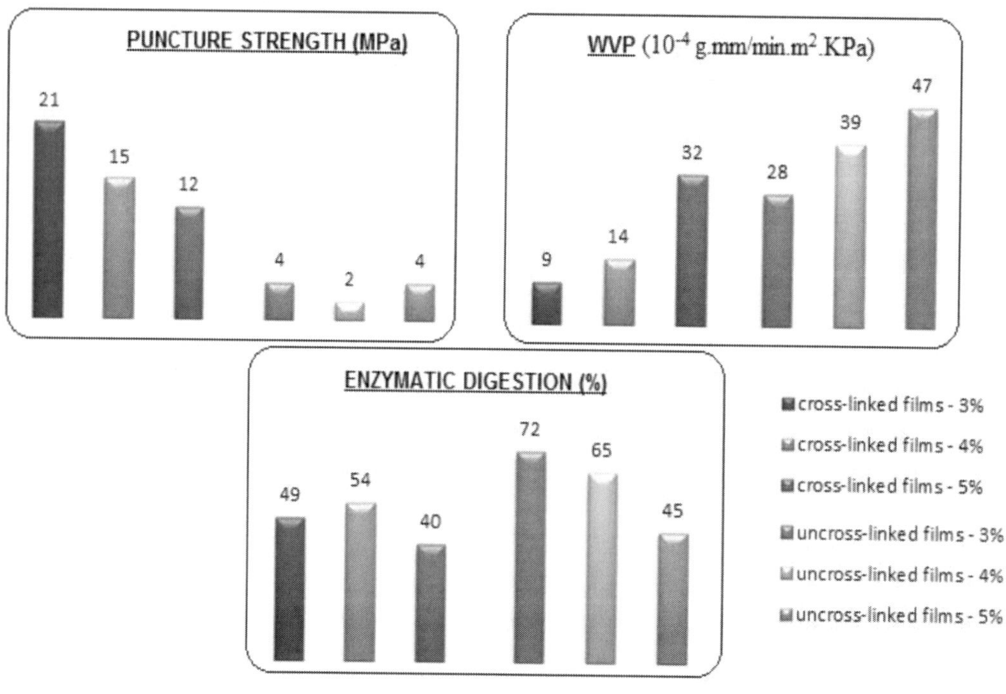

Figure 2. Properties of cross-linked pectin-HAS free films.

Cross-linked films with lower polymer content (3 and 4%) presented the lowest WVP values (Figure 2), indicating, once more, that cross-linking process promotes the building of a tighter and more rigid polymer network, that difficult the water vapor penetration once it restricts molecular motions.

The cross-linking reaction with STMP could reduce the WVP of the films up to 67%, while increased the mechanical strength up to 5.7 times in comparison to films prepared with uncross-linked polymers blends. These data demonstrate that both blend with other polysaccharide and chemical modification by cross-linking process allow the modulation of pectin properties as solubility and hydrophilicity according to specific goals.

Multiparticulated and Compacted Systems

Blends of pectin and HAS at different ratios were cross-linked with STMP in alkaline media for preparing multiparticulate matrix systems [65]. The cross-linking reaction allowed the building of covalent gels with higher thermal stability than uncross-linked samples, making this material a promising excipient for the design of controlled drug release systems based on hydrogel matrices.

Rheological studies by dynamic oscillatory measurements represent a reliable tool to evaluate the gel structure and, according to entangled networks, gels can be classified as covalently or physically cross-linked [59, 66, 67]. The study revealed that the rising of pectin proportion in cross-linked blends led to the formation of weaker gel structures and this behavior was evidenced by the lowest critical stress values supported by these samples (Figure 3). Furthermore, the mechanical spectrum of samples demonstrated the predominance of G' values over G'' in the whole frequency range, indicating an elastic behavior [68]. The creep-recovery tests showed that when pectin proportion was increased, the recovery ability of the samples was disfavored, indicating again the formation of a weaker gel structure.

In the diffractograms of polymer blends treated in alkaline medium without cross-linker, a decay of crystallinity degree was observed as result of some structural reorganization. For cross-linked blends, at both 2 and 4% of alkali, events as the occurrence of new predominant peaks, reduction of the intensity of some peaks or even the disappearance of peculiar peaks of the original polymers were observed. These features should be attributed to significant changes of the tridimensional network due to the cross-linking process [65].

According to nuclear magnetic resonance analysis (NMR), the same characteristic peaks of pectin were displaced or even disappeared from the NMR spectra of cross-linked samples and those submitted to alkaline treatments, pointing again to a structural rearrangement [65].

After characterization of blends of pectin-HAS cross-linked with STMP in alkaline medium by Carbinatto and coworkers (2012), these materials were evaluated for the application as excipient in matrix tablets and the influence of cross-linking degree and polymers ratio on the drug release patterns and mechanisms was evaluated [69].

The increasing of pectin proportion promoted the rising of LU ability when polymer blends were cross-linked at 2% of NaOH, since the most hydrophilic polymer (pectin) is in higher proportion, favoring the hydrophilicity of the system. Otherwise, blends cross-linked at 4% of NaOH presented the lowest LU ability because in this high cross-linking degree, the reduced mesh size of polymer network limits the water entrance in the polymer structure [69].

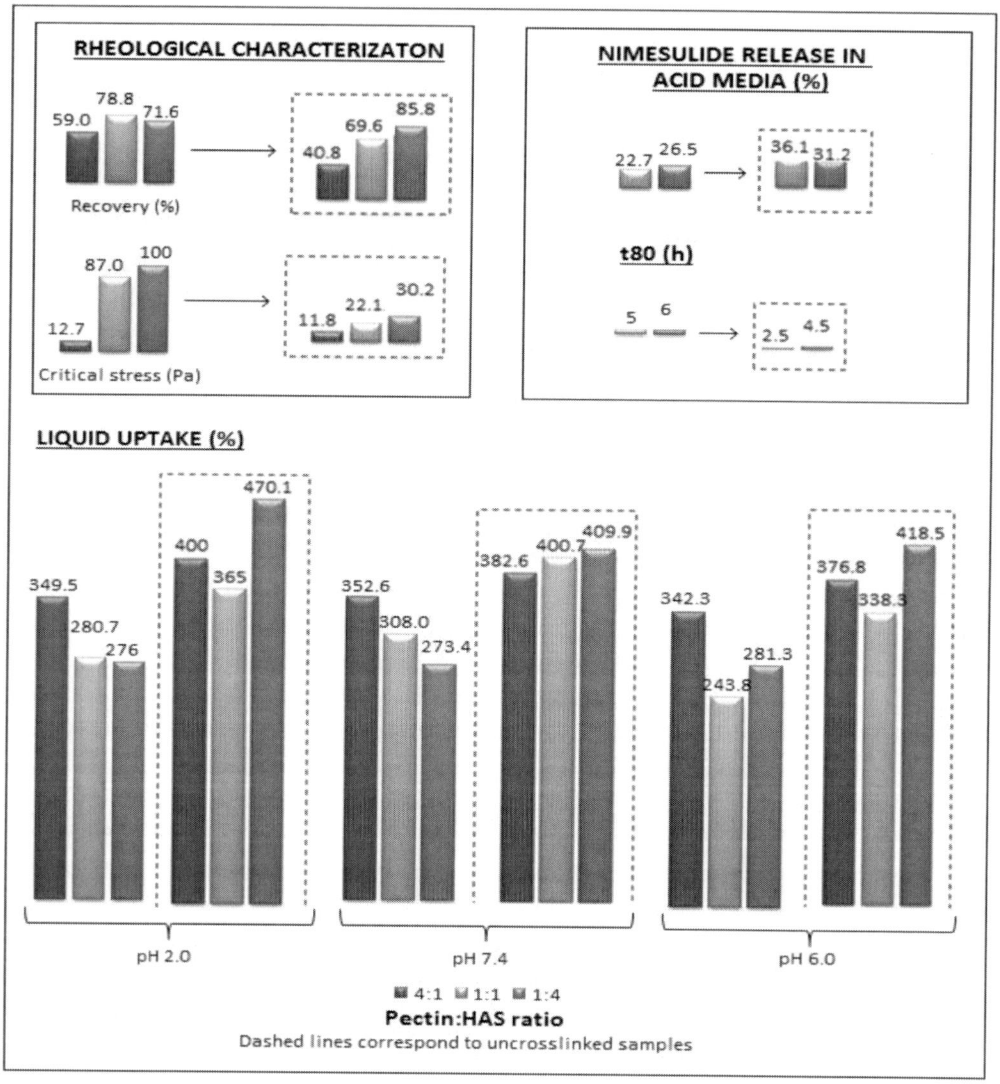

Figure 3. Properties of cross-linked pectin-HAS microparticles.

Flow and density of powdered cross-linked polymers was favored by the cross-linking of all polymer blends, since this process allow a more packed polymer network and their higher densities favored the flow ability [69].

In vitro dissolution of tablets prepared with cross-linked polymer blends containing nimesulide evidenced the reduction of drug release rates in acid medium and the lowest drug release (%) (Figure 3) occurred when higher pectin proportion was used because a more viscous and thicker gel layer may have been built, which represents a more resistant barrier against drug diffusion [63, 70].

In higher pH value (7.4), increasing pectin proportion promoted an acceleration of drug release rates because this hydrophilic polysaccharide contributed to the dissolution of polymer matrix, resulting in its erosion [69].

The change of pectin ratio in the polymer blends promoted an important change in drug release mechanism so that its increasing made the erosion take a place in the release process, probably by favoring the dissolution of the matrix. When pectin was in lower proportion, the drug release occurred according to the anomalous transport, in which this process is driven by both swelling of the matrix and diffusion of the drug.

In another study of our research group, microparticles were obtained with the pectin-HAS blends at different ratios (4:1, 1:1 and 4:1) cross-linked with STMP and loaded with SD [35].

The rheological characterization of hydrogels prepared from the aqueous dispersions (5%, w/v) of blends of cross-linked pectin-HAS was performed by dynamic oscillatory and creep-recovery tests. The mechanical spectra demonstrated that all hydrogels have storage modulus (G') higher than loss modulus (G") within the whole frequency range, indicating a behavior of elastic gel, peculiar of covalently cross-linked networks.

Pectin favored the building of stronger structures, since samples with higher proportion of this polysaccharide (4:1) exhibited the highest *S* values (Figure 4), which is a coefficient related to the cross-linking density inside the gel. So, higher *S* values have been related to more cross-linked and stronger gels. Inversely, the *n* viscoelastic exponent value decreases with the increase of cross-linking density. Both *S* and *n* values can be calculated by a "Power Law" (Eq.1) [71], given by:

$$G' = S\omega^n \qquad \text{Equation 1}$$

where G' is the storage modulus; *S* is the gel strength, ω the oscillation frequency and *n* is the viscoelastic exponent.

Likewise, the creep-recovery tests (Figure 4) performed to provide more information about internal structure of systems that presented an elastic behavior, indicated that the highest pectin ratio led to formation of more elastic gels, verified by the higher value of recovery (*R%*), corroborating the G' data.

Microparticles prepared with these blends of cross-linked pectin-HAS presented high circularity (0.704-0,756) and shape regularity, which are important features that contribute to provide more even drug release patterns. Furthermore, microparticles had high SD level (92-98%) and the pectin-HAS proportion did not influence this parameter.

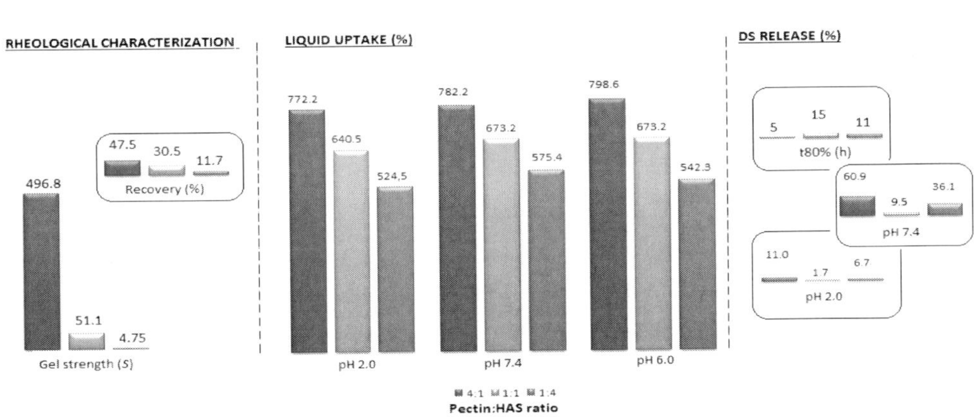

Figure 4. Properties of pectin-HAS microparticles containing DS.

Thermal analysis of microparticles showed events of mass loss between 40°C and 110°C related to the moisture evaporation, and around 180-400°C due to simultaneous degradation of drug and both polysaccharides. The presence of a slight shoulder between 210°C and 220°C without changes in overall termoanalytical profile was associated to physicochemical interactions between drug and polymer [72]. The lack of peculiar peaks of drug in the DSC curves of microparticles indicated that the SD was molecularly dispersed within the polymer matrix, building a solid solution [73].

Well defined peaks of pectin proper of its crystallinity and some others related to HAS were not preserved in the diffractograms of microparticles, indicating the amorphization of polysaccharides due to the structural reorganization caused by cross-linking and liophilization processes involved in the synthesis of microparticles. Likewise, peculiar peaks of SD were not observed, indicating that the drug molecules interact with polymers, dispersing inside of polymer matrix and its original crystalline structure becomes deformed [74].

The LU ability of microparticles was evaluated in media with different pH values (Figure 4), simulating those of the different segments of the GIT and exhibited the lowest values in acid medium (pH 2.0). This property was improved when the pH was raised, because in high pH the carboxylic groups of the anionic polymers are ionized and the network is expanded, since the polymers chains are apart.

The increasing of pectin proportion in the polymer blends enhanced the LU ability of microparticles, so that this hydrophilic polymer should contribute to the water entrance [75, 76].

In vitro dissolution tests for determining the release profile of tablets containing SD and cross-linked polymers blends as excipient showed that the release rate of SD in acid medium (pH 2.0) was lower (at least 4.5 x) than at pH 7.4 and 6.0 (Figure 4), demonstrating the pH-responsive behavior of these systems, as observed for LU studies. Besides of the low LU ability in acid media, the protonation of carboxylic and phosphate groups of the polymeric matrix in this pH should restrict the motion and/or relaxation of the chains, hindering the diffusion and release of the drug [77].

Pectin-Gellan Gum Blends

Beads

Besides of other favorable properties of pectin, its mucoadhesiveness has been well reported and it is an important feature that makes pectin a promising polysaccharide to be used in drug release systems with bioadhesive properties [21, 78-81].

Beads of pectin and gellan gum mixtures were successfully prepared by ionotropic gelation technique using $AlCl_3$ as cross-linking agent and ketoprofen as model drug. Entrapment efficiency (EE%) up to 89% was reached and no influence of pectin:gellan gum ratio on this parameter was observed, although higher polymer and drug concentrations improved the encapsulation process (unpublished data).

The increase of polymer and cross-linker concentrations led to an increase in particle size and circularity because a higher amount of cross-linker can react with more sites of the polymers, building a more branched and packed structure.

Beads without drug exhibited a homogeneous polymeric matrix, which was disturbed and expanded in the presence of drug. FTIR analysis revealed that the drug was physically trapped within the polymer chains.

LU ability of beads (Figure 5) was strongly dependent on the pH because in high pH values the anionic polysaccharides pectin and gellan gum remain in ionized form and have their network expanded due to electrostatic repulsion, increasing the hydrophilicity and favoring the penetration of liquid in the system. This property was not significantly influenced by polymer ratio, demonstrating that cross-linking process was able to restrict the hydrophilicity of the systems, even when pectin was in increased amounts.

The high mucoadhesive ability of such beads was evidenced by *in vitro* mucin adsorption tests that exhibited values (Figure 5) higher than those reported in the literature for chitosan beads [82], as well as by *ex vivo* test, in which all beads were kept strongly attached to the intestinal mucosa of porcine.

Pectin-gellan gum beads were able to reduce the release of ketoprofen in acid medium (0.1N HCl pH 1.2) and control the drug release in phosphate buffer (pH 7.4) up to 6 h.

Beads of pectin-gellan gum were able to decrease the drug released in acid medium by half in comparison with beads prepared with only gellan gum (Figure 5). In acid media, pectin can remain as aggregates of macromolecules hindering the drug diffusion. These results show the important role played by pectin in controlling drug release rates in acid pHs.

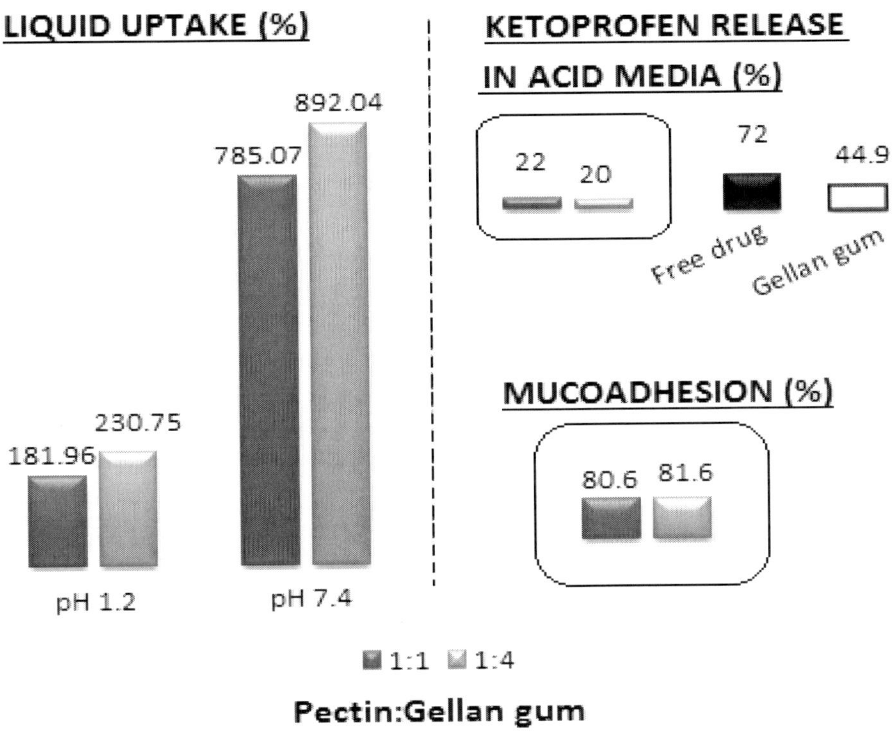

Figure 5. Properties of pectin:gellan gum beads.

Conclusion

The purpose of this chapter was to compile the main findings of our research group about the use of pectin blended with other polysaccharides for designing different drug release systems intended to release the drug in specific organs of the GIT, mainly to the colon. The covalent or ionic cross-linking of these materials was also evaluated as an additional strategy to improve the performance of these drug release systems. All systems were systematically characterized according to their physicochemical properties that revealed important changes of them in relation to isolated polymers, so that the polymer blends and cross-linking process showed to be useful tools to modulate the pectin properties for specific purposes. These changes of properties allowed the control of drug release rates and, so that, generally, the release was restricted in acid media, demonstrating the potential of these materials in protecting the drug in upper portions of the GIT and targeting drugs to the colon.

Acknowledgment

The financial support of the FAPESP, CAPES and CNPq made the preparation of this entry possible.

References

[1] Mignani, S; El Kazzouli, S; Bousmina, M; Majoral, J-P. Expand classical drug administration ways by emerging routes using dendrimer drug delivery systems: A concise overview. *Advanced Drug Delivery Reviews,* 2013. 65, 1316–1330.

[2] Sadeghi, AM-M; Dorkoosh, FA; Avadi, MR; Saadat, P; Rafiee-Tehrani, M; Junginger, HE. Preparation, characterization and antibacterial activities of chitosan,N-trimethyl chitosan (TMC) and N-diethylmethyl chitosan (DEMC) nanoparticles loaded with insulin using both the ionotropic gelation and polyelectrolyte complexation methods. *International Journal of Pharmaceutics*, 2008. 355(1), 299-306.

[3] Lam, PL & Gambari, R. Advanced progress of microencapsulation technologies: in vivo and in vitro models for studying oral and transdermal drug deliveries. *Journal of Controlled Release,* 2014. 178, 25-45.

[4] Maior, JFAS; Reis, AV; Muniz, EC; Cavalcanti, OA. Reaction of pectin and glycidyl methacrylate and ulterior formation of free films by reticulation. *International Journal of Pharmaceutics,* 2008. 355(1–2), 184-194.

[5] Sinha, VR & Kumria, R. Microbially triggered drug delivery to the colon. *European Journal of Pharmaceutical Sciences,* 2003. 18(1), 3-18.

[6] Li, X; Tang, T; Zhou, Y; Zhang, Y; Sun, Y. Applicability of enzyme-responsive mesoporous silica supports capped with bridged silsesquioxane for colon-specific drug delivery. *Microporous and Mesoporous Materials,* 2014. 184, 83-89.

[7] McConnell, RE & Tyska, MJ. Myosin-1a powers the sliding of apical membrane along microvillar actin bundles. *The Journal of Cell Biology,* 2007. 177(4), 671-681.

[8] Chourasia, MK & Jain, SK. Pharmaceutical approaches to colon targeted drug delivery systems. *Journal of Pharmacy & Pharmaceutical Sciences,* 2003. 6(1), 33-66.

[9] Liu, LS; Fishman, ML; Kost, J; Hicks, KB. Pectin-based systems for colon-specific drug delivery via oral route. *Biomaterials,* 2003. 24(19), 3333-3343.

[10] Yang, L. Biorelevant dissolution testing of colon-specific delivery systems activated by colonic microflora. *Journal of Controlled Release,* 2008. 125(2), 77-86.

[11] Sinha, VR & Kumria, R. Polysaccharides in colon-specific drug delivery. *International Journal of Pharmaceutics,* 2001. 224(1–2), 19-38.

[12] Friend, DR. New oral delivery systems for treatment of inflammatory bowel disease. *Advanced Drug Delivery Reviews,* 2005. 57(2), 247-265.

[13] Orlu, M; Cevher, E; Araman, A. Design and evaluation of colon specific drug delivery system containing flurbiprofen microsponges. *International Journal of Pharmaceutics,* 2006. 318(1–2), 103-117.

[14] Ahmed, IS & Ayres, JW. Comparison of in vitro and in vivo performance of a colonic delivery system. *International Journal of Pharmaceutics,* 2011. 409(1), 169-177.

[15] Schols, HA & Voragen, AGJ. The chemical structure of pectins. In: Seymour GB & Knox JP, editors. Pectins and their manipulation. *Boca Raton, FL: CRC Press LLC;* 2002; 1-29.

[16] Silva, JAL & Rao, MA. Pectins: Structure, functionality, and uses. In: Stephen AM, Phillips GO, Williams PA, editors. *Food polysaccharides and their applications.* Boca Raton, FL: Taylor and Francis; 2006; 353-411.

[17] Schols, HA & Voragen, AGJ. Complex Pectins: Structure elucidation using enzymes. In: Visser J & Voragen AGJ, editors. *Pectins and Pectinases.* Amsterdam: Elsevier; 1996; 3-20.

[18] Morris, GA; Castile, J; Smith, A; Adams, GG; Harding, SE. The effect of different storage temperatures on the physical properties of pectin solutions and gels. *Polymer Degradation and Stability,* 2010. 95(12), 2670-2673.

[19] Vincken, J-P; Schols, HA; Oomen, RJFJ; McCann, MC; Ulvskov, P; Voragen, AGJ; Visser, RGF. If Homogalacturonan Were a Side Chain of Rhamnogalacturonan I. Implications for Cell Wall Architecture. *Plant Physiology,* 2003. 132(4), 1781-1789.

[20] Jung, J; Arnold, RD; Wicker, L. Pectin and charge modified pectin hydrogel beads as a colon-targeted drug delivery carrier. *Colloids and Surfaces B: Biointerfaces, 2013.* 104, 116-121.

[21] Thirawong, N; Nunthanid, J; Puttipipatkhachorn, S; Sriamornsak, P. Mucoadhesive properties of various pectins on gastrointestinal mucosa: An in vitro evaluation using texture analyzer. *European Journal of Pharmaceutics and Biopharmaceutics, 2007.* 67(1), 132-140.

[22] Thakur, BR; Singh, RK; Handa, AK; Rao, MA. Chemistry and uses of pectin - a review. *Critical Reviews in Food Science & Nutrition,* 1997. 37(1), 47-73.

[23] Tombs, M & Harding, SE. Structural polysaccharides. In: Tombs M & Harding SE, editors. *An introduction to polysaccharide biotechnology.* London, UK: Taylor and Francis; 1998; 47-94.

[24] Axelos, MAV & Thibault, JF. The chemistry of low-methoxyl pectin gelation. In: Walter RH, Editor. *The chemistry and technology of pectin.* San Diego, CA: Academic Press, INC 1991; 109-118.

[25] Morris, GA; de al Torre, JG; Ortega, A; Castile, J; Smith, A; Harding, SE. Molecular flexibility of citrus pectins by combined sedimentation and viscosity analysis. *Food Hydrocolloids,* 2008. 22(8), 1435-1442.

[26] Lapasin, R & Pricl, S, *Rheology of industrial polysaccharides: theory and applications.* 1995, London: Blackie Academic & Professional London.

[27] Munarin, F; Tanzi, MC; Petrini, P. Advances in biomedical applications of pectin gels. *International Journal of Biological Macromolecules,* 2012. 51(4), 681-689.

[28] Gilsenan, PM; Richardson, RK; Morris, ER. Thermally reversible acid-induced gelation of low-methoxy pectin. *Carbohydrate Polymers,* 2000. 41(4), 339-349.

[29] Akhtar, MJ; Jacquot, M; Jasniewski, J; Jacquot, C; Imran, M; Jamshidian, M; Paris, C; Desobry, S. Antioxidant capacity and light-aging study of HPMC films functionalized with natural plant extract. *Carbohydrate Polymers,* 2012. 89(4), 1150-1158.

[30] Khurana, R; Singh, K; Sapra, B; Tiwary, AK; Rana, V. Tamarindus indica pectin blend film composition for coating tablets with enhanced adhesive force strength. *Carbohydrate Polymers,* 2014. 102, 55-65.

[31] Evageliou, V; Richardson, RK; Morris, ER. Effect of pH, sugar type and thermal annealing on high-methoxy pectin gels. *Carbohydrate Polymers,* 2000. 42(3), 245-259.

[32] Löfgren, C & Hermansson, A-M. Synergistic rheological behaviour of mixed HM/LM pectin gels. *Food Hydrocolloids,* 2007. 21(3), 480-486.

[33] Willats, WGT; Knox, JP; Mikkelsen, JD. Pectin: new insights into an old polymer are starting to gel. *Trends in Food Science & Technology,* 2006. 17(3), 97-104.

[34] Itoh, K; Hirayama, T; Takahashi, A; Kubo, W; Miyazaki, S; Dairaku, M; Togashi, M; Mikami, R; Attwood, D. In situ gelling pectin formulations for oral drug delivery at high gastric pH. *International Journal of Pharmaceutics,* 2007. 335(1), 90-96.

[35] Soares, GA; Castro, ADd; Cury, BSF; Evangelista, RC. Blends of cross-linked high amylose starch/pectin loaded with diclofenac. *Carbohydrate Polymers,* 2013. 91(1), 135-142.

[36] Prezotti, FG; Meneguin, AB; Evangelista, RC; Cury, BSF. Preparation and characterization of free films of high amylose/pectin mixtures cross-linked with sodium trimetaphosphate. *Drug Development and Industrial Pharmacy,* 2012. 38(11), 1354-1359.

[37] Kim, TH; Park, YH; Kim, KJ; Cho, CS. Release of albumin from chitosan-coated pectin beads in vitro. *International Journal of Pharmaceutics,* 2003. 250(2), 371-383.

[38] Thirawong, N; Thongborisute, J; Takeuchi, H; Sriamornsak, P. Improved intestinal absorption of calcitonin by mucoadhesive delivery of novel pectin–liposome nanocomplexes. *Journal of Controlled Release,* 2008. 125(3), 236-245.

[39] Chambin, O; Dupuis, G; Champion, D; Voilley, A; Pourcelot, Y. Colon-specific drug delivery: Influence of solution reticulation properties upon pectin beads performance. *International Journal of Pharmaceutics,* 2006. 321(1–2), 86-93.

[40] Bigucci, F; Luppi, B; Monaco, L; Cerchiara, T; Zecchi, V. Pectin-based microspheres for colon-specific delivery of vancomycin. *Journal of Pharmacy and Pharmacology,* 2009. 61(1), 41-46.

[41] Sriamornsak, P; Thirawong, N; Weerapol, Y; Nunthanid, J; Sungthongjeen, S. Swelling and erosion of pectin matrix tablets and their impact on drug release behavior. *European Journal of Pharmaceutics and Biopharmaceutics,* 2007. 67(1), 211-219.

[42] Das, S & Ng, K-Y. Colon-specific delivery of resveratrol: Optimization of multi-particulate calcium-pectinate carrier. *International Journal of Pharmaceutics,* 2010. 385(1–2), 20-28.

[43] Vandamme, TF; Lenourry, A; Charrueau, C; Chaumeil, JC. The use of polysaccharides to target drugs to the colon. *Carbohydrate Polymers,* 2002. 48(3), 219-231.

[44] Mishra, RK; Banthia, AK; Majeed, ABA. Pectin based formulations for biomedical applications: a review. *Asian Journal of Pharmaceutical & Clinical Research,* 2012. 5(4), 1-7.

[45] Maestrelli, F; Cirri, M; Corti, G; Mennini, N; Mura, P. Development of enteric-coated calcium pectinate microspheres intended for colonic drug delivery. *European Journal of Pharmaceutics and Biopharmaceutics,* 2008. 69(2), 508-518.

[46] Singh, J; Kaur, L; McCarthy, OJ. Factors influencing the physico-chemical, morphological, thermal and rheological properties of some chemically modified starches for food applications - A review. *Food Hydrocolloids,* 2007. 21(1), 1-22.

[47] Mocanu, G; Souguir, Z; Picton, L; Le Cerf, D. Multi-responsive carboxymethyl polysaccharide crosslinked hydrogels containing Jeffamine side-chains. *Carbohydrate Polymers,* 2012. 89(2), 578-585.

[48] Bajpai, AK; Shukla, SK; Bhanu, S; Kankane, S. Responsive polymers in controlled drug delivery. *Progress in Polymer Science,* 2008. 33(11), 1088-1118.

[49] Luo, Y; Zhu, J; Ma, Y; Zhang, H. Dry coating, a novel coating technology for solid pharmaceutical dosage forms. *International Journal of Pharmaceutics,* 2008. 358(1), 16-22.

[50] Miro, A; d'Angelo, I; Nappi, A; La Manna, P; Biondi, M; Mayol, L; Musto, P; Russo, R; Rotonda, MIL; Ungaro, F. Engineering poly (ethylene oxide) buccal films with cyclodextrin: A novel role for an old excipient? *International Journal of Pharmaceutics,* 2013. 452(1), 283-291.

[51] Meneguin, AB; Cury, BSF; Evangelista, RC. Films from resistant starch-pectin dispersions intended for colonic drug delivery. *Carbohydrate Polymers,* 2014. 99(0), 140-149.

[52] Yuan, RC; Thompson, DB; Boyer, CD. Fine structure of amylopectin in relation to gelatinization and retrogradation behavior of maize starches from three wx-containing genotypes in two inbred lines. *Cereal Chemistry,* 1993. 70, 81-89.

[53] Thompson, DB. Strategies for the manufacture of resistant starch. *Trends in Food Science & Technology,* 2000. 11(7), 245-253.

[54] Chung, H-J; Lim, HS; Lim, S-T. Effect of partial gelatinization and retrogradation on the enzymatic digestion of waxy rice starch. *Journal of Cereal Science,* 2006. 43(3), 353-359.

[55] Htoon, AK; Uthayakumaran, S; Piyasiri, U; Appelqvist, IAM; López-Rubio, A; Gilbert, EP; Mulder, RJ. The effect of acid dextrinisation on enzyme-resistant starch content in extruded maize starch. *Food Chemistry,* 2010. 120(1), 140-149.

[56] Felton, LA. Characterization of coating systems. *AAPS Pharm SciTech,* 2007. 8(4), 258-266.

[57] López, OV; García, MA; Zaritzky, NE. Film forming capacity of chemically modified corn starches. *Carbohydrate Polymers*, 2008. 73(4), 573-581.

[58] Xie, F; Halley, PJ; Avérous, L. Rheology to understand and optimize processibility, structures and properties of starch polymeric materials. *Progress in Polymer Science,* 2012. 37(4), 595-623.

[59] Khondkar, D; Tester, RF; Hudson, N; Karkalas, J; Morrow, J. Rheological behaviour of uncross-linked and cross-linked gelatinised waxy maize starch with pectin gels. *Food Hydrocolloids,* 2007. 21(8), 1296-1301.

[60] Lawal, OS; Lapasin, R; Bellich, B; Olayiwola, TO; Cesàro, A; Yoshimura, M; Nishinari, K. Rheology and functional properties of starches isolated from five improved rice varieties from West Africa. *Food Hydrocolloids,* 2011. 25(7), 1785-1792.

[61] Carbinatto, FM, *Matrizes poliméricas reticuladas de alta amilose e pectina para liberação controlada de fármacos,* 2010, Universidade Estadual Paulista "Júlio de Mesquita Filho", UNESP: Araraquara. p. 109f.

[62] Sriamornsak, P & Wattanakorn, N. Rheological synergy in aqueous mixtures of pectin and mucin. *Carbohydrate Polymers,* 2008. 74(3), 474-481.

[63] Bigucci, F; Luppi, B; Cerchiara, T; Sorrenti, M; Bettinetti, G; Rodriguez, L; Zecchi, V. Chitosan/pectin polyelectrolyte complexes: Selection of suitable preparative conditions for colon-specific delivery of vancomycin. *European Journal of Pharmaceutical Sciences,* 2008. 35(5), 435-441.

[64] Espitia, PJP; Du, W-X; Avena-Bustillos, RdJ; Soares, NdFF; McHugh, TH. Edible films from pectin: Physical-mechanical and antimicrobial properties - A review. *Food Hydrocolloids,* 2014. 35(0), 287-296.

[65] Carbinatto, FM; de Castro, AD; Cury, BSF; Magalhães, A; Evangelista, RC. Physical properties of pectin–high amylose starch mixtures cross-linked with sodium trimetaphosphate. *International Journal of Pharmaceutics,* 2012. 423(2), 281-288.

[66] Clark, AH & Ross-Murphy, SB. Structural and mechanical properties of biopolymer gels. *Biopolymers.* Berlin Heidelberg Springer; 1987; 57-192.

[67] Doucet, D; Gauthier, SF; Foegeding, EA. Rheological characterization of a gel formed during extensive enzymatic hydrolysis. *Journal of Food Science,* 2001. 66(5), 711-715.

[68] O'Brien, S; Wang, Y-J; Vervaet, C; Remon, JP. Starch phosphates prepared by reactive extrusion as a sustained release agent. *Carbohydrate Polymers,* 2009. 76(4), 557-566.

[69] Carbinatto, FM; de Castro, AD; Evangelista, RC; Cury, BSF. Insights into the swelling process and drug release mechanisms from cross-linked pectin/high amylose starch matrices. *Asian Journal of Pharmaceutical Sciences,* 2014. 9(1), 27-34.

[70] Luppi, B; Bigucci, F; Abruzzo, A; Corace, G; Cerchiara, T; Zecchi, V. Freeze-dried chitosan/pectin nasal inserts for antipsychotic drug delivery. *European Journal of Pharmaceutics and Biopharmaceutics,* 2010. 75(3), 381-387.

[71] Saxena, A; Kaloti, M; Bohidar, H. Rheological properties of binary and ternary protein–polysaccharide co-hydrogels and comparative release kinetics of salbutamol sulphate from their matrices. *International Journal of Biological Macromolecules,* 2011. 48(2), 263-270.

[72] Mora, MJ; Longhi, MR; Granero, GE. Synthesis and characterization of binary and ternary complexes of diclofenac with a methyl-β-CD and monoethanolamine and in vitro transdermal evaluation. *European Journal of Medicinal Chemistry,* 2010. 45(9), 4079-4088.

[73] Sipos, P; Szűcs, M; Szabó, A; Erős, I; Szabó-Révész, P. An assessment of the interactions between diclofenac sodium and ammonio methacrylate copolymer using thermal analysis and Raman spectroscopy. *Journal of Pharmaceutical and Biomedical Analysis,* 2008. 46(2), 288-294.

[74] Fini, A; Moyano, JR; Ginés, JM; Perez-Martinez, JI; Rabasco, AM. Diclofenac salts, II. Solid dispersions in PEG6000 and Gelucire 50/13. *European Journal of Pharmaceutics and Biopharmaceutics,* 2005. 60(1), 99-111.

[75] Guimarães, F; Oliveira, C; Sequeiros, E; Torres, M; Susano, M; Henriques, M; Oliveira, R; Escobar Galindo, R; Carvalho, S; Parreira, NMG. Structural and Mechanical properties of Ti–Si–C–ON for biomedical applications. *Surface and Coatings Technology,* 2008. 202(11), 2403-2407.

[76] Sriamornsak, P & Kennedy, RA. Swelling and diffusion studies of calcium polysaccharide gels intended for film coating. *International Journal of Pharmaceutics,* 2008. 358(1), 205-213.

[77] Souto-Maior, JFA; Reis, AV; Pedreiro, LN; Cavalcanti, OA. Phosphated crosslinked pectin as a potential excipient for specific drug delivery: preparation and physicochemical characterization. *Polymer International,* 2010. 59(1), 127-135.

[78] Thirawong, N; Kennedy, RA; Sriamornsak, P. Viscometric study of pectin–mucin interaction and its mucoadhesive bond strength. *Carbohydrate Polymers,* 2008. 71(2), 170-179.

[79] Liu, L; Fishman, ML; Hicks, KB; Kende, M. Interaction of various pectin formulations with porcine colonic tissues. *Biomaterials,* 2005. 26(29), 5907-5916.

[80] Li, Y; Zhao, H; Duan, L-R; Li, H; Yang, Q; Tu, H-H; Cao, W; Wang, S-W. Preparation, characterization and evaluation of bufalin liposomes coated with citrus pectin. *Colloids and Surfaces A: Physicochemical and Engineering Aspects,* 2014. 444, 54-62.

[81] Sriamornsak, P; Wattanakorn, N; Nunthanid, J; Puttipipatkhachorn, S. Mucoadhesion of pectin as evidence by wettability and chain interpenetration. *Carbohydrate Polymers,* 2008. 74(3), 458-467.

[82] Dhawan, S; Singla, A; Sinha, V. Evaluation of mucoadhesive properties of chitosan microspheres prepared by different methods. *AAPS PharmSciTech,* 2004. 5(4), 122-128.

In: Pectin: Chemical Properties, Uses and Health Benefits ISBN: 978-1-63321-438-5
Editor: Phillip L. Bush

Chapter 6

PECTIN: STRUCTURE, MODIFICATION AND THE HUMAN DISTAL GUT MICROBIOTA

D. W. Abbott*[*], *B. Farnell and J. W. Yamashita
Agriculture and Agri-Food Canada, Lethbridge Research Centre,
Lethbridge, Alberta, Canada

ABSTRACT

Homogalacturonan (HG), rhamnogalacturonan-I (RG-I) and rhamnogalacturonan-II (RG-II) are structural pectic polysaccharides (i.e. pectin) found within the cell wall of terrestrial plants, and common sources of dietary fibre. The human genome does not contain any enzymes predicted to be involved in pectin digestion; therefore, in order to extract nutritional value from HG, RG-I, and RG-II humans rely on a consortium of symbiotic intestinal bacteria, commonly referred to as the distal gut microbiota (DGM), to deconstruct and ferment pectins and other complex carbohydrates into host-absorbable products. Currently, intestinal applications for bioactive pectins, such as HG, are under intensive investigation as nutraceuticals, prebiotics, and drug delivery systems. In this light, elucidating the incremental process of HG recognition and deconstruction by intestinal pectinolytic bacteria will provide new insights into the dynamic relationship between diet, human intestinal health, and DGM community structure. This chapter will define the different types of pectin structure, review mechanisms of pectinase function, provide insights into pectinolytic genes present within the genomes of intestinal pectinolytic bacteria, such as *Bacteroides thetaiotaomicron*, and summarize key functions of pectin in the maintenance of intestinal health.

PECTIC POLYSACCHARIDES AND CELL WALL STRUCTURE

Unlike animals which possess a skeletal system, plants rely on an extracellular matrix called the plant cell wall to regulate development and support plant architecture. This network

[*] To whom correspondence should be addressed: wade.abbott@agr.gc.ca, Agriculture and Agri-Food Canada, Lethbridge Research Centre, 5403-1st Avenue South, Lethbridge, Alberta, Canada, T1J4B1.

must be rigid enough to endure immense weight, yet pliable to allow for transformation and expansion during various growth stages. The plant cell wall is divided into two main components referred to as the 'primary' and 'secondary' wall [1]. Both walls are predominantly composed of different compositions of structural polysaccharides, which include cellulose, hemicellulose, and pectin [2]. Cellulose is an unbranched and unmodified β1,4-glucan. Individual homopolymers can interact to form higher-order structures called microfibrils and fibrils that are primarily crystalline due to dense, water excluding intermolecular and intramolecular hydrogen bonding [1]. Cellulose is synthesized at the plasma membrane from nucleotide sugar substrates and is a component of both the primary and secondary wall [3].

Hemicelluloses differ in composition from cellulose but are structurally similar in that they are connected through β1,4-linkages. The main classes of hemicelluloses include xylans, mannans and glucans, which are homopolymers of xylose, mannose, and glucose respectively; however, variations in structure do exist. Glucomannans for example, have a backbone of randomly dispersed β1,4-linked glucose and mannose [2]. These backbone sugars can be extensively decorated with a variety of sugars and acetyl groups which account for the non-crystalline nature of these polymers. Much like cellulose, hemicelluloses are present in both the primary and secondary wall; however they are synthesized in the golgi from the corresponding nucleotide sugar substrates [3].

Pectin is a plant cell wall structural polysaccharide within the primary cell wall and the middle lamella, which punctuates the junctions between primary walls of neighboring cells and participates in intercellular connections [4]. In addition it is found in the cell walls of some freshwater and marine algae [5, 6]. Pectin is the most complex carbohydrate found in nature due to the diversity of stereochemical glycosidic bonds that link a variety of common and rare carbohydrate subunits. A defining feature of all pectins is that they display a high D-galacturonic acid (GalA) content [7] (FIG 1). GalA adopts a $^4C_1$ conformation and is structurally analogous to D-galactose (Gal) with an equatorial C6 that has been oxidized into an uronic acid (FIG 2A-C). Each GalA moiety therefore contains an inherent negative charge at physiologic pH.

Pectin is divided into three classes of distinct pectic polysaccharides: homogalacturonan (HG), rhamnogalacturonan-I (RG-I) and rhamnogalacturonan-II (RG-II) [8] that vary in size, branching, and function. HG, RG-I, and RG-II are believed to be found as an extensive interconnected network within the plant cell wall (FIG 1A-C) [9]. There appears to be multiple levels of covalent crosslinking that contribute to this network, which include but are not limited to, backbone glycosidic linkages, calcium crosslinking [10], borate ester coordination [11] and covalent linkages to phenols (lignin), proteins, and possibly other compounds yet to be discovered [12]. Elucidating the specific assemblies and the degrees of polymerization of each pectic domain remains a difficult task as the 'native' pectin structure is disrupted by chemical and enzymatic treatments required to extract it from the primary cell wall. Despite these limitations, distinct functions have been correlated with pectin, including, cell-cell adhesion via HG cross-linking and RG-II dimerization [12, 13], chemical signaling [12], growth and development, fruit development (ripening) [14, 15] and plant defenses [16-18].

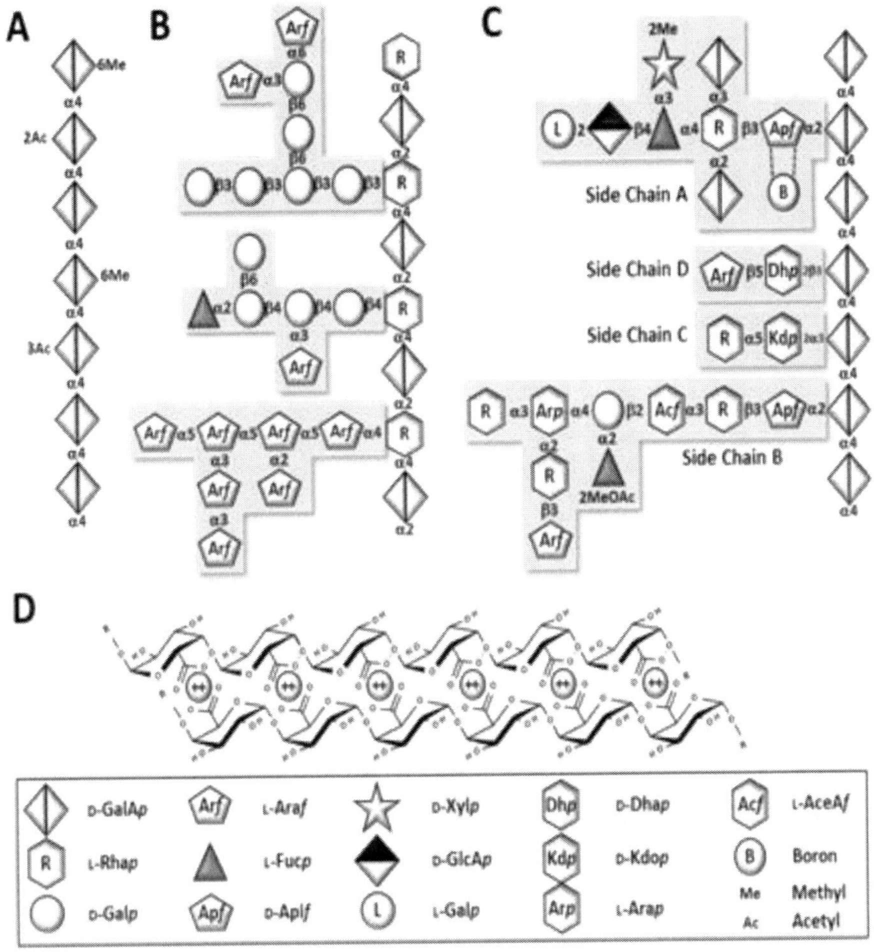

Figure 1. Pectin structure. Schematic representations of the three main pectic polysaccharides (A) HG, (B) RG-I and (C) RG-II. Sugars are represented using the standard symbol nomenclature. Enantiomers are indicated by D and L followed by *p* (pyranose) or *f* (furanose) to indicate ring configuration. Glycosidic linkages are labeled between residues. Carbohydrate nomenclature: GalA: Galacturonic acid; Rha: Rhamnose; Gal: Galactose; Ara: Arabinose; Fuc: Fucose; Api: Apiose; Xyl: Xylose; GlcA: Glucuronic acid; Dha: 2-keto-3-deoxy-D-lyxoheptulosaric acid; Kdo: 2-keto-3-deoxy-D-manno-octulosonic acid; and AceA: Aceric acid. (D) Egg-box model. The intermolecular coordination of a calcium divalent cation (++), by the C5 uronic acid group of proximal GalA moieties creates a tightly packed structure, wherein the calcium represents the egg, 'boxed' in between two HG polymers.

Homogalacturonan

HG is synthesized by α1,4-galacturonosyltransferases, such as GAUT1 [19], which create highly polymerized fibres of α1,4-linked GalA (also referred to as polygalacturonic acid and pectate) [16]. The glycosidic bonds of HG are connected through a C1 axial - C4 axial 'accordion-like' structure that is uncommon in other polysaccharides (FIG 2M-N). For instance, galactans (neutral galactose homopolysaccharides with an axial C4) are often found in β1-4 linkages, which reflect the stereochemical linkages within α1,4 glucans (neutral glucose homopolysaccharides with axial C1), and give rise to a characteristic helical structure in polymers [20]. HG is the backbone of pectin, accounting for greater than 60% of the total pectin assembly [16, 21]. This backbone can be esterified by either methyl groups (at the C6-O, FIG 2G-I) and / or acetyl groups (at the C2-O or C3-O, FIG 2J-L) [16]. The pattern of esterifications varies between plant species, suggesting that the degree of chemical modification is related to developmental and tissue-specific phytophysiology [22]. Contiguous regions of HG (>10 residues) lacking these chemical modifications are capable of forming Ca^{2+} salt bridges between the negative charges of uronate groups, stabilizing a defined higher order structure, referred to as the 'egg-box model' (FIG 1D) [10, 12]. ^{13}C NMR experiments have shown that the gelatinous HG (egg-box model) adopts a 2_1 helical confirmation (two residues per turn), whereas dried HG adopts a 3_1 helical confirmation [23]. This model contributes to dense packing of HG into pectic gels, with ~70% of pectate adopting this gel form.

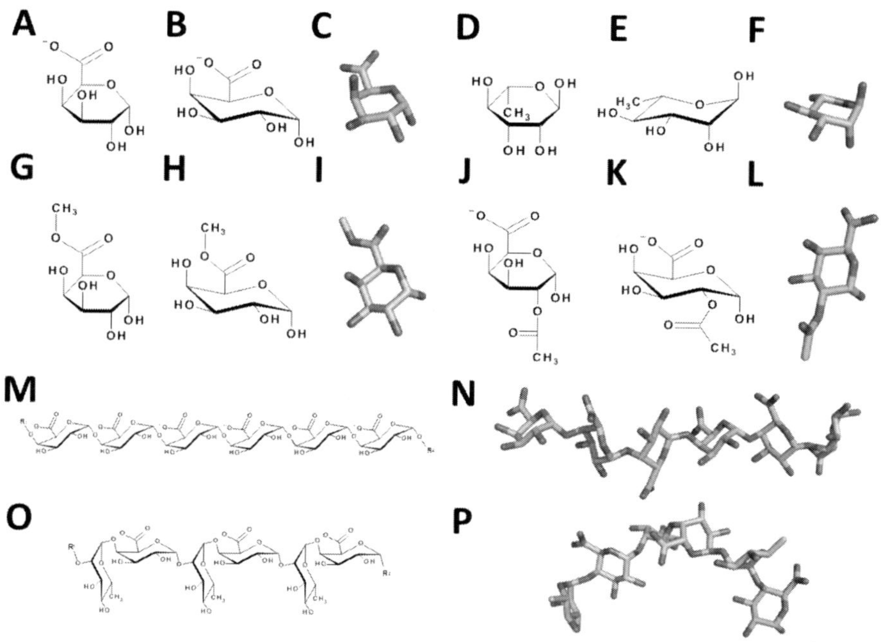

Figure 2. Chemical structure of HG and RG-I. α-D-GalA displayed as a Haworth (A) and chair (B) projection. (C) Three-dimensional structure of α-D-GalA extracted from the PL1 structure from *E. chrysanthemi EC16* (PDB ID: 2ewe) [152]. α-L-Rha displayed as a Haworth (D) and chair (E) projection. (F) Three-dimensional structure of α-L-Rha extracted from the PL4 structure from *A. aculeatus KSM 510EC16* (PDB ID: 3njv) [60]. Methylesterified α-D-GalA displayed as a Haworth (G) and chair (H) projection. (I) Three-dimensional structure of methylesterified α-D-GalA extracted from the CE8 structure from *D. dadantii 3937* (PDB ID: 2nst) [91]. Acetylesterified α-D-GalA displayed as a Haworth (J) and chair (K) projection. (L) Three-dimensional model of acetylesterified α-D-GalA built from an β-N-acetylgalactosamine scaffold and validated for bond angles and distances using COOT [153]. (M) Schematic representation of HG. (N) Three-dimensional structure of a HG hexasaccharide extracted from the PL1 structure of *E. chrysanthemi EC16* (PDB ID: 2ewe). (O) Schematic representation of RG-I. (P) Three-dimensional structure of a RG-I hexasaccharide extracted from the PL4 structure of *A. aculeatus KSM 510EC16* (PDB ID: 3njv).

Rhamnogalacturonan-I

RG-I is unique amongst the pectic polysaccharides as its backbone is comprised of a repeating disaccharide of GalA and α-L-rhamnose (Rha) [4)-α-D-GalA-(1,2)-α-L-Rha-(1,]$_n$ [21, 24, 25] (FIG 2D-F, O-P). Rha is a stereomimic of 6-deoxy-L-mannose, which displays a C2 axial hydroxyl in the $^1C_4$ conformation. The axial-axial linkage of GalA (O4) to Rha (O2) results in the three-dimensional structure of RG-I adopting a curved helix (FIG 2O-P), which is strikingly different than the twisting linear structure of HG (FIG 2M-N). Similar to HG, the backbone GalA residues may be hyperacetylated at the O-2 and O-3 positions and 25-80% of the Rha residues are decorated by branching at the O-4 position [8]. These decorations include linear or branched patterns of defined polysaccharides: α1,5 arabinans, and β1,3 or β1,4 galactans (FIG 1B) [16, 26, 27]. RG-I side-chains can contain further branching at certain positions (O-2 and O-3 for arabinans and O-3 and O-6 for galactans) by the following sugars: arabinose, arabinan, galactan, arabinogalactan [26]. Secondary substitutions of the linear or branched polysaccharides from the main GalA-Rha backbone increase its complexity and lead to a wide diversity of possible RG-I structures. Due to these extensive decorations, RG-I is commonly referred to as the 'hairy-region' of pectin. Interestingly, the structure of RG-I is not strictly conserved between tissues and species, but rather its side-chains appear to be developmentally and differentially regulated [14, 28, 29].

Rhamnogalacturonan-II

RG-II, which accounts for ~10% of pectin [21], is comprised of an HG backbone (~7-9 α1,4-linked GalA) with four (A-D) well-defined side-chains [12] (FIG 1C). Side-chains A (an octasaccharide) and B (a nonasaccharide) are linked to the HG backbone at the O-2 position. Side-chains C and D are both disaccharides and are linked to the HG backbone at the O-3 position. These four defined and well conserved side chains add to the complexity of RG-II molecule by presenting 12 different types of monosaccharides, including the following rare sugars: 2-O-methylxylose, 2-O-methylfucose [30], aceric acid [31], 2-keto-3-deoxy-D-lyxoheptulosaric acid (Dha) [32], and 2-keto-3-deoxy-D-manno-octulosonic acid (Kdo) (FIG 1C) [33]. In addition to the abundance of different carbohydrate subunits, RG-II displays 21 different linkages. Despite this structural diversity (i.e. sugars and linkages), RG-II is highly conserved between plant species [8, 11]. RG-II generally exists as a RG-II dimer that is crosslinked by a bidentate borate diester between apiose residues in side-chain A [13], which covalently crosslinks two distinct RG-II molecules and fortifies the pectin network [13]. These conserved features of RG-II structure play a critical function in plant growth and development, as minor modifications of the RG-II structure have shown near fatal effects on plant growth [34].

Cell Wall Dynamics

Though a common characteristic of plants, cell wall structure varies by source and can be altered over time. Lignification is the hardening process that occurs in the secondary wall which renders the wall resistant to compressive forces, while restricting passage of small molecules [1]. Ripening (softening of the cell wall, with other associated chemical changes) on the other hand occurs due to the regulated modification of polysaccharides found within the primary cell wall and middle lamella. The majority of these polysaccharide modifications are the result of secreted carbohydrate active enzymes (CAZymes) [15, 35]. Initially, these enzymes target the HG rich middle lamella, which results in a loss of intercellular connections [4]. This is followed by enzymatic modifications to the fibrous polysaccharides within the cell wall, resulting in a weakened cell wall structure and network [36]. These modifications allow for the hard and acidic unripe tissue to transform into a sweet, fragrant, and soft fruit [15]. Over-ripening is a limiting factor in the distribution of fruit worldwide, and thus an area of financial significance to the fruit industry.

Table 1. Functions and first structures of pectinase families [35]

FAMILY	ACTIVITY	PROTEIN	PDB	SPECIES	FOLD	REF
Polysaccharide Lyases						
PL1	EC 4.2.2.2	PelE	1pcl	*E. chrysanthemi EC16*	β-helix	[63]
PL2	EC 4.2.2.2	YePL2A	2v8i	*Y. enterocolitica subsp. enterocolitica 8081*	α/α_7-barrel	[56]
PL3	EC 4.2.2.2	Pel-15	1ee6	*Bacillus sp. KSM-P15*	β-helix	[155
PL4	EC 4.2.2.23	RghB	1nkg	*A. aculeatus KSM 510*	β-sandwich	[67]
PL9	EC 4.2.2.2	PelL	1ru4	*D. dadantii 3937*	β-helix	[66]
PL10	EC 4.2.2.2	PelA	1gxn	*Cellvibrio japonicus Ueda107*	α/α_3-barrel	[54]
PL11	EC 4.2.2.23	YesW	2z8r	*B. subtilis subsp. subtilis str. 168*	β_8-propeller	[49]
PL22	*ND*[1]	VPA0088	3c5m	*Vibrio parahaemolyticus RIMD 2210633*	β_7-propeller	*TBP*[2]
Glycoside Hydrolases						
GH28	3.2.1.171	RhgA	1rmg	*A. aculeatus KSM 510 / CBS 115.80*	β-helix	[74]
GH105	3.2.1.172	YteR	1nc5	*B. subtilis subsp. subtilis str. 168*	α/α_6-barrel	[85]
Carbohydrate Esterases						
CE8	3.1.1.11	PemA	1qjv	*E. chrysanthemi B374 /B364*	β-helix	[90]
CE12	3.1.1.86	Rha1	1deo	*A. aculeatus KSM 510*	α/β/α-sandwich	[101]

[1]*ND* – not determined.
[2]*TBP* – to be published.

Dynamic modification of the plant cell wall during infectious disease is catalyzed by CAZymes produced by various organisms. Bacterial, fungal, and insect pathogens are known to contain enzymes, such as pectin methylesterases (PMEs), that degrade the plant cell wall, a debilitating process that can lead to disease and even plant death (soft-rot) [37]. In order to defend against invading pathogens, plants deploy a coordinated immune cascade. A primary line of defense involves an immune protein, referred to as pectin methylesterase inhibitor (PMEI), which disrupts the function of PMEs [38, 39]. PMEIs, bind to the active site of pectin methylesterases (PME) secreted by phytopathogens, which renders the enzyme inactive [40]. PMEs are a primary virulence factor during infection and function upstream of depolymerases such as pectate lyases and polygalacturonases. If the pectic network becomes compromised, plants have an innate immune response that is incited by the release of oligogalacturonides (OGs), which includes accumulation of reactive oxygen species [41] and pathogenesis-related proteins [42, 43]. Recently the existence of a pectin integrity monitoring system (PIMS) has been proposed [18], which are regions of HG that when cleaved by invading species act as a signal to activate plant innate immunity.

ENZYMATIC MODIFCATION OF PECTIN

Despite its ubiquity in nature, cleavage of glycosidic linkages within the pectic backbones of HG and RG-I are catalyzed by relatively few enzyme families. Whether this observation is due to the properties of the substrate (e.g. stability, steric constraints, and charge potential) or selective stringency on the convergent evolution of pectinases is not known. Pectinases are often found in multiple copies within the genomes of pectinolytic bacteria, which suggests that individual genes are differentially regulated or have preferential activities on microheterogeneous substrates that may vary in their degree of esterification, polymerization, or carbohydrate composition (FIG 2) [44, 45]. Currently, there are two different enzyme classes known to cleave the glycosidic linkages within pectic sugars that operate with distinct mechanisms (Table 1): polysaccharide lyases (PLs) and glycoside hydrolases (GHs). Deesterification of methyl and acetyl esters is catalyzed by a third class of enzymes, called carbohydrate esterases (CEs), from families 8 (CE8: PMEs) and 12 (CE12: acetylesterases).

Polysaccharide Lyases

PLs that are active on pectin are commonly referred to as pectate lyases; however, pectate lyases are defined by having exclusive activity on HG (i.e. pectate). Variations do exist in nature that are active on methylated HG (i.e. pectin lyase) [46, 47] and RG-I (rhamnogalacturonan lyase, RG-lyase) [48, 49]. PLs active on other uronic acid polysaccharides, such as heparin [50] and alginate [51], deploy unrelated mechanisms, involving unique catalytic residues (e.g. isoleucine and tyrosine) and except for PL6s do not require metals for catalysis [52, 54]. For the purposes of this discussion, the use of the PL acronym below will refer to polysaccharide lyases.

PL Mechanism – PLs utilize a β-elimination mechanism to cleave glycosyl bonds in uronic acid containing pectic sugars. The $^4C_1$ chair conformation of GalA within HG presents an ideal geometry for anti-periplanar β-elimination with the C5 hydrogen and C4 hydroxyl group positioned in opposing axial configurations (FIG 2A-C). The reaction progresses through an e1cb pathway (H-Cα cleavage) [54], in which the rate-limiting step is C5 proton abstraction by a Brønstead base (FIG 3A). Proton acidification is facilitated by the C5 uronate, coordination of catalytic divalent metals, which draw charge from the C5 carbon, and localized basic residues within the active site that participate to alter the local pKa environment. Most commonly the Brønstead base is a catalytic arginine (PLs 1, 2, 3, and 10) or lysine (family 9) [52, 53], which explains the high pH optima observed across the PL landscape. This observation also reveals a dichotomy in PL function; however, as the common pH optima (8.0-9.5) [55] is more alkaline than the biological environments these enzymes are secreted into and are active in. The lowered optima for intracellular lyases suggest that this pH effect is primarily related to secreted PL function [56-58]. The transition state proceeds through an enolate-enolate intermediate, which was recently reported to be resonance stabilized by hydrogen bonding or donation to the oxyanion from a dedicated lysine in family 1 PLs and asparagine in family 9 [59]. Decomposition of the intermediate occurs by protonation of the scissile glycosyl oxygen, through a yet to be determined mechanism, and elimination of the axial O4 creating an unsaturation between C4 and C5. The unsaturated product (△GalA), distorts the pyranosyl GalA conformation from a $^4C_1$ into a trigonal planar geometry, which is unstable for cyclized monosaccharide products.

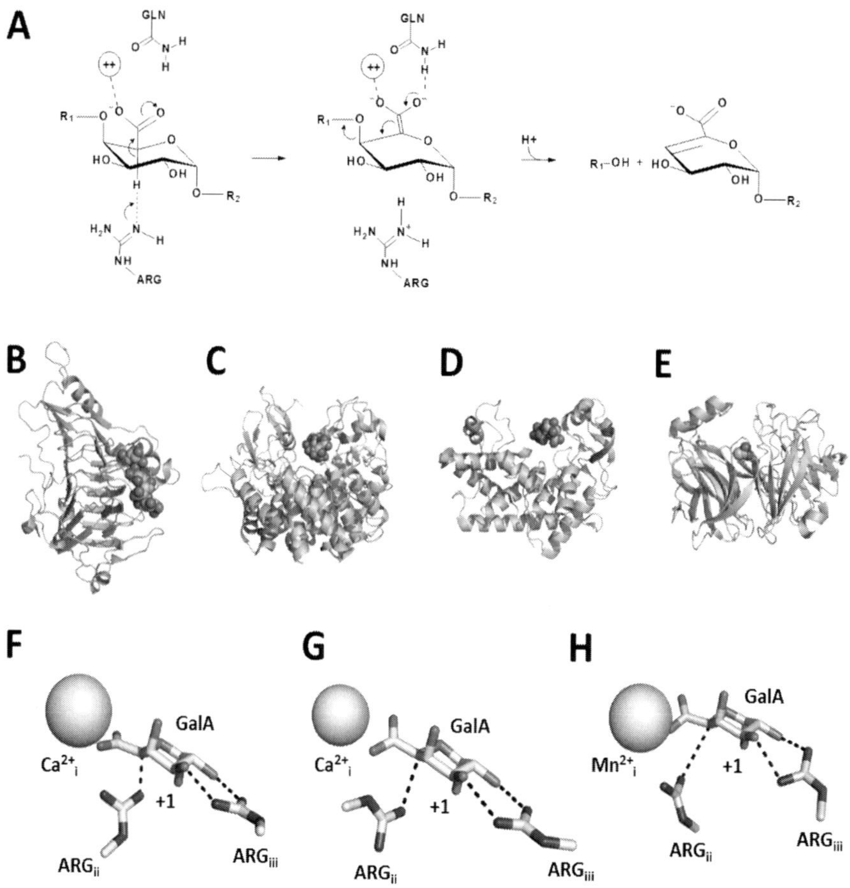

Figure 3. Structure and function of PLs active on HG. (A) β-elimination of an α-D-GalA configured substrate. The metal cofactor is delineated as (++). (B-E) Three-dimensional structures of polysaccharide lyase-complexes active on HG and RG-I shown in cartoon representation with ligands as spheres. (B) PL1 β-helix from *E. chrysanthemi EC16* in complex with $GalA_6$ (PDB ID: 2ewe) [152]. (C) PL2 α/$α_7$-barrel from *Y. enterocolitica subsp. enterocolitica 8081* in complex with $GalA_3$ (PDB ID: 2v8k) [56]. (D) PL10 α/$α_3$-barrel from *C. japonicus Ueda107* in complex with $GalA_3$ (PDB ID: 1gxo). (E) PL22 $β_7$-propeller from *Y. enterocolitica subsp. enterocolitica 8081* in complex with acetate (PDB ID: 3pe7). (F-H) Evolutionary convergence of the +1 subsite and β-elimination in pectate lyases [54, 56, 57]. The metal (i), Brønstead base (ii), and stabilizing Arg (iii) are shown for PL1 (F), PL9 (G), and PL2 (H) respectively.

RG-lyases (PL4 and PL11) deploy different mechanisms than pectate lyases [60]. In PL4s, β-hydrogen abstraction is catalyzed by a lysine, which is similar to PL9s; however, that is where the similarities end. The report of a catalytic mutant in complex with a hexasaccharide defined six subsites (-3 to +3), and following superimposition of the wild-type lysine residue, the function of two unique aspects of catalysis were illuminated (FIG 4B). Firstly, PL4s do not require a metal cofactor and secondly, they deploy a histidine as a catalytic acid that protonates the scissle glycosidic linkage (FIG 4A-B). By comparison, a detailed understanding of the PL11 mechanism is still lacking; however, based upon a $GalA_2$ complex with YesW and some structural convergence with PL4s, a model has been proposed [49, 61]. Intriguingly, PL11s may harness a metal cofactor [49], and either a histidine [49] or aspartate [61] as a catalytic base. Further research is required to clarify the mechanism of this family, which displays preferential activity on RG-I, but also has described activity on HG [49], and therefore, may represent a lyase with a hybrid mechanism on structurally distinct pectins.

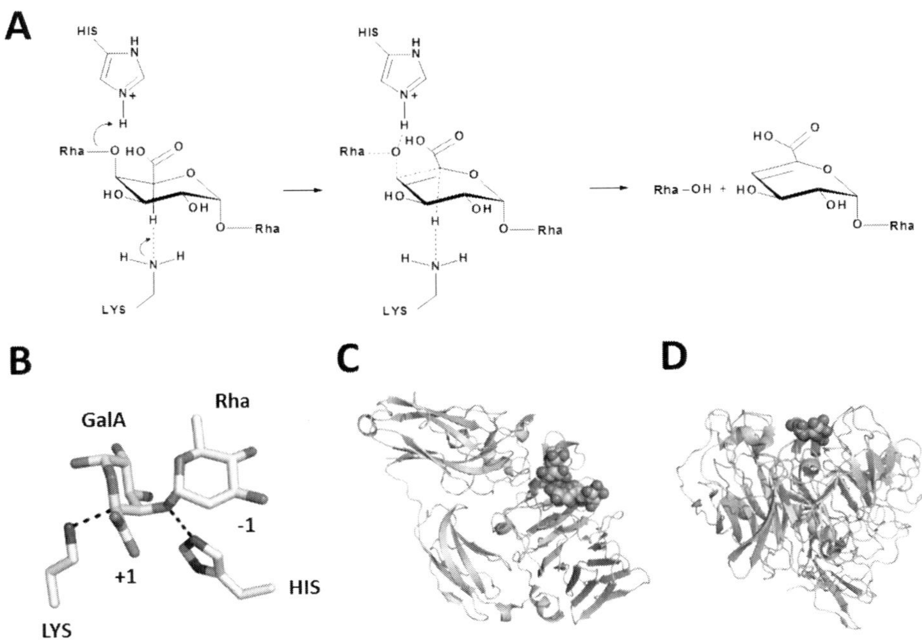

Figure 4. Structure and function of PLs active on RG-I. (A) Proposed mechanism for β-elimination of RG-I by PL4s [60]. (B) Catalytic residues involved in β–elimination by PL4s [60]. (C) PL4 β-sandwich fold from *A. aculeatus KSM 510EC16* in complex with a RG-I hexasaccharide (PDB ID: 3njv). (D) β_8-propeller PL11 from *Bacillus subtilis subsp. subtilis str. 168* in complex with a $GalA_2$ (PDB ID: 2z8s).

PL Activities – Endolytic and exolytic HG and RG-I PLs have been described in the literature (Table 1) [35]. Pectinolytic microorganisms will often contain multiple copies of PLs from the same family within their genomes [52]. In some cases, such as *Bacteroides* spp. [35], this redundancy is explained by segregated regulation of unique catabolic pathways (HG, RG-I, RG-II) or the existence of differential activities within a common family. For example, PL1 and PL2 isoforms display different activities and are likely active within the same pathway at different stages of pectin processing. PL2 represents an interesting example as the majority of sequences within this family are found as two paralogous copies, which cluster into two distinct subfamilies associated with either secreted endolytic (subfamily 1) or cytoplasmic exolytic (subfamily 2) activities [35, 58]. Across most families with characterized exolytic PL activities, the product generated is △GalA-α1,4-GalA (△$GalA_2$, α-galacturonate-4-[4-deoxygalact-4-uronosyl]). The exception to this rule is PL22s (EC 4.2.2.6), which are cytoplasmic enzymes that are preferentially active on △$GalA_2$ and $GalA_2$ and can produce GalA and △GalA depending upon the substrate [57-62].

PL Structural Highlights – The first pectinase structure solved was PelC from *Erwinia chrysanthemi EC16* [63] (FIG 3B). PelC adopts a right-handed β-helix that coils into three parallel β-sheets that are stabilized by aromatic stacks that run longitudinally through the protein core. At the time of its discovery, the structure of PelC defined a novel fold family. Since, it has proven to represent a plastic scaffold with utility for diverse pectinase activities [64], including several PL families (PL1, 3, and 9), and more surprisingly, distinct pectinase enzyme classes (PLs, GH28s, and CE8s) [45]. The structural conservation in β-helix enzymes are believed to be a product of fold stability [45, 64, 65], as pectinases are commonly secreted into harsh and competitive environments, such as the gastrointestinal tract of animals, soil, and plant cell walls. More recently, several new fold families have been described for PLs (Table 1), including the PL2 α/α_7-barrel (FIG 3C) [56], PL10 α/α_3-barrel (FIG 3D) [54], and PL22 β_7-propeller (FIG 3E) [57]. Despite this structural diversity, however, a common theme has emerged from the analysis of these enzymes. Investigation into the catalytic residues, metal cofactors, and substrates within active sites of these fold families has revealed a functional convergence of three key substructures (i-iii) that appear to be perquisites for β-elimination (FIG 3F-H) [57, 66]. These substructures include a (i) metal coordination pocket, (ii) Brønstead base, and (iii) stabilizing arginine. There is plasticity in two of these substructures as the metal binding pocket displays tailored chemistries for Ca^{2+}, Mn^{2+}, or Mg^{2+} [56-58], and the catalytic base has been determined to be most commonly an arginine, lysine [66] and perhaps histidine [57]. The stabilizing arginine, however, is invariant which suggests that it may be essential for catalysis [57].

Rhamnogalacturonan lyases from PL4 and PL11 display unrelated folds to PLs active on HG (Table 1). The PL4 from *A. aculeatus* KSM 510 adopts a β–sandwich fold arranged as three distinct modular domains each with structural homology to carbohydrate binding domains found in other carbohydrate active enzymes (FIG 4C) [67]. The first PL11 structure solved was YesW, an endolytic enzyme from *Bacillus subtilis subsp. subtilis str. 168* (FIG 4D) [49]. The fold consists of a β_8-propeller scaffold that houses a deep active site cleft, flanked by catalytic arms rich in β–structure. The structure of the exolytic PL11 homolog (i.e. YesX) has illuminated the structural basis of exolytic activity within this family [61]. YesX contains a specific loop that interacts with the terminal saccharide molecule and restricts access of the RG-I polymer into the active site cleft. Deletion of this loop transforms YesX into an exolytic enzyme.

Glycoside Hydrolases

GH28 and GH105 have emerged within the literature as the primary agents of HG and RG-I hydrolysis. In both cases, these families have been shown to cleave linkages with different chemistries. GH28 in particular is a well-characterized family that is differentially active on substructures within HG and RG-I. The mechanism, activity, and structures of these families will be discussed in further detail below.

GH Mechanism – Despite recognizing diverse pectic substrates, GH28s operate by a conserved single-displacement mechanism resulting in inversion of stereochemistry of the anomeric carbon from C1-α to C1-β (FIG 5A). The first detailed description of the GH28 mechanism was from endopolygalaturonases I and II (*Aspergillus niger* and *Aspergillus tubingensis*) using reduced substrates [68]. The reaction is catalyzed by a triad of aspartates clustered on the same side of the active site cleft (sometimes referred to as the 'syn' conformation) (FIG 5E,F). This architecture differs from the canonical tandem general acid and general base orientation observed in other inverting GHs, in which the residues are ~10 Å apart and opposed on either side of the substrate [69]. Asp201 [68] (Asp223 in *Pectobacterium carotovorum* PehA [70] / Asp402 in exoGH28 in *Y. enterocolitica* [71]) is believed to function as the general acid by donating a proton to the glycosidic oxygen. The other two aspartates, 180 (202/381) and 202 (224/403) have been identified as general bases by interacting with the nucleophilic water and charging it for attack of the anomeric carbon. Although not as well understood, the GH28 α-L-rhamnohydrolase releases β-rhamnose from the non-reducing termini of α1-2-rhamnosyl substrates within RG-I [72, 73]. Conservation of the catalytic aspartates [74] underpins that distinct anomeric chemistries can be accommodated within the active sites of GH28s.

GH105s use a unique mechanism that differs from the canonical inverting and retaining mechanisms of most GHs (FIG 6A). The structure of YteR (FIG 6B), an unsaturated rhamnogalacturonyl hydrolase, in complex with unsaturated △GlcA-GalNAc [75] and △GalA-Rha [76] (△GlcA and △GalA are sterically identical) identified Asp143 as the general acid, and His189 as a general base that activated a catalytic water (FIG 6C). Significantly, the scissle bond in the hydrolytic reaction is the C4 and C5 alkene as opposed to the glycosidic bond. Asp143 donates a proton to the C4 atom and the activated water attacks the C5 creating an unstable hemiacetal. This compound decomposes into the linear DKI aldehyde, releasing the glycone leaving group.

GH Activity – GH28s and GH105s contain diverse activities for enzymes that are active on functionally related (i.e. pectins) but chemically distinct substrates (e.g. α1,4 galacturonosyl and α1,2 rhamnosyl) (Table 1). Characterized members from GH28 include polygalacturonase (EC 3.2.1.15) [77], exopolygalacturonase (EC 3.2.1.67) [78], exopolygalacturonosidase (EC 3.2.1.82) [78], rhamnogalacturonase (EC 3.2.1.171) [48], rhamnogalacturonan α-1,2-galacturonohydrolase (EC 3.2.1.173) [72], and rhamnogalacturonan α-L-rhamnopyranohydrolase (EC 3.2.1.174) [79]. These coordinated activities have the capacity to saccharify complex polymerized HG and RG-I into GalA and Rha monosaccharides, and many of these activities have been harnessed for food processing applications [80, 81].

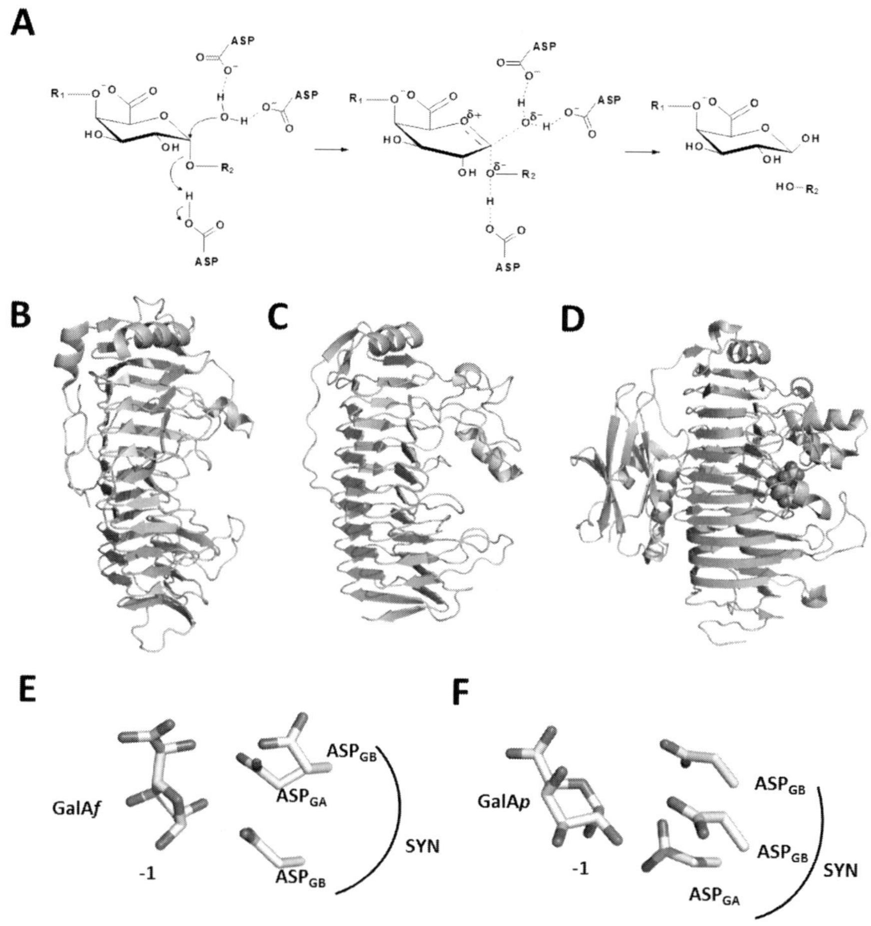

Figure 5. Structure and function of GH28s on pectic carbohydrates. (A) Inverting hydrolysis mechanism performed by GH28s. Three dimensional structures of (B) *A. aculeatus KSM 510 / CBS 115.80* endorhamnogalacturonase (PDB ID: 1rmg) [74], (C) *P. carotovorum SCC3193* endopolygalacturonase (PDB ID: 1bhe) [70], and (D) *Y. enterocolitica spp. enterocolitica 8081* exopolygalacturonase (PDB ID: 2uvf) [71]. Comparison of the three catalytic aspartates (i.e. 'syn' conformation) in GH28s from the -1 subsite of endopolygalacturonase I from *Stereum purpureum* with α-D-galacturofuranose (PDB ID: 1kcd) [84] (E) and exopolygalacturonase from *Y. enterocolitica spp. enterocolitica 8081* with α-D-galacturonopyranose (PDB ID: 2uvf) [71] (F).

GH105s are exolytic enzymes that remove unsaturated GalA from the non-reducing end of PL products (EC 3.2.1.172). The first activity was observed in a *Bacillus subtilis* subsp. *subtilis* str. 168 enzyme that was specifically active on RG lyase products [75, 76]. More recently a GH105 homolog has been described from the green macroalgae *Nonlabens ulvanivorans* (NuGH105) [82]. This enzyme harnesses a similar vinyl-ether hydration mechanism to cleave a unique β-△GlcA-Rha-(sulphate)$_3$ linkage, indicating that this mechanism is not exclusive to a defined stereochemistry of the anomeric carbon. NuGH105 is active on the β-△GlcA moiety present within the algal cell wall polysaccharide ulvan (3-sulfated rhamnose, glucuronic acid, iduronic acid, and small amounts of xylose) that had been treated with ulvan lyases. The determination that GH105s are active on both α– and β–configured glycosidic linkages highlights the plasticity between active sites, and suggests the possibility that a wide spectrum of activities on lyase products may yet be discovered within this family.

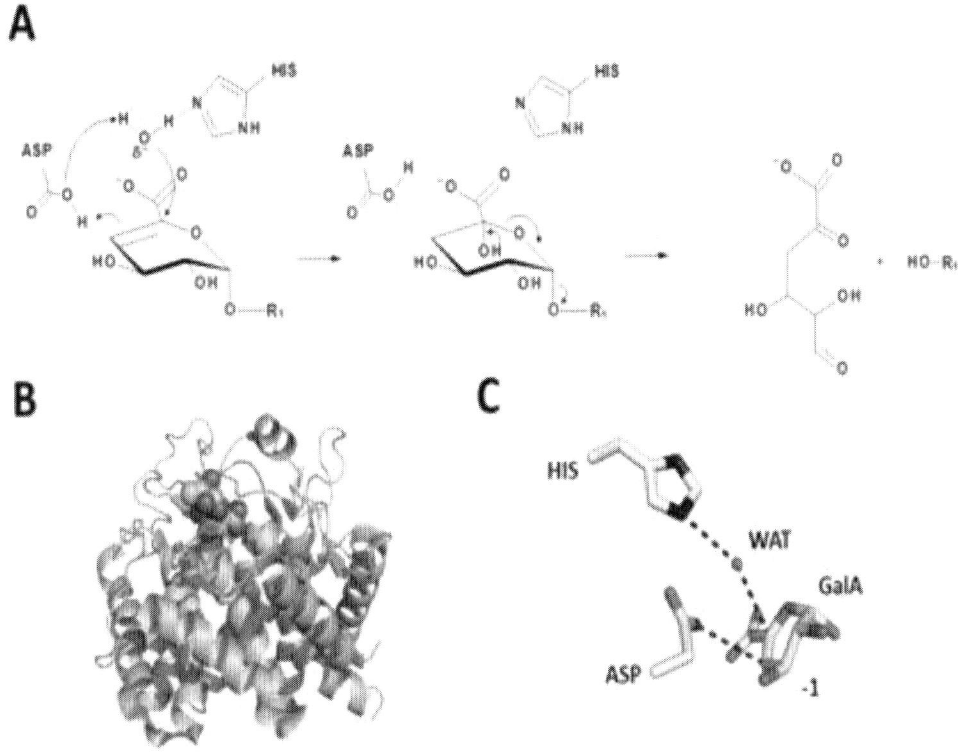

Figure 6. Structure and function of GH105s on unsaturated pectic carbohydrates. (A) Vinyl ether hydrolysis mechanism catalyzed by GH105s. The catalytic Asp and His involved in hydration of the △C4-C5 bond in GalA are shown. (B) Cartoon representation of the *Bacillus subtilis subsp. subtilis str. 168* in complex with α-D-4-deoxy-GlcpA-(1-2)-α-L-Rhap (PDB ID: 2gh4) [154]. Catalytic residues within the -1 subsite involved in vinyl ether hydrolysis.

GH Structural Highlights – GH28 and GH105 adopt distinct folds (FIG 5&6, Table 1). GH28s have right-handed parallel β-helixes that differ from the β-helix PLs by displaying a four-sided β-sheet architecture as opposed to three. The first family structure was the rhamnogalacturonase from *Aspergillus aculeatus* (FIG 5B) [74], which was followed by a bacterial endopolygalacturonase from *Pectobacterium carotovorum SCC3193* (FIG 5C) [70]. Structural superimposition of these two enzymes revealed that they are very similar in overall structure; however, the bacterial endopolygalacturonase had a shortened topology with one less β-helix turn and a unique C-terminus cap [70]. Elucidating the structural basis of exopolygalacturonase activity took nearly a decade after these seminal insights. YeGH28, which is a disaccharide releasing exopolygalacturonosidase (EC 3.2.1.82), provides an example of large-scale changes to the active site cleft transforming endolytic to exolytic activity (FIG 5D). Four loop insertions converge to form a blind canyon, which restricts access of substrate from one direction and results in exclusive production of disaccharide products [71]. The structure of a monosaccharide releasing exopolygalacturonase (EC 3.2.1.67) from *Thermotoga maratima* revealed that GH28s can oligomerize to increase stability under thermophilic conditions [83]. Product complexes for an endopolygalacturonase (-1 and +1 subsites) [84] and an exopolygalacturonase (-1 and -2 subsites) have helped to define the subsite architecture in GH28s (FIG 5E,F) [71].

Several structures have been deposited for bacterial homologs of GH105. YteR exhibits an α/α double-toroid structure with six α-hairpins arranged in a double α-helical barrel (FIG 6B, Table 1) [85]. Overlays of YteR complexes with α-△GlcA-GalNAc [75] and α-GalA-Rha [76] revealed that the anhydro subunit is positioned in nearly identical orientations. Superimpositions with a GH88 complex, which are related β-△glucuronyl hydrolases active on the products of chondroitin lyases, confirmed that the catalytic machinery is conserved between these families, and suggests a distant relatedness and conserved mechanism (FIG 6C) [75].

Carbohydrate Esterases

HG can be modified with methyl esters at C6 (methoxylation) and acetyl esters at C2 or C3 (acetylation) (FIG 2G-L), which alter the structure, packing, and solubility of pectin fibres. Methoxylation in particular induces significant changes in pectin solubility by neutralizing the charge of GalA, and occluding a hallmark recognition motif of many different families of pectinases. Enzymatic removal of methylesters and acetylations results in the production of methanol and acetate respectively [86]. In general, esterified pectin provides protection against cell wall deconstruction as removal of these substituent modifications is a preliminary reaction for saprophyte and phytopathogen metabolism [87, 88]. Pectin specific esterases belong to two different sequence related families, which encompasses the methylesterases (CE8) and acetylesterases (CE12).

CE Mechanism – Most serine proteases, lipases, and carbohydrate esterases, including CE12s contain a conserved Asp-His-Ser catalytic triad. CE8 pectin methylesterases (PMEs; EC 3.1.1.11) are distinct in this regard as they contain two catalytic Asp residues [89, 90] and an oxyanion stabilizing Gln residue [91] (FIG 7A). A breakthrough paper in 2007 from the laboratory of Pickersgill elucidated the mechanism of the phytopathogen *Dickeya dadantii 3937* CE8 (renamed from *Erwinia chrysanthemi* 3937), using X-ray crystallography, mutagenesis, kinetics, and a complex synthetic library of differentially methylated oligogalacturonides [91]. Within the active site, the general acid Asp178 is buried within a hydrophobic environment and likely protonated, whereas the nucleophile Asp199 is solvent accessible and negatively charged at physiological pH. The ester is activated by protonation of the carbohyl oxygen by Asp178. The Asp199 nucleophile attacks the C6 (carboxylate carbon), forming a tetrahedral intermediate and the carbonyl oxygen evolves into an oxyanion, stabilized by a protonated Asp178 and Gln177. Methanol departs, forming an anhydride intermediate and a charged water attacks the C6, generating a titratable uronic acid group. In comparison to PMEs, very little is known about the mechanism of CE12s and a detailed study of their mechanism remains to be reported.

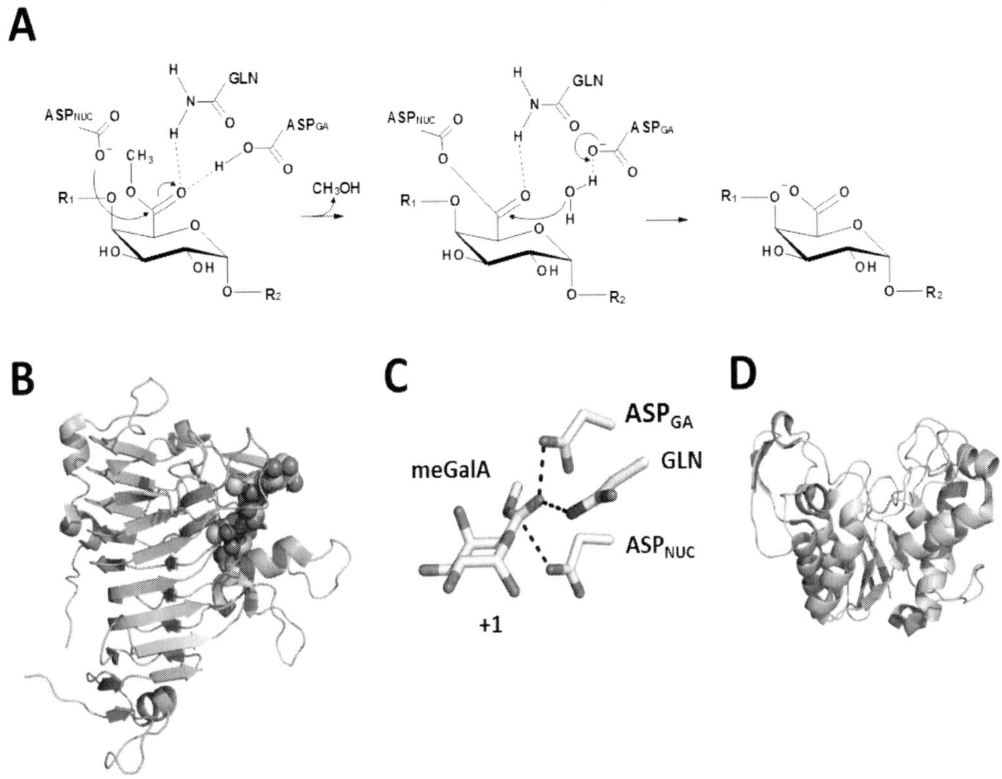

Figure 7. Structure and function of CEs. (A) Methanol producing demethylesterification mechanism catalyzed by CE8s. (B) Three-dimensional structure of PemA from *D. dadantii* 3937 in complex with α-D-methylesterified hexagalacturonide (Substrate II, PDB ID: 2nst) [91]. (C) Constellation of catalytic residues involved in the hydrolysis of methylesters within CE8s. (D) Three-dimensional structure of CE12 acetylesterase from *A. aculeatus KSM 510* (PDB ID: 1deo).

CE activity – A variety of CE12 acetylesterase activities have been described (Table 1). These include pectin acetylesterase (EC 3.1.1.-) [92], which displays preferential activity after pectin is demethoxylated; rhamnogalacturonan acetylesterase (EC 3.1.1-), which catalyzes the removal of acetate from the RG-I backbone facilitating the action of depolymerases [93]; and acetyl xylan esterase (EC 3.1.1.72), which play analogous roles in the deacteylation of xylan but are not active on pectins [94]. Contrastingly, CE8s appear to have strict specificity for methylesterified HG. Often, both CE8 and CE12 genes are found within the genomes of the same microorganism, and there appears to be a hierarchy in preferential activities as acetylesterase efficiency is increased when esterified substrates are pretreated with methylesterase [92, 95]. Deesterification is believed to occur through three distinct patterns: 'single-chain', in which a processive enzyme removes all esters contiguously; 'multiple-chain', in which the enzyme dissociates after each reaction; and 'multiple-attack', where multiple reactions are catalyzed before dissociation [91].

Comparative genomics of pectinolytic microorganisms has revealed that multiple copies of CE families can exist in within the same organism. For example, the bacterial phytopathogen *Dickeya dadantii 3937* has two CE8s, PmeA and PmeB, which are associated with distinct activities. PmeA is closely related to plant PMEs and preferentially active on polymeric methylesterified pectin; whereas, PmeB is specific for methylesterified oligogalactuonides [96-99]. Commonly, these complimentary activities represent different stages along the pectinolytic cascade and can be compartmentalized within different regions of the cell (e.g. outer membrane, periplasm) [45, 100]. Interestingly, a group of enteric pathogens, including *Klebsiella*, *Salmonella*, *Shigella*, and *Escherichia coli*, contain a CE8 homolog (YbhC) of unknown function, which appears to have diverged from the PmeB subfamily [98]. YbhC has lost its ability to bind to or modify pectin, and displays notable structural signatures, including a lipidation motif, remodelled active-site and an Asn residue as a general acid catalyst [98]. Significantly, the CE8 from *Y. enterocolitica* displays more similarity to PmeA than YbhC [99], which suggests that pectinolysis is a more important metabolic signature in the *Yersinia* genus compared to other enteric pathogens.

CE Structural Highlights – Structures have been solved for both bacterial and eukaryotic CE8s and CE12s (Table 1). CE8s adopt a right-handed parallel β-helix, similar to polygalacturonases, rhamnogalacturonases and several PL families. The first structure reported was from *D. dadantii* 3937 (previously *E. chrysanthemi 3937*) (FIG 7B) [90], and followed by a homolog from carrot [89]. The systematic investigation of the active site of the *D. dadantii* 3937 using a series of synthesized methylesterified oligogalacturonides helped to define the role of charge neutralization and GalA recognition in ten subsites, including five in the positive and negative directions [91]. In addition, a catalytic Asp mutant in combination with other active site mutations, were performed to elucidate the tetrahedral-intermediate mechanism of methanol production and oxidation of C6 to a uronic acid (FIG 7A-C).

Comparing the global folds of PmeA and PmeB revealed that the active site of PmeB was sealed off at one end by loop insertions, which provides a structural basis for the distinct activities [98, 99]. Further transformations to the topography of the active site in the YbhC subclade is believed to be responsible for its loss of PmeB-like pectinolytic activity and gain of its unknown function [98, 99].

The first RG-acetylesterase structure solved was from *A. aculeatus* (FIG 7D) [101]. Similar to the majority other esterase classes, CE12s adopt an α/β hydrolase fold and position the Ser-His-Asp catalytic triad within an open active site cleft. Despite very low primary structure homology, CE12 represents one subfamily within SGNH-hydrolase family [101]. Insights into the molecular basis of substrate specificity and catalytic mechanism of CE12s would benefit from the solution of enzyme-complex.

PECTINASES AND THE DISTAL GUT MICROBIOTA

The majority of our understanding of pectinase structure-function relationships has been derived from characterizing recombinant and purified enzymes from phytopathogenic and saprophytic microorganisms [52, 53]. Correspondingly, the field has several entrenched model systems, including soft-rot pathogens from Enterobacteriaceae (i.e. *Erwinia spp.*, *Dickeya spp.*, and *Pectobacterium spp.*) [45, 102] and pectinolytic fungi (e.g. *Fusarium graminearum*, *Aspergillus sp.*) [103, 104]. These pectinophiles are well-adapted to deconstruct intact plant cell walls, and correspondingly, they possess extensive arsenals of pectinolytic enzymes within their genomes that vary in regulation, cellular targeting, and activity. In comparison, pectinolytic intestinal bacteria appear to have more limited and specialized pathways for pectin utilization. Of the annotated genomes in the CAZy database (i.e. CAZomes) only the *Bacteroides* genus display extensive levels of genes belonging to defined pectinase families (Table 2). However, there are several examples of pectinolytic strains of bacteria, including *Eubacterium eligens ATCC27750* (16 pectinase genes) and *Klebsiella oxytoca E718* (11 pectinase genes), with augmented levels of predicted pectinases when compared to other strains within their own species (Table 2). This observation highlights the functional diversity that exists in the metabolic potential of intestinal bacteria at the strain level, and underpins the importance of next-generation metagenomic initiatives, such as the Human Microbiome Project [105], that are providing the sequencing depth required to define these relationships.

Table 2. Pectinases within intestinal bacteria that contain multiple genes with predicted or characterized activity on pectic substrates [35]

SPECIES	PL1	PL2	PL3	PL4	PL9	PL10	PL11	PL22	GH28	GH105	CE8	CE12	Total
BACTEROIDES													
B. salanitronis DSM 18170[1]	3					3	2		11	6	5	4	**34**
B. thetaiotaomicron VPI-5482	5				2	1	1		9	7	3	4	**32**
B. vulgatus ATCC 8482	2					2	3		13	7	4	5	**36**
B. xylanisolvens XB1A	5				1	1	4		9	5	2	2	**29**
ENTEROBACTER													
E. aerogenes EA1509E								1	1	2	1		**5**
E. aerogenes KCTC 2190									1	2	1		**4**
Enterobacter sp. 638		1			1			1	1	1	1		**6**
ENTEROCOCCUS													
E. casseliflavus EC20									1	2	1	1	**5**
E. faecium Aus0004					1				2	2	1	1	**7**
E. mundtii QU 25 QU25									1	3	1	1	**6**
KLEBSIELLA													
K. oxytoca E718		1			1			2	3	2	2		**11**
K. pneumoniae 342[2]								1	1	2	1		**5**
SPECIES	PL1	PL2	PL3	PL4	PL9	PL10	PL11	PL22	GH28	GH105	CE8	CE12	**Total**
ROSEBURIUM													
R. hominis A2-183									2	1	1		**4**
R. intestinalis M50/1									1	1	1		**3**
YERSINIA													
Y. enterocolitica 8081[2]		2						1	1				**4**
Y. pseudotuberculosis YPIII[2]		2						1		1	1		**5**
OUTLIERS													
Eubacter. eligens ATCC27750	3				4				3	2	2	2	**16**
Faecalibacter. prausnitzii SL3/3									3	1	1		**5**
Salmonella bongori N268-08[2]									1	1	1		**3**

[1] *B. salanitronis DSM 18170* was isolated from poultry.

[2] Human pectinolytic pathogens.

The human genome does not contain requisite enzymes required to digest HG, RG-I, or RG-II, and therefore, we rely on symbiotic bacteria to unlock the energy contained within pectic glycosidic linkages [106]. These enzymatic reactions occur in an anaerobic environment and result in the formation of fermented by-products (e.g. acetate, propionate, and butyrate) that are utilized by the host [107]. Pectic substrates metabolized by intestinal bacteria are 'pretreated' by the upper stages of digestion, such as mastication and acid-hydrolysis, and therefore represent a more accessible substrate than what is presented to saprophytes and phytopathogens in the environment and during plant disease.

Bacteroides is a genus of ubiquitous glycophilic bacteria found in terrestrial, marine, and host-intestinal habitats. Next-generation sequencing and functional genomics are beginning to unravel the mechanism by which intestinal *Bacteroides* sp. saccharify indigestible complex carbohydrates derived from the diet, host glycans, and microbial capsular polysaccharides. Comparison of four different species reveals distinct metabolic signatures for pectinolysis (Table 2). For example, *B. vulgatus* has an enrichment of hydrolases, whereas *B. xylanisolvens* displays the most PLs. Metabolic pathways in *Bacteroides* spp. are organized into dedicated gene clusters, referred to as Polysaccharide Utilization Loci (PULs). PULs are 'self-contained' and respond to discrete carbohydrate signals by deploying specific enzymes for deconstructing detected polysaccharides and transport proteins to ensure passage of the products into the cell for energy harvest [108]. In this regard, *B. thetaiotaomicron* was recently shown to metabolize HG, RG-I, and RG-II as sole carbon sources [109]. Correspondingly, three distinct PULs were correlated with the catabolism of these substrates, complete with predicted pectinolytic enzymes, regulators (HTCSs), and transport machinery (SusCDE-like proteins) [109]. The HG PUL for example, spans from Bt4108-Bt4124, and contains three PL1s (Bt4115, Bt4116, and Bt4119), one GH28 (Bt4123), one GH105 (Bt4108), two CE8s (Bt4109 and Bt4110), and one CE12 (Bt4110, which is a bimodular enzyme). These observations underpin that searching for annotated pectinase gene families can be a potent tool for predicting new pectinolytic PULs or pathways in *Bacteroides* spp. and other bacterial genomes.

In contrast to *B. thetaiotaomicron*, the role for abridged pectinolytic pathways observed within enteric pathogens is intriguing (Table 2). The presence of hallmark pectinase genes and pathways within enteropathogenic Enterobacteriacea has been known for over a decade [110]; however, their biological significance remains elusive. Potentially, these catabolic pathways enable enteric pathogens to contaminate food crops, which would provide a vector for transmission, or to compete for dietary plant cell wall polysaccharides as a nutrient source and colonize the host intestine [45, 57, 99]. Alternatively, they may simply represent orphaned catabolic remnants from ancestral microorganisms that targeted pectin as a primary nutrient source. Further investigation into the microbiology and symbiosis of pectinolysis in these microorganisms is required to determine its significance if any for enteric pathogen lifecycles and the onset of foodborne illness.

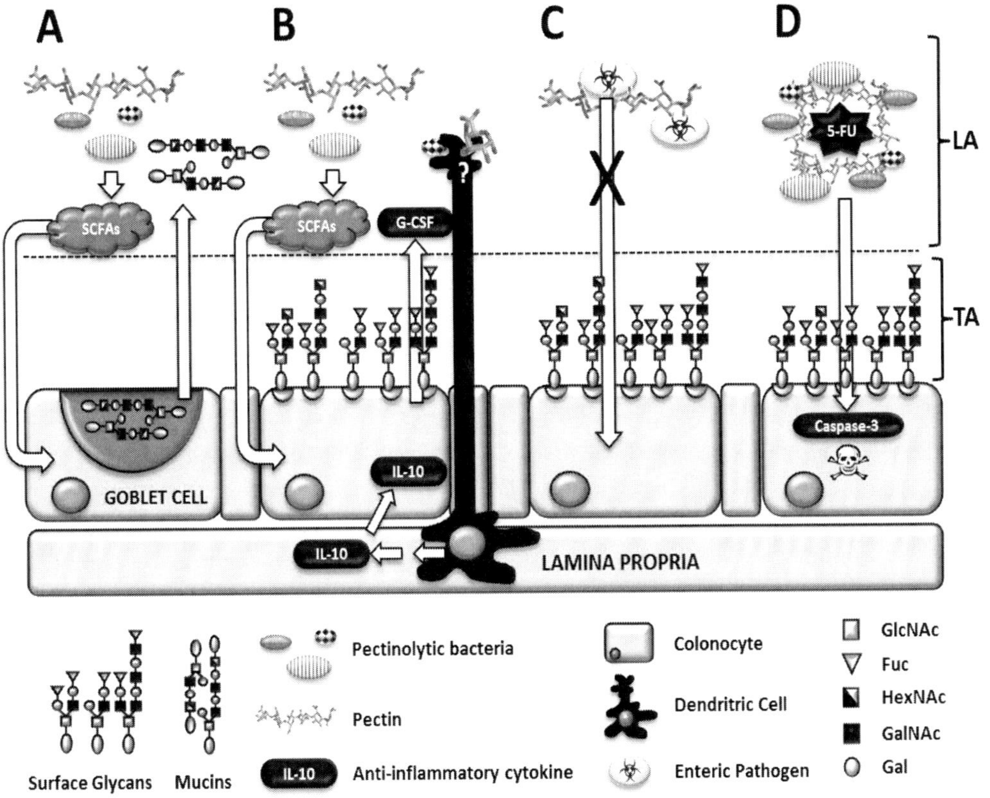

Figure 8. Beneficial roles of dietary pectin in intestinal health. (A) Enteric pectinophilic bacteria degrade pectin producing SCFA by-products and stimulate mucin secretion from goblet cells, which results in the formation of the loosely adherent (LA) mucus layer. (B) Dendritic cell sampling of luminal bacteria results in increasing anti-inflammatory cytokine production (IL-10) from immune cells located in the lamina propria, as well as, increasing beneficial immune molecules (e.g. granulocyte colony-stimulating factor (G-CSF). (C) Pectin and pectic oligosaccharides also serve as receptor mimics and prevent enteric pathogens from binding host glycans that comprise the tightly adherent (TA) layer. D) Pectin capsules are degraded by pectinolytic microflora in the intestine, which releases bioactive compounds (i.e. 5-FU) at the site of disease.

PECTIN, THE DGM AND INTESTINAL HEALTH

Within the DGM, members of the microbiota interact in a syntrophic (i.e. cross-feeding) network in which the metabolic by-products of primary feeders are used as nutrients for downstream feeders. As a result, dietary pectin can function as a prebiotic by directly modulating pectinolytic bacterial populations, but can also have indirect effects on secondary feeders. Indirect effects on community structure also exist between commensal bacteria and intestinal mucus production. *B. thetaiotaomicron* and *F. prausnitzii* have been shown to influence the production of mucus glycans (i.e. mucins) from goblet cells (Figure 8A) [111, 112]. Acetate formed by *B. thetaiotaomicron* during pectin metabolism is utilized by *F. prausnitzii* to produce butyrate [113, 114], a short chain fatty acid (SCFA) by-product that has been shown to modulate mucin (*MUC*) gene expression in goblet cells [115]. The mucus composition likewise has an effect on the DGM community. Colonocytes have their apical surfaces decorated by integral membrane mucins forming a tightly adherent sterile layer, which is itself coated by loosely adherent gel-forming mucins (i.e. mucus) [116, 117] (FIG 8A). The loosely adherent layer is readily colonized by the DGM and mucins are a nutrient source for many species, including *B. thetaiotaomicron*. In this capacity, dietary pectin and its metabolism by intestinal bacteria can directly or indirectly induce changes in DGM community structure [111]. In addition, SCFA production resulting from pectin fermentation lowers the colonic pH, which has been determined to have a protective effect against colon cancer [118]. SCFAs can also stimulate pro-apoptotic pathways (e.g. caspase-3) within intestinal cells [119, 120]. Induction of apoptotic cascades result in growth inhibition and attrition of colon cancer cells, suggesting that pectin rich foods may have a dietary protection role against colon cancer [121].

Pectin is abundant in the diets of most humans. Beyond providing basic nutritional benefits it has been shown to reduce the risk and morbidity of chronic diseases. In this capacity, pectin and pectic oligosaccharides (POS) generated by pectinases within the DGM can be classified as functional foods. While the primary role of food is to meet the basic nutritional requirements of the host for maintenance and growth, some food components surpass this role by providing additional benefits for organismal health [122]. These roles include immunomodulation, reducing damage inflicted by toxic materials, and preventing infection by enteric pathogens (FIG 8) [123]. In addition, chemically modified pectin and POS represent an emerging field of nutraceutical applications, which bridges the disciplines of nutrition and pharmaceuticals to complement conventional medicine [124].

Immunomodulation

Immunomodulatory properties have been reported for pectins, although whether these observations are due to direct interactions of the carbohydrate with the host or indirect effects resulting from DGM-host interactions remains to be determined. Pectin has been reported to have a protective, anti-inflammatory effect on inflammatory bowel disorder (IBD); however the exact mechanisms remain unknown. Using IL-$10^{-/-}$ mice, a murine model for IBD, pectin treatment reduced expression of pro-inflammatory TNF-α and reduced activity of the GATA-3 transcription factor that features heavily in a Th_2 immune response [125]. Therefore pectin was shown to downregulate the colonic inflammatory response by moderating the production of potentially harmful pro-inflammatory cytokines and immunoglobulins, such as the modified citrus pectin (MCP) dose-dependent inhibition of pro-inflammatory cytokine IL-1β and increased secretion of the anti-inflammatory cytokines IL-1ra and IL-10 using human peripheral blood cells [126]. Similarly, in a study investigating bupleuran 2IIc, a pectic polysaccharide isolated from the roots of *Bupleurum falcatum* L., increased granulocyte colony-stimulating factor (G-CSF) secretion in cultured colonic epithelial cells was observed [127]. G-CSF is characterized as both a cytokine and hormone, with pleiotropic effects on hematopoiesis, and innate and adaptive immune responses (FIG 8B). The pectic polysaccharides as well as its enzymatically degraded fractions from the vegetative parts of the medicinal plant *Biophytum petersianum* have been shown to stimulate activation of nitric oxide synthesis of macrophages as well as the MHC expression of dendritic cells [128]. Further, the β-D-(1,4)-galactan-containing side chains, in the pectic polysaccharide fragments was suggested to have an important immunomodulatory role in both Peyer's patch immunocompetent cells and macrophages [129].

Detoxification

The consumption of pectin can potentially play a role in the detoxification of deleterious chemicals, toxins, and metals within the body. This property makes pectin and POS attractive options for treating heavy metal poisoning and bacterial toxin exposure by reducing inflammation. In chelation therapy, heavy metals are removed from the body using chemical chelators introduced intravenously with multiple possible side effects. Pectin cannot cross into the blood stream due to its size; however MCP, in contrast is divided into lower molecular weight fractions allowing for absorption and chelation. MCP can be useful for ongoing detoxification without the harsh side effects such as leaching essential nutrients from the body [130]. Pectin is also of interest in cancer research given its involvement in carcinogen detoxification for toxins produced by the human body resulting from deregulated metabolism or environmental, foodborne, and waterborne toxin exposure [131]. Cell or tissue exposure to toxins can produce free radicals, which damage organ function over time and DNA maintenance.

Food Safety and Enteric Pathogens

Pectin has been shown to protect host tissues from enteric pathogen adherence (FIG 8C). In undifferentiated and differentiated Caco-2 human colonic cells, POS inhibits the invasion of pathogenic *Campylobacter jejuni* [132]. During infection, pathogenic bacteria commonly bind surface glycans on host cells with adhesins, carbohydrate binding proteins (i.e. lectins) that contain multivalent binding sites [133]. These glycans are generally oligosaccharides, linked to the host membrane through proteins or lipids to form glycoproteins/glycolipids. In this capacity, free oligosaccharides have been proposed to prevent attachment to epithelial cell surfaces by competing for bind sites within bacterial adhesions [134, 135]. To date, using free oligosaccharides, such as POS, to inhibit bacterial attachment has been successful for a number of food-borne pathogens, including: *Helicobacter pylori*, *E. coli*, and *C. jejuni* [121, 133, 136, 137]. By acting as a receptor mimic, dietary polysaccharides (e.g. pectin/POS) can prevent several infectious bacterial pathologies [138] (FIG 8C). Dietary pectin and POS have further functions related to food safety as they been shown to protect colonic cells against bacterial toxins such as the Shiga-like toxin from *Escherichia coli* O157:H7 [121].

Pectin Encapsulation

Providing localized health benefits within the colon depends upon several factors, including the presence and persistence of pectinophiles, and accessibility and solubility of pectin and POS during transit. In this light, pectin is under investigation in the biotechnology sector as an encapsulating agent to deliver biologically active compounds to the intestine because of its low toxicity, biocompatibility, biodegradability, and ability to form gels [139].

Bioactive cargo (Table 3) encapsulated within indigestible polysaccharides prevents degradation during passage through the upstream stages of digestion. In this capacity, pectin is an acid sugar, physiochemical properties which decrease solubility within the low pH of the stomach. In addition, its recalcitrance to human digestive enzymes prevents its modification within the upper gastrointestinal tract and absorption within the small intestine. Upon transit to the colon, pectin capsules are readily dismantled by pectinolytic members of the DGM. Pectin encapsulation therefore protects cargo molecules, prolongs the absorption of encapsulated bioactive components, and provides targeted delivery of therapeutics to the colon by depending on DGM catalyzed release (FIG 8D) [140].

Table 3. Compounds for bioactive supplement delivery in the human colon, using pectin encapsulation

Intestinal Disorder	Compound	Ref
IBD	Sulphasalazine, olsalazine, mesalazine, fludrocortisone, budesonide,	[156, 157]
Colon Cancer	5-fluorouracil (5-FU), doxorubicin, methotrexate	[158-160]
Immunomodulation	Peptide drugs delivered to lymphoid tissues (growth hormones, insulin, interleukins, interferon, and erythropoietin)	[150, 168]
Metabolic deficiency	Lipophilic vitamins, bile salts, and some steroids	[161]
Live Cells	Human PICs, Human FLSCs, Human umbilical cord blood cells, Human parathyroid tissue, human erythroleukemia cells, Human genetically engineered kidney cells, various commensal bacteria	[141, 162]

Cargo is encapsulated by a variety of chemical, physical, and physicochemical methods (e.g. matrix polymerization, spray drying, and ionotropic gelation) that coat bioactive molecules with pectic polysaccharides to form core-shell biopolymer particles. Described cargo molecules include lipids, enzymes, essential nutrients, peptides, and other lipophilic materials (Table 3) [141, 142]. Live bacterial encapsulation helps to deliver viable commensals through the harsh upper gastrointestinal tract conditions, which have been proven successful in treating renal failure, CVD, and colonic disorders [141, 143-145]. Pectin is an inexpensive starting material, which enables microbeads to be produced economically and on a large scale. In addition this technique is versatile, as different sources of pectin can be utilized including beet, apple or citrus, which provides distinct chemistries and solubilities, resulting in microbeads with different molecular properties and cargo release profiles. Cross-linking agents, such as Ca^{2+} salts and protons [139], in addition to chemical and enzymatic modification of pectin may help to tune methylesterification levels by altering the potential for divalent metal coordination and regulating solubility of pectate/pectinate (salts containing partially methylesterified HG)-metal gels [146]. Furthermore, pectins with low levels of methoxylation can also undergo amidation of their uronic acid groups. Amidated pectic acids are more prone to form rigid gel structures with divalent cations (pectates) than non-amidated pectin [147]. Reports suggest that zinc-pectin salts outperform calcium-pectin salts in delivery to the colon in encapsulation efficiency and substrate retention when used as formulated pectinate beads [148].

The implications of pectin as an enteric delivery technology are far-reaching and may improve current treatment options for colonic diseases such as Irritable Bowel Disease (i.e. ulcerative colitis and Crohn's disease) and colon cancer. The colonic region is the preferred target for drug absorption due to the less acidic conditions compared with those found in the stomach and small intestine. Similarly, drugs susceptible to digestive or pancreatic enzyme catabolism are minimally affected in the colon [149]. Additionally, providing higher drug concentrations at disease sites can increase drug efficacy and minimize systemic side effects at non-target regions (FIG 8D & Table 3). Due to the lymphoid tissue density in the colon, enteric epithelial cells can present antigens to the underlying immune cells to stimulate the rapid production of antibodies [150, 151]. This presents an elegant route for the delivery of immunomodulatory compounds, such as vaccines, growth hormones, insulin, interleukins, interferon, erythropoietin, some lipophilic vitamins, bile salts, and steroids undergoing enterohepatic circulation (Table 3).

Future of Research in Pectin Modification

In recent years, significant advances in pectin research have led to an understanding of pectin structure, pectinase mechanisms, interactions between dietary pectin and the DGM, and targeted delivery of beneficial compounds to the intestine using pectin encapsulation. Despite these successes, however, the field would still benefit from several keystone discoveries. These include developing a small molecule inhibitor of a glycoside hydrolase or pectate lyase, determining how the glycosidic bond is protonated during β-elimination of HG, elucidating the function of the conserved stabilizing arginine in PLs (FIG 3F-H), and defining the mechanism of CE12s, which would be facilitated by a substrate or product complex. Although the general biology of microorganisms involved in saprophytic, phytopathogenic, and intestinal pectin utilization has been well studied; the role for pectinases in the lifecycles of human enteric pathogens remains a mystery. Chromosomal mutagenesis will help illuminate these relationships. The role of pectin as a functional food (e.g. nutraceutical, prebiotic) and in encapsulation appears to have great potential for improving intestinal health. Further developments towards pectin synbiotics (i.e. the joint administration of pectin and intestinal pectinolytic bacteria) will help to advance these technologies.

References

[1] Burton, R. A., Gidley, M. J., and Fincher, G. B. (2010) Heterogeneity in the chemistry, structure and function of plant cell walls, *Nat Chem Biol 6*, 724-732.

[2] Gilbert, H. J. (2010) The Biochemistry and Structural Biology of Plant Cell Wall Deconstruction, *Plant Physiology 153*, 444-455.

[3] Mohnen, D., Bar-Peled, M., and Somerville, C. (2009) Cell Wall Polysaccharide Synthesis, In *Biomass Recalcitrance*, pp 94-187, Blackwell Publishing Ltd.

[4] Jarvis, M. C., Briggs, S. P. H., and Knox, J. P. (2003) Intercellular adhesion and cell separation in plants, *Plant, Cell & Environment 26*, 977-989.

[5] Popper, Z. A., Michel, G., Hervé, C., Domozych, D. S., Willats, W. G. T., Tuohy, M. G., Kloareg, B., and Stengel, D. B. (2011) Evolution and Diversity of Plant Cell Walls: From Algae to Flowering Plants, *Annual Review of Plant Biology 62*, 567-590.

[6] Domozych, D. S. (2001) Algal Cell Walls, In *eLS*, John Wiley & Sons, Ltd.

[7] Selvendran, R. R., and O'Neill, M. A. (1987) Isolation and analysis of cell walls from plant material, *Methods of biochemical analysis 32*, 25-153.

[8] Atmodjo, M. A., Hao, Z., and Mohnen, D. (2013) Evolving Views of Pectin Biosynthesis, *Annual Review of Plant Biology 64*, 747-779.

[9] O'Neill, M. A., Warrenfeltz, D., Kates, K., Pellerin, P., Doco, T., Darvill, A. G., and Albersheim, P. (1996) Rhamnogalacturonan-II, a pectic polysaccharide in the walls of growing plant cell, forms a dimer that is covalently cross-linked by a borate ester. In vitro conditions for the formation and hydrolysis of the dimer, *J Biol Chem 271*, 22923-22930.

[10] Liners, F., Letesson, J. J., Didembourg, C., and Van Cutsem, P. (1989) Monoclonal Antibodies against Pectin: Recognition of a Conformation Induced by Calcium, *Plant Physiol 91*, 1419-1424.

[11] O'Neill, M. A., Ishii, T., Albersheim, P., and Darvill, A. G. (2004) Rhamnogalacturonan II: structure and function of a borate cross-linked cell wall pectic polysaccharide, *Annu Rev Plant Biol 55*, 109-139.

[12] Caffall, K. H., and Mohnen, D. (2009) The structure, function, and biosynthesis of plant cell wall pectic polysaccharides, *Carbohydr Res 344*, 1879-1900.

[13] Bar-Peled, M., Urbanowicz, B. R., and O'Neill, M. A. (2012) The Synthesis and Origin of the Pectic Polysaccharide Rhamnogalacturonan II - Insights from Nucleotide Sugar Formation and Diversity, *Frontiers in plant science 3*, 92.

[14] Willats, W. G., McCartney, L., Mackie, W., and Knox, J. P. (2001) Pectin: cell biology and prospects for functional analysis, *Plant molecular biology 47*, 9-27.

[15] Brummell, D. A. (2006) Cell wall disassembly in ripening fruit, *Functional Plant Biology 33*, 103.

[16] Ridley, B. L., O'Neill, M. A., and Mohnen, D. (2001) Pectins: structure, biosynthesis, and oligogalacturonide-related signaling, *Phytochemistry 57*, 929-967.

[17] Savatin, D. V., Ferrari, S., Sicilia, F., and De Lorenzo, G. (2011) Oligogalacturonide-auxin antagonism does not require posttranscriptional gene silencing or stabilization of auxin response repressors in Arabidopsis, *Plant Physiol 157*, 1163-1174.

[18] Ferrari, S., Savatin, D. V., Sicilia, F., Gramegna, G., Cervone, F., and Lorenzo, G. D. (2013) Oligogalacturonides: plant damage-associated molecular patterns and regulators of growth and development, *Frontiers in plant science 4*, 49.

[19] Sterling, J. D., Atmodjo, M. A., Inwood, S. E., Kumar Kolli, V. S., Quigley, H. F., Hahn, M. G., and Mohnen, D. (2006) Functional identification of an Arabidopsis pectin biosynthetic homogalacturonan galacturonosyltransferase, *Proceedings of the National Academy of Sciences of the United States of America 103*, 5236-5241.

[20] Cid, M., Pedersen, H. L., Kaneko, S., Coutinho, P. M., Henrissat, B., Willats, W. G., and Boraston, A. B. (2010) Recognition of the helical structure of beta-1,4-galactan by a new family of carbohydrate-binding modules, *J Biol Chem 285*, 35999-36009.

[21] Mohnen, D. (2008) Pectin structure and biosynthesis, *Curr Opin Plant Biol 11*, 266-277.
[22] Willats, W. G., McCartney, L., and Knox, J. P. (2001) In-situ analysis of pectic polysaccharides in seed mucilage and at the root surface of Arabidopsis thaliana, *Planta 213*, 37-44.
[23] Jarvis, M. C., and Apperley, D. C. (1995) Chain conformation in concentrated pectic gels: evidence from 13C NMR, *Carbohydr Res 275*, 131-145.
[24] Nakamura, A., Furuta, H., Maeda, H., Takao, T., and Nagamatsu, Y. (2002) Structural studies by stepwise enzymatic degradation of the main backbone of soybean soluble polysaccharides consisting of galacturonan and rhamnogalacturonan, *Bioscience, biotechnology, and biochemistry 66*, 1301-1313.
[25] Lau, J. M., McNeil, M., Darvill, A. G., and Albersheim, P. (1985) Structure of the backbone of rhamnogalacturonan I, a pectic polysaccharide in the primary cell walls of plants, *Carbohydr Res 137*, 111-125.
[26] Atmodjo, M. A., Hao, Z., and Mohnen, D. (2013) Evolving views of pectin biosynthesis, *Annu Rev Plant Biol 64*, 747-779.
[27] J, O. B., Harholt, J., Scheller, H. V., and Orfila, C. (2004) Rhamnogalacturonan I in Solanum tuberosum tubers contains complex arabinogalactan structures, *Phytochemistry 65*, 1429-1438.
[28] Freshour, G., Clay, R. P., Fuller, M. S., Albersheim, P., Darvill, A. G., and Hahn, M. G. (1996) Developmental and Tissue-Specific Structural Alterations of the Cell-Wall Polysaccharides of Arabidopsis thaliana Roots, *Plant Physiol 110*, 1413-1429.
[29] Guillemin, F., Guillon, F., Bonnin, E., Devaux, M. F., Chevalier, T., Knox, J. P., Liners, F., and Thibault, J. F. (2005) Distribution of pectic epitopes in cell walls of the sugar beet root, *Planta 222*, 355-371.
[30] Thomas, J. R., McNeil, M., Darvill, A. G., and Albersheim, P. (1987) Structure of Plant Cell Walls : XIX. Isolation and Characterization of Wall Polysaccharides from Suspension-Cultured Douglas Fir Cells, *Plant Physiol 83*, 659-671.
[31] Spellman, M. W., McNeil, M., Darvill, A. G., Albersheim, P., and Dell, A. (1983) Characterization of a structurally complex heptasaccharide isolated from the pectic polysaccharide rhamnogalacturonan II, *Carbohydr Res 122*, 131-153.
[32] Stevenson, T. T., Darvill, A. G., and Albersheim, P. (1988) 3-deoxy-d-lyxo-2-heptulosaric acid, a component of the plant cell-wall polysaccharide rhamnogalacturonan-II, *Carbohydr Res 179*, 269-288.
[33] York, W. S., Darvill, A. G., McNeil, M., and Albersheim, P. (1985) 3-deoxy-d-manno-2-octulosonic acid (KDO) is a component of rhamnogalacturonan II, a pectic polysaccharide in the primary cell walls of plants, *Carbohydr Res 138*, 109-126.
[34] O'Neill, M. A., Eberhard, S., Albersheim, P., and Darvill, A. G. (2001) Requirement of borate cross-linking of cell wall rhamnogalacturonan II for Arabidopsis growth, *Science 294*, 846-849.
[35] Cantarel, B. L., Coutinho, P. M., Rancurel, C., Bernard, T., Lombard, V., and Henrissat, B. (2009) The Carbohydrate-Active EnZymes database (CAZy): an expert resource for Glycogenomics, *Nucleic Acids Res 37*, D233-238.
[36] Brummell, D. A., and Harpster, M. H. (2001) Cell wall metabolism in fruit softening and quality and its manipulation in transgenic plants, *Plant molecular biology 47*, 311-340.

[37] Toth, I. K., Bell, K. S., Holeva, M. C., and Birch, P. R. J. (2003) Soft rot erwiniae: from genes to genomes, *Molecular Plant Pathology 4*, 17-30.

[38] Giovane, A., Servillo, L., Balestrieri, C., Raiola, A., D'Avino, R., Tamburrini, M., Ciardiello, M. A., and Camardella, L. (2004) Pectin methylesterase inhibitor, *Biochimica et biophysica acta 1696*, 245-252.

[39] Pelloux, J., Rusterucci, C., and Mellerowicz, E. J. (2007) New insights into pectin methylesterase structure and function, *Trends in plant science 12*, 267-277.

[40] Di Matteo, A., Giovane, A., Raiola, A., Camardella, L., Bonivento, D., De Lorenzo, G., Cervone, F., Bellincampi, D., and Tsernoglou, D. (2005) Structural basis for the interaction between pectin methylesterase and a specific inhibitor protein, *The Plant cell 17*, 849-858.

[41] Bellincampi, D., Dipierro, N., Salvi, G., Cervone, F., and De Lorenzo, G. (2000) Extracellular H(2)O(2) induced by oligogalacturonides is not involved in the inhibition of the auxin-regulated rolB gene expression in tobacco leaf explants, *Plant Physiol 122*, 1379-1385.

[42] Davis, K. R., and Hahlbrock, K. (1987) Induction of defense responses in cultured parsley cells by plant cell wall fragments, *Plant Physiol 84*, 1286-1290.

[43] Broekaert, W. F., Lambrechts, D., Verbelen, J. P., and Peumans, W. J. (1988) Datura stramonium Agglutinin : Location in the Seed and Release upon Imbibition, *Plant Physiol 86*, 569-574.

[44] Herron, S. R., Benen, J. A., Scavetta, R. D., Visser, J., and Jurnak, F. (2000) Structure and function of pectic enzymes: virulence factors of plant pathogens, *Proc Natl Acad Sci U S A 97*, 8762-8769.

[45] Abbott, D. W., and Boraston, A. B. (2008) Structural biology of pectin degradation by Enterobacteriaceae, *Microbiol Mol Biol Rev 72*, 301-316, table of contents.

[46] Schlemmer, A. F., Ware, C. F., and Keen, N. T. (1987) Purification and characterization of a pectin lyase produced by Pseudomonas fluorescens W51, *J Bacteriol 169*, 4493-4498.

[47] Mayans, O., Scott, M., Connerton, I., Gravesen, T., Benen, J., Visser, J., Pickersgill, R., and Jenkins, J. (1997) Two crystal structures of pectin lyase A from Aspergillus reveal a pH driven conformational change and striking divergence in the substrate-binding clefts of pectin and pectate lyases, *Structure 5*, 677-689.

[48] Kofod, L. V., Kauppinen, S., Christgau, S., Andersen, L. N., Heldt-Hansen, H. P., Dorreich, K., and Dalboge, H. (1994) Cloning and characterization of two structurally and functionally divergent rhamnogalacturonases from Aspergillus aculeatus, *J Biol Chem 269*, 29182-29189.

[49] Ochiai, A., Itoh, T., Maruyama, Y., Kawamata, A., Mikami, B., Hashimoto, W., and Murata, K. (2007) A novel structural fold in polysaccharide lyases: Bacillus subtilis family 11 rhamnogalacturonan lyase YesW with an eight-bladed beta-propeller, *J Biol Chem 282*, 37134-37145.

[50] Tripathi, C. K., Banga, J., and Mishra, V. (2012) Microbial heparin/heparan sulphate lyases: potential and applications, *Appl Microbiol Biotechnol 94*, 307-321.

[51] Wong, T. Y., Preston, L. A., and Schiller, N. L. (2000) ALGINATE LYASE: review of major sources and enzyme characteristics, structure-function analysis, biological roles, and applications, *Annu Rev Microbiol 54*, 289-340.

[52] Lombard, V., Bernard, T., Rancurel, C., Brumer, H., Coutinho, P. M., and Henrissat, B. (2010) A hierarchical classification of polysaccharide lyases for glycogenomics, *Biochem J 432*, 437-444.

[53] Garron, M. L., and Cygler, M. (2010) Structural and mechanistic classification of uronic acid-containing polysaccharide lyases, *Glycobiology 20*, 1547-1573.

[54] Charnock, S. J., Brown, I. E., Turkenburg, J. P., Black, G. W., and Davies, G. J. (2002) Convergent evolution sheds light on the anti-beta -elimination mechanism common to family 1 and 10 polysaccharide lyases, *Proc Natl Acad Sci U S A 99*, 12067-12072.

[55] Tardy, F., Nasser, W., Robert-Baudouy, J., and Hugouvieux-Cotte-Pattat, N. (1997) Comparative analysis of the five major Erwinia chrysanthemi pectate lyases: enzyme characteristics and potential inhibitors, *J Bacteriol 179*, 2503-2511.

[56] Abbott, D. W., and Boraston, A. B. (2007) A family 2 pectate lyase displays a rare fold and transition metal-assisted beta-elimination, *J Biol Chem 282*, 35328-35336.

[57] Abbott, D. W., Gilbert, H. J., and Boraston, A. B. (2010) The active site of oligogalacturonate lyase provides unique insights into cytoplasmic oligogalacturonate beta-elimination, *J Biol Chem 285*, 39029-39038.

[58] Abbott, D. W., Thomas, D., Pluvinage, B., and Boraston, A. B. (2013) An ancestral member of the polysaccharide lyase family 2 displays endolytic activity and magnesium dependence, *Appl Biochem Biotechnol 171*, 1911-1923.

[59] Seyedarabi, A., To, T. T., Ali, S., Hussain, S., Fries, M., Madsen, R., Clausen, M. H., Teixteira, S., Brocklehurst, K., and Pickersgill, R. W. (2010) Structural insights into substrate specificity and the anti beta-elimination mechanism of pectate lyase, *Biochemistry 49*, 539-546.

[60] Jensen, M. H., Otten, H., Christensen, U., Borchert, T. V., Christensen, L. L., Larsen, S., and Leggio, L. L. (2010) Structural and biochemical studies elucidate the mechanism of rhamnogalacturonan lyase from Aspergillus aculeatus, *J Mol Biol 404*, 100-111.

[61] Ochiai, A., Itoh, T., Mikami, B., Hashimoto, W., and Murata, K. (2009) Structural determinants responsible for substrate recognition and mode of action in family 11 polysaccharide lyases, *J Biol Chem 284*, 10181-10189.

[62] Shevchik, V. E., Condemine, G., Robert-Baudouy, J., and Hugouvieux-Cotte-Pattat, N. (1999) The exopolygalacturonate lyase PelW and the oligogalacturonate lyase Ogl, two cytoplasmic enzymes of pectin catabolism in Erwinia chrysanthemi 3937, *J Bacteriol 181*, 3912-3919.

[63] Yoder, M. D., Keen, N. T., and Jurnak, F. (1993) New domain motif: the structure of pectate lyase C, a secreted plant virulence factor, *Science 260*, 1503-1507.

[64] Jenkins, J., and Pickersgill, R. (2001) The architecture of parallel beta-helices and related folds, *Prog Biophys Mol Biol 77*, 111-175.

[65] Jenkins, J., Mayans, O., and Pickersgill, R. (1998) Structure and evolution of parallel beta-helix proteins, *J Struct Biol 122*, 236-246.

[66] Jenkins, J., Shevchik, V. E., Hugouvieux-Cotte-Pattat, N., and Pickersgill, R. W. (2004) The crystal structure of pectate lyase Pel9A from Erwinia chrysanthemi, *J Biol Chem 279*, 9139-9145.

[67] McDonough, M. A., Kadirvelraj, R., Harris, P., Poulsen, J. C., and Larsen, S. (2004) Rhamnogalacturonan lyase reveals a unique three-domain modular structure for polysaccharide lyase family 4, *FEBS Lett 565*, 188-194.

[68] Biely, P., Benen, J., Heinrichova, K., Kester, H. C., and Visser, J. (1996) Inversion of configuration during hydrolysis of alpha-1,4-galacturonidic linkage by three Aspergillus polygalacturonases, *FEBS Lett 382*, 249-255.

[69] Vasella, A., Davies, G. J., and Bohm, M. (2002) Glycosidase mechanisms, *Curr Opin Chem Biol 6*, 619-629.

[70] Pickersgill, R., Smith, D., Worboys, K., and Jenkins, J. (1998) Crystal structure of polygalacturonase from Erwinia carotovora ssp. carotovora, *J Biol Chem 273*, 24660-24664.

[71] Abbott, D. W., and Boraston, A. B. (2007) The structural basis for exopolygalacturonase activity in a family 28 glycoside hydrolase, *J Mol Biol 368*, 1215-1222.

[72] Mutter, M., Beldman, G., Pitson, S. M., Schols, H. A., and Voragen, A. G. (1998) Rhamnogalacturonan alpha-d-galactopyranosyluronohydrolase. An enzyme that specifically removes the terminal nonreducing galacturonosyl residue in rhamnogalacturonan regions of pectin, *Plant Physiol 117*, 153-163.

[73] Mutter, M., Beldman, G., Schols, H. A., and Voragen, A. G. (1994) Rhamnogalacturonan alpha-L-rhamnopyranohydrolase. A novel enzyme specific for the terminal nonreducing rhamnosyl unit in rhamnogalacturonan regions of pectin, *Plant Physiol 106*, 241-250.

[74] Petersen, T. N., Kauppinen, S., and Larsen, S. (1997) The crystal structure of rhamnogalacturonase A from Aspergillus aculeatus: a right-handed parallel beta helix, *Structure 5*, 533-544.

[75] Itoh, T., Ochiai, A., Mikami, B., Hashimoto, W., and Murata, K. (2006) A novel glycoside hydrolase family 105: the structure of family 105 unsaturated rhamnogalacturonyl hydrolase complexed with a disaccharide in comparison with family 88 enzyme complexed with the disaccharide, *J Mol Biol 360*, 573-585.

[76] Itoh, T., Ochiai, A., Mikami, B., Hashimoto, W., and Murata, K. (2006) Structure of unsaturated rhamnogalacturonyl hydrolase complexed with substrate, *Biochem Biophys Res Commun 347*, 1021-1029.

[77] Mill, P. J., and Tuttobello, R. (1961) The pectic enzymes of Aspergillus niger. 2. Endopolygalacturonase, *Biochem J 79*, 57-64.

[78] Hasegawa, S., and Nagel, C. W. (1968) Isolation of an oligogalacturonate hydrolase from a Bacillus specie, *Archives of biochemistry and biophysics 124*, 513-520.

[79] Pitson, S. M., Mutter, M., van den Broek, L. A., Voragen, A. G., and Beldman, G. (1998) Stereochemical course of hydrolysis catalysed by alpha-L-rhamnosyl and alpha-D-galacturonosyl hydrolases from Aspergillus aculeatus, *Biochem Biophys Res Commun 242*, 552-559.

[80] Khan, M., Nakkeeran, E., and Umesh-Kumar, S. (2013) Potential application of pectinase in developing functional foods, *Annual review of food science and technology 4*, 21-34.

[81] Lang, C., and Dornenburg, H. (2000) Perspectives in the biological function and the technological application of polygalacturonases, *Appl Microbiol Biotechnol 53*, 366-375.

[82] Nyvall-Collen, P., Jeudy, A., Sassi, J. F., Groisillier, A., Czjzek, M., Coutinho, P. M., and Helbert, W. (2014) A novel unsaturated beta-glucuronyl hydrolase involved in

ulvan degradation unveils the versatility of stereochemistry requirements in family GH105, *J Biol Chem.*

[83] Pijning, T., van Pouderoyen, G., Kluskens, L., van der Oost, J., and Dijkstra, B. W. (2009) The crystal structure of a hyperthermoactive exopolygalacturonase from Thermotoga maritima reveals a unique tetramer, *FEBS Lett 583*, 3665-3670.

[84] Shimizu, T., Nakatsu, T., Miyairi, K., Okuno, T., and Kato, H. (2002) Active-site architecture of endopolygalacturonase I from Stereum purpureum revealed by crystal structures in native and ligand-bound forms at atomic resolution, *Biochemistry 41*, 6651-6659.

[85] Zhang, R., Minh, T., Lezondra, L., Korolev, S., Moy, S. F., Collart, F., and Joachimiak, A. (2005) 1.6 A crystal structure of YteR protein from Bacillus subtilis, a predicted lyase, *Proteins 60*, 561-565.

[86] Biely, P. (2012) Microbial carbohydrate esterases deacetylating plant polysaccharides, *Biotechnol Adv 30*, 1575-1588.

[87] Lionetti, V., Cervone, F., and Bellincampi, D. (2012) Methyl esterification of pectin plays a role during plant-pathogen interactions and affects plant resistance to diseases, *J Plant Physiol 169*, 1623-1630.

[88] Jolie, R. P., Duvetter, T., Van Loey, A. M., and Hendrickx, M. E. (2010) Pectin methylesterase and its proteinaceous inhibitor: a review, *Carbohydr Res 345*, 2583-2595.

[89] Johansson, K., El-Ahmad, M., Friemann, R., Jornvall, H., Markovic, O., and Eklund, H. (2002) Crystal structure of plant pectin methylesterase, *FEBS Lett 514*, 243-249.

[90] Jenkins, J., Mayans, O., Smith, D., Worboys, K., and Pickersgill, R. W. (2001) Three-dimensional structure of Erwinia chrysanthemi pectin methylesterase reveals a novel esterase active site, *J Mol Biol 305*, 951-960.

[91] Fries, M., Ihrig, J., Brocklehurst, K., Shevchik, V. E., and Pickersgill, R. W. (2007) Molecular basis of the activity of the phytopathogen pectin methylesterase, *Embo J 26*, 3879-3887.

[92] Shevchik, V. E., and Hugouvieux-Cotte-Pattat, N. (1997) Identification of a bacterial pectin acetyl esterase in Erwinia chrysanthemi 3937, *Mol Microbiol 24*, 1285-1301.

[93] Kauppinen, S., Christgau, S., Kofod, L. V., Halkier, T., Dorreich, K., and Dalboge, H. (1995) Molecular cloning and characterization of a rhamnogalacturonan acetylesterase from Aspergillus aculeatus. Synergism between rhamnogalacturonan degrading enzymes, *J Biol Chem 270*, 27172-27178.

[94] Christov, L. P., and Prior, B. A. (1993) Esterases of xylan-degrading microorganisms: production, properties, and significance, *Enzyme Microb Technol 15*, 460-475.

[95] Shevchik, V. E., and Hugouvieux-Cotte-Pattat, N. (2003) PaeX, a second pectin acetylesterase of Erwinia chrysanthemi 3937, *J Bacteriol 185*, 3091-3100.

[96] Shevchik, V. E., Condemine, G., Hugouvieux-Cotte-Pattat, N., and Robert-Baudouy, J. (1996) Characterization of pectin methylesterase B, an outer membrane lipoprotein of Erwinia chrysanthemi 3937, *Mol Microbiol 19*, 455-466.

[97] Kazemi-Pour, N., Condemine, G., and Hugouvieux-Cotte-Pattat, N. (2004) The secretome of the plant pathogenic bacterium Erwinia chrysanthemi, *Proteomics 4*, 3177-3186.

[98] Eklof, J. M., Tan, T. C., Divne, C., and Brumer, H. (2009) The crystal structure of the outer membrane lipoprotein YbhC from Escherichia coli sheds new light on the phylogeny of carbohydrate esterase family 8, *Proteins 76*, 1029-1036.

[99] Boraston, A. B., and Abbott, D. W. (2012) Structure of a pectin methylesterase from Yersinia enterocolitica, *Acta Crystallogr Sect F Struct Biol Cryst Commun 68*, 129-133.

[100] Hugouvieux-Cotte-Pattat, N., Condemine, G., Nasser, W., and Reverchon, S. (1996) Regulation of pectinolysis in Erwinia chrysanthemi, *Annu Rev Microbiol 50*, 213-257.

[101] Molgaard, A., Kauppinen, S., and Larsen, S. (2000) Rhamnogalacturonan acetylesterase elucidates the structure and function of a new family of hydrolases, *Structure 8*, 373-383.

[102] James, V., and Hugouvieux-Cotte-Pattat, N. (1996) Regulatory systems modulating the transcription of the pectinase genes of Erwinia chrysanthemi are conserved in Escherichia coli, *Microbiology 142 (Pt 9)*, 2613-2619.

[103] Lara-Marquez, A., Zavala-Paramo, M. G., Lopez-Romero, E., and Camacho, H. C. (2011) Biotechnological potential of pectinolytic complexes of fungi, *Biotechnol Lett 33*, 859-868.

[104] Kikot, G. E., Hours, R. A., and Alconada, T. M. (2009) Contribution of cell wall degrading enzymes to pathogenesis of Fusarium graminearum: a review, *J Basic Microbiol 49*, 231-241.

[105] *http://commonfund.nih.gov/hmp/index*. Human Microbiome Project.

[106] Koropatkin, N. M., Cameron, E. A., and Martens, E. C. (2012) How glycan metabolism shapes the human gut microbiota, *Nature reviews. Microbiology 10*, 323-335.

[107] McNeil, N. I. (1984) The contribution of the large intestine to energy supplies in man, *The American journal of clinical nutrition 39*, 338-342.

[108] Martens, E. C., Koropatkin, N. M., Smith, T. J., and Gordon, J. I. (2009) Complex glycan catabolism by the human gut microbiota: the Bacteroidetes Sus-like paradigm, *J Biol Chem 284*, 24673-24677.

[109] Martens, E. C., Lowe, E. C., Chiang, H., Pudlo, N. A., Wu, M., McNulty, N. P., Abbott, D. W., Henrissat, B., Gilbert, H. J., Bolam, D. N., and Gordon, J. I. (2011) Recognition and degradation of plant cell wall polysaccharides by two human gut symbionts, *PLoS Biol 9*, e1001221.

[110] Rodionov, D. A., Gelfand, M. S., and Hugouvieux-Cotte-Pattat, N. (2004) Comparative genomics of the KdgR regulon in Erwinia chrysanthemi 3937 and other gamma-proteobacteria, *Microbiology 150*, 3571-3590.

[111] Wrzosek, L., Miquel, S., Noordine, M. L., Bouet, S., Joncquel Chevalier-Curt, M., Robert, V., Philippe, C., Bridonneau, C., Cherbuy, C., Robbe-Masselot, C., Langella, P., and Thomas, M. (2013) Bacteroides thetaiotaomicron and Faecalibacterium prausnitzii influence the production of mucus glycans and the development of goblet cells in the colonic epithelium of a gnotobiotic model rodent, *BMC Biol 11*, 61.

[112] Kim, Y. S., and Ho, S. B. (2010) Intestinal goblet cells and mucins in health and disease: recent insights and progress, *Curr Gastroenterol Rep 12*, 319-330.

[113] Duncan, S. H. (2002) Growth requirements and fermentation products of Fusobacterium prausnitzii, and a proposal to reclassify it as Faecalibacterium prausnitzii gen. nov., comb. nov, *International Journal of Systematic and Evolutionary Microbiology 52*, 2141-2146.

[114] Dongowski, G., Lorenz, A., and Anger, H. (2000) Degradation of pectins with different degrees of esterification by Bacteroides thetaiotaomicron isolated from human gut flora, *Appl Environ Microbiol 66*, 1321-1327.

[115] Gaudier, E., Jarry, A., Blottière, H. M., de Coppet, P., Buisine, M. P., Aubert, J. P., Laboisse, C., Cherbut, C., and Hoebler, C. (2004) Butyrate specifically modulates MUC gene expression in intestinal epithelial goblet cells deprived of glucose, *American Journal of Physiology - Gastrointestinal and Liver Physiology 287*, G1168-G1174.

[116] Holmen Larsson, J. M., Thomsson, K. A., Rodriguez-Pineiro, A. M., Karlsson, H., and Hansson, G. C. (2013) Studies of mucus in mouse stomach, small intestine, and colon. III. Gastrointestinal Muc5ac and Muc2 mucin O-glycan patterns reveal a regiospecific distribution, *Am J Physiol Gastrointest Liver Physiol 305*, G357-363.

[117] Johansson, M. E., Sjovall, H., and Hansson, G. C. (2013) The gastrointestinal mucus system in health and disease, *Nat Rev Gastroenterol Hepatol 10*, 352-361.

[118] Fung, K. Y., Ooi, C. C., Zucker, M. H., Lockett, T., Williams, D. B., Cosgrove, L. J., and Topping, D. L. (2013) Colorectal carcinogenesis: a cellular response to sustained risk environment, *Int J Mol Sci 14*, 13525-13541.

[119] Jackson, C. L., Dreaden, T. M., Theobald, L. K., Tran, N. M., Beal, T. L., Eid, M., Gao, M. Y., Shirley, R. B., Stoffel, M. T., Kumar, M. V., and Mohnen, D. (2007) Pectin induces apoptosis in human prostate cancer cells: correlation of apoptotic function with pectin structure, *Glycobiology 17*, 805-819.

[120] Vayssade, M., Sengkhamparn, N., Verhoef, R., Delaigue, C., Goundiam, O., Vigneron, P., Voragen, A. G., Schols, H. A., and Nagel, M. D. (2010) Antiproliferative and proapoptotic actions of okra pectin on B16F10 melanoma cells, *Phytotherapy research : PTR 24*, 982-989.

[121] Olano-Martin, E., Rimbach, G. H., Gibson, G. R., and Rastall, R. A. (2003) Pectin and pectic-oligosaccharides induce apoptosis in in vitro human colonic adenocarcinoma cells, *Anticancer Res 23*, 341-346.

[122] Roberfroid, M. B. (2000) Prebiotics and probiotics: are they functional foods?, *The American Journal of Clinical Nutrition 71*, 1682S-1687S.

[123] Chirdo, F. G., Millington, O. R., Beacock-Sharp, H., and Mowat, A. M. (2005) Immunomodulatory dendritic cells in intestinal lamina propria, *European Journal of Immunology 35*, 1831-1840.

[124] Keservani, R. K., Kesharwani, R. K., Vyas, N., Jain, S., Ramsaneh, R., and Sharma, A. K. (2010) Nutraceutical and Functional Food as Future Food: A Review, *Der Pharmacia Lettre 2*, 106-116.

[125] Ye, M. B., and Lim, B. O. (2010) Dietary pectin regulates the levels of inflammatory cytokines and immunoglobulins in interleukin-10 knockout mice, *J Agric Food Chem 58*, 11281-11286.

[126] Salman, H., Bergman, M., Djaldetti, M., Orlin, J., and Bessler, H. (2008) Citrus pectin affects cytokine production by human peripheral blood mononuclear cells, *Biomedicine & Pharmacotherapy 62*, 579-582.

[127] Matsumoto, T., Moriya, M., Sakurai, M. H., Kiyohara, H., Tabuchi, Y., and Yamada, H. (2008) Stimulatory effect of a pectic polysaccharide from a medicinal herb, the roots of Bupleurum falcatum L., on G-CSF secretion from intestinal epithelial cells, *Int Immunopharmacol 8*, 581-588.

[128] Inngjerdingen, M., Inngjerdingen, K. T., Patel, T. R., Allen, S., Chen, X., Rolstad, B., Morris, G. A., Harding, S. E., Michaelsen, T. E., Diallo, D., and Paulsen, B. S. (2008) Pectic polysaccharides from Biophytum petersianum Klotzsch, and their activation of macrophages and dendritic cells, *Glycobiology 18*, 1074-1084.

[129] Gronhaug, T. E., Kiyohara, H., Sveaass, A., Diallo, D., Yamada, H., and Paulsen, B. S. (2011) Beta-D-(1-->4)-galactan-containing side chains in RG-I regions of pectic polysaccharides from Biophytum petersianum Klotzsch. contribute to expression of immunomodulating activity against intestinal Peyer's patch cells and macrophages, *Phytochemistry 72*, 2139-2147.

[130] Niture, S. K., and Refai, L. (2013) Plant Pectin: A Potential Source for Cancer Suppression, *American Journal of Pharmacology and Toxicology 8*, 9-19.

[131] Eliaz, I., Hotchkiss, A. T., Fishman, M. L., and Rode, D. (2006) The effect of modified citrus pectin on urinary excretion of toxic elements, *Phytotherapy research : PTR 20*, 859-864.

[132] Ganan, M., Collins, M., Rastall, R., Hotchkiss, A. T., Chau, H. K., Carrascosa, A. V., and Martinez-Rodriguez, A. J. (2010) Inhibition by pectic oligosaccharides of the invasion of undifferentiated and differentiated Caco-2 cells by Campylobacter jejuni, *Int J Food Microbiol 137*, 181-185.

[133] Gustafsson, A., Hultberg, A., Sjostrom, R., Kacskovics, I., Breimer, M. E., Boren, T., Hammarstrom, L., and Holgersson, J. (2006) Carbohydrate-dependent inhibition of Helicobacter pylori colonization using porcine milk, *Glycobiology 16*, 1-10.

[134] Sharon, N., and Ofek, I. (2002) Fighting infectious diseases with inhibitors of microbial adhesion to host tissues, *Crit Rev Food Sci Nutr 42*, 267-272.

[135] Sharon, N. (2006) Carbohydrates as future anti-adhesion drugs for infectious diseases, *Biochimica et biophysica acta 1760*, 527-537.

[136] Cundell, D. R., and Tuomanen, E. I. (1994) Receptor specificity of adherence of Streptococcus pneumoniae to human type-II pneumocytes and vascular endothelial cells in vitro, *Microb Pathog 17*, 361-374.

[137] Simon, P. M., Goode, P. L., Mobasseri, A., and Zopf, D. (1997) Inhibition of Helicobacter pylori binding to gastrointestinal epithelial cells by sialic acid-containing oligosaccharides, *Infect Immun 65*, 750-757.

[138] Thomas, R. J. (2010) Receptor mimicry as novel therapeutic treatment for biothreat agents, *Bioengineered Bugs 1*, 17-30.

[139] Ogonczyk, D., Siek, M., and Garstecki, P. (2011) Microfluidic formulation of pectin microbeads for encapsulation and controlled release of nanoparticles, *Biomicrofluidics 5*, 13405.

[140] de Vos, P., Faas, M. M., Spasojevic, M., and Sikkema, J. (2010) Encapsulation for preservation of functionality and targeted delivery of bioactive food components, *International Dairy Journal 20*, 292-302.

[141] Tomaro-Duchesneau, C., Saha, S., Malhotra, M., Kahouli, I., and Prakash, S. (2013) Microencapsulation for the Therapeutic Delivery of Drugs, Live Mammalian and Bacterial Cells, and Other Biopharmaceutics: Current Status and Future Directions, *Journal of Pharmaceutics 2013*, 1-19.

[142] Santipanichwong, R., Suphantharika, M., Weiss, J., and McClements, D. J. (2008) Core-shell biopolymer nanoparticles produced by electrostatic deposition of beet pectin onto heat-denatured beta-lactoglobulin aggregates, *J Food Sci 73*, N23-30.

[143] Coussa, R., Martoni, C., Bhathena, J., Urbanska, A. M., and Prakash, S. (2010) Oral microencapsulated live Saccharomyces cerevisiae cells for use in renal failure uremia: preparation and in vivo analysis, *Journal of biomedicine & biotechnology 2010*, 7.

[144] Jones, M. L., Chen, H., Ouyang, W., Metz, T., and Prakash, S. (2004) Microencapsulated Genetically Engineered Lactobacillus plantarum 80 (pCBH1) for Bile Acid Deconjugation and Its Implication in Lowering Cholesterol, *Journal of biomedicine & biotechnology 2004*, 61-69.

[145] Urbanska, A. M., Paul, A., Bhathena, J., and Prakash, S. (2010) Suppression of tumorigenesis: modulation of inflammatory cytokines by oral administration of microencapsulated probiotic yogurt formulation, *Int J Inflam 2010*, 894972.

[146] Liu, L., Fishman, M. L., Kost, J., and Hicks, K. B. (2003) Pectin-based systems for colon-specific drug delivery via oral route, *Biomaterials 24*, 3333-3343.

[147] Sundar, R. A. A., Rubila, S., Jayabalan, R., and Ranganathan, T. V. (2012) A Review on Pectin: Chemistry due to General Properties of Pectin and its Pharmaceutical Uses, *1*.

[148] Wong, T. W., Colombo, G., and Sonvico, F. (2011) Pectin matrix as oral drug delivery vehicle for colon cancer treatment, *AAPS PharmSciTech 12*, 201-214.

[149] Kinget, R., Kalala, W., Vervoort, L., and van den Mooter, G. (1998) Colonic drug targeting, *J Drug Target 6*, 129-149.

[150] Sarasija, S., and Hota, A. (2000) Colon specific drug delivery systems, *Ind J Pharm Sci 62*, 1-8.

[151] Campeau, J. L., Salim, S. Y., Albert, E. J., Hotte, N., and Madsen, K. L. (2012) Intestinal Epithelial Cells Modulate Antigen-Presenting Cell Responses to Bacterial DNA, *Infection and Immunity 80*, 2632-2644.

[152] Scavetta, R. D., Herron, S. R., Hotchkiss, A. T., Kita, N., Keen, N. T., Benen, J. A., Kester, H. C., Visser, J., and Jurnak, F. (1999) Structure of a plant cell wall fragment complexed to pectate lyase C, *Plant Cell 11*, 1081-1092.

[153] Emsley, P., Lohkamp, B., Scott, W. G., and Cowtan, K. (2010) Features and development of Coot, *Acta Crystallogr D Biol Crystallogr 66*, 486-501.

[154] Itoh, T., Ochiai, A., Mikami, B., Hashimoto, W., and Murata, K. (2006) Structure of unsaturated rhamnogalacturonyl hydrolase complexed with substrate, *Biochem Biophys Res Commun 347*, 1021-1029.

[155] Akita, M., Suzuki, A., Kobayashi, T., Ito, S., and Yamane, T. (2001) The first structure of pectate lyase belonging to polysaccharide lyase family 3, *Acta Crystallogr D Biol Crystallogr 57*, 1786-1792.

[156] Patel, P. K., Satwara, R.S., and Pandya, S.S. (2012) Bacteria aided biopolymers as carriers for colon specific drug delivery system: A Review, *International Journal of PharmTech Research 4*, 1191-1214.

[157] Tomaro-Duchesneau, C., Saha, S., Malhotra, M., Kahouli, I., and Prakash, S. (2013) Microencapsulation for the Therapeutic Delivery of Drugs, Live Mammalian and Bacterial Cells, and Other Biopharmaceutics: Current Status and Future Directions, *Journal of Pharmaceutics* 1-19.

[158] Ferreira, T. M., Leonel, A. J., Melo, M. A., Santos, R. R., Cara, D. C., Cardoso, V. N., Correia, M. I., and Alvarez-Leite, J. I. (2012) Oral supplementation of butyrate reduces mucositis and intestinal permeability associated with 5-Fluorouracil administration, *Lipids 47*, 669-678.

[159] Cho, K., Shin, H. W., Cho, C. H., Chun, Y. S., Kim, T. Y., and Park, J. W. (2013) Mad1 mediates hypoxia-induced doxorubicin resistance in colon cancer cells by inhibiting mitochondrial function, *Free Radic Biol Med 60*, 201-220.

[160] Zweibaum, A., Lesuffleur, T., Barbat, A., Dussaulx, E., Chantret, I., Mahraoui, L., Chevalier, G., Brot-Laroche, E., and Rousset, M. (1993) *Colon cancer cell differentiation as related to methotrexate and 5-fluorouracil resistance*, Vol. 75, Springer Berlin Heidelberg.

[161] Kararli, T. T. (1995) Comparison of the gastrointestinal anatomy, physiology, and biochemistry of humans and commonly used laboratory animals, *Biopharmaceutics & Drug Disposition 16*, 351-380.

[162] Prakash, S., Tomaro-Duchesneau, C., Saha, S., and Cantor, A. (2011) The gut microbiota and human health with an emphasis on the use of microencapsulated bacterial cells, *Journal of biomedicine & biotechnology 2011*, 981214.

In: Pectin: Chemical Properties, Uses and Health Benefits
Editor: Phillip L. Bush
ISBN: 978-1-63321-438-5

Chapter 7

NOVEL USES OF PECTINS AS HEALTH INGREDIENTS FOR FOOD AND PHARMACEUTICAL APPLICATIONS

Dongxiao Sun-Waterhouse[1,2*], Geoffrey I. N. Waterhouse[2], Mouming Zhao[1] and Qiangzhong Zhao[1]

[1]College of Light Industry and Food Sciences,
South China University of Technology, Guangzhou, China
[2]School of Chemical Sciences, The University of Auckland, Auckland, New Zealand

ABSTRACT

Pectic polysaccharides, commonly termed pectins, are a group of natural polymers containing (1→4)-linked α-D-galacturonosyl residues such as homogalacturonan, arabinan-rich rhamnogalacturonan and xylogalacturonan. Pectins are abundant in many fruits and higher plants such as citrus, apples, pears and carrots, and have long been used as food additives or as active/stabilising components of pharmaceutical and cosmetic products. Global annual use of pectins is estimated at around 45 million kilograms, with a market value exceeding €400 million. The positive physiological effect of pectins (mainly as soluble dietary fibers) on humans stimulates manufacturing opportunities for novel pectin fibre ingredients and derived functional foods. However, ensuring that the desirable biological functionality and food processing properties of pectins are retained during product manufacturing remains a challenge. This chapter explores novel approaches for manipulating and optimising the positive role(s) of pectins in food processing and digestion based on pectins' sensitivity to pH, temperature, enzymes and other matrix factors. Several case studies, such as pectins' inhibitory effects on the activity of lipase enzymes and pectin's role in preservation of ascorbic acid and other phytochemicals during food processing, will be used to highlight key concepts. Plant origin, extraction method and degree of methyl esterification, acetylation and amidation of pectin ingredients determine finished food quality attributes as well as bioactive content and activities. The importance of advanced characterisation techniques, such as SEM, FT-IR, HPLC, GC, rheometry, cyclic voltammetry and solid-state NMR spectroscopy, in addition to conventional chemical analysis assays, for evaluating the suitability of a pectin ingredient for a specific functional food application, is

* E-mail address: dxsun72@hotmail.com.

demonstrated. The future outlook suggests that pectins will be increasingly utilised as soluble fibers for conventional food systems, and also as encapsulants for bioactive delivery systems. It is expected that 'designer pectins' such as those enriched in uronic acid or oligosaccharide fractions, or having specific methylation or branching degree to confer desirable health-promoting functionality (e.g. preserving antioxidants and enhancing prebiotic effects) will gain increased market share in the food ingredient sector.

INTRODUCTION

Dietary fiber (DF) plays a critical role in human health and well-being, though its exact definition has changed many times over the past 50-60 years. Our knowledge of the chemistry and nutrition of DF has increased considerably over the past 20 years, stimulated by debate around its definition and method of analysis. The recently established DF definition in the *Codex Alimentarius* [1] is as "carbohydrate polymers with 10 or more monomeric units, which are not hydrolysed by the endogenous enzymes in the small intestine of humans". This definition recognizes DF's unique chemical and physical properties. The rising consumer awareness of the link between diet and health along with the accumulated scientific evidence on DF's positive roles in human health, leads to the promotion of DF in the rapidly evolving functional food market. Food laws have efficiently kept up with advancements in DF understanding, e.g., DFs are listed as a nutrient in Nutrition Fact Panels, as a food ingredient or additive in the list of specific ingredients, and as a bioactive substance for validated claims such as "good source of fibre" and "excellent source of fibre" allowed by the *Code of Federal Regulations* (Title 21, Part 101.54). In February 2014, the US Food and Drug Administration (FDA) proposed to update the Nutrition Fact labels for packaged foods to recognize the latest discoveries on the effect of human diet on chronic diseases (http://www.fda.gov/NewsEvents/Newsroom/PressAnnouncements/ucm387418.htm, access on 22 March 2014). In light of this proposal, the recommended daily amount of DF intake is being increased from 25 to 28 grams.

Both soluble and insoluble fibers contribute to the conferred health benefits of DFs. Indeed, soluble and insoluble fibers complement each other in terms of chemical and physical properties as well as physiological effects [2, 3]. A balanced proportion of soluble and insoluble fibers is recommended e.g., an insoluble-to-soluble fiber ratio of 75:25 by the *Code of Federal Regulations*. Amongst the soluble fibers, pectins have attracted most attention. Pectins possess distinct physical and chemical structure that imparts their marked physicochemical properties and nutritional value. Persuasive evidence for pectin's protective effects against chronic diseases justifies its inclusion into foods at significant quantities to confer specific health benefits. This stimulates interest from consumers and food manufacturers.

There exist numerous studies and publications on the chemistry and nutrition of pectins prepared by health organizations, research scientists and industrial specialists. This chapter focuses on pectins as ingredients for food and pharmaceutical use. The molecular chemical structure, polymer arrangement, and properties that directly influence product quality and bioactive profiles of a pectin-rich functional food are discussed, including pectins in plant cell wall preparations and commercially isolated pectins.

It is highlighted that pectins are structural materials and their versatile physicochemical properties can be further manipulated to aid the delivery of novel functional foods. The resultant "designer pectin-fortified functional foods" may convey synergistic health-promoting properties beyond the well-known nutritional value of pectins as DFs. The positive effects of pectins on the digestive enzymes and the delivery of desired nutrients such as proteins and other bioactive component such as polyphenol and vitamin antioxidants are also explored herein.

PECTINS AS DIETARY FIBERS

Epidemiological research has shown that regular or increased consumption of fruits, vegetables and cereals results in decreased risk of chronic diseases [4-6]. DF is a collective term established for a group of polysaccharide polymers and non-polysaccharides [7]. DFs can be classified into cellulose and non-cellulosic polysaccharides such as pectins, xyloglucans, arabinoxylans, glucomannans and galactoglucomannans [8]. Its heterogeneity is attributed to its chemical composition (i.e. biopolymers with different side chains) and physical arrangement (i.e. cross-linked network matrix). DFs contain carboxyl, hydroxyl, ester, ether and cyclic acetal functional groups that can participate in various chemical reactions [9]. These chemical reactions and physical interactions render DFs' water-holding capacity, bulk volume and viscosity properties which vary with pH and ionic environment during gastrointestinal transit [10, 11].

DF's positive physiological effects on human health are well documented and include the promotion of energy regulation and balance, digestive health, satiety and appetitie control, and reduced incidence of cancer, heart, obesity and diabetes [9, 12-15]. Covalent linkages within DFs that are retained after food processing (such as cooking), can be cleaved by appropriate enzymes generated by the gut bacterial flora during food digestion and colonic fermentation. Soluble fibers possess positive physiological effects such as significant serum cholesterol-lowering, prebiotic, hypocholesterolemic and anti-cancer effects [16-19]. Soluble DFs considerably influence nutrient digestion and assimilation in the digestive tract through changing the physicochemical properties of the stomach and small intestine media. Nutrients in food matrices have to be released first into the surrounding digestive fluid prior to absorption through the gut and/or intestinal walls. There are many factors that control the rate of such a release including the nature of food solids (e.g. particle size, shape, microstructure, charge and surface characteristics) and physical state of the digesta medium or solute (e.g. viscosity, colloidal attributes and pH).

Humans consume on average around 5 grams pectin per day [20]. Pectins are fermentable and viscous DFs, and have been well studied as functional components in cell walls of fruits, vegetables and cereals, and as isolated ingredients or additives in a wide range of manufactured foods and pharmaceutical products. Pectins possess prebiotic effects and stimulate intestinal microbial growth and improve immune function [21]. They can suppress some cancers and degenerative diseases like diabetes and coronary heart disease, decrease blood glucose concentration, improve insulin response and lower plasma LDL cholesterol concentrations [6, 20, 22, 23]. At gastric pHs, pectins remain partially un-ionised and can shed their bound counter ions. Pectin's tight site-binding of counterions in the surrounding matrix is attributed to its ordered structure. In the small intestine where the pH is high, pectins are highly charged and facilitate electrostatic binding. These properties confer some of the important health benefits of pectin including increased sensations of satiety [24, 25]. Pectin can also affect the metabolism of nutrients such as protein digestion [26] and the activity of digestive enzymes such as amylase and lipase [27, 28].

Pectin can be extracted from all plant material. Commercially available pectins are manufactured through utilizing by-products of fruit and vegetable processing as raw materials, including apple pomace, citrus peels, tropical fruit waste, potato skins, onion skins and green pea peel. Industrial manufacturing of pectins often includes extraction (i.e. using acidified water at pH 1.5-3.0 and 60-100°C for 0.5-6 h), separation (i.e. using alcoholic precipitation, filtration and/or centrifugation) and modification or optimization (i.e. tailoring degree of methylation, acetylation, amidation and branching). The obtained protopectins have reduced branching and chain length after prolonged extraction. Washing is first required to leach out organic acids, sugar and pigments to prevent discoloration or browning and increase the purity of pectin product. During extraction, methylation and acetylation are possible and the chain length of polymer can also be reduced. Monitoring the viscosity of extracts to reasonably low values is required to ensure efficient subsequent filtration. Commercially available pectin ingredients are mostly homogalacturonans.

The versatile functionality of pectins comes from their intrinsic molecular structure and polymeric arrangement. The most important chemical indices for describing pectins include monosaccharide composition (i.e. uronic acid and neutral sugar contents), degree of methyl esterification (DM), acetylation and amidation, as well as total, soluble and insoluble fiber, protein and ash contents [29]. Based on the DM value, pectins are divided into high methoxyl pectins (HM, DM > 50%) and low methoxyl pectins (LM, DM < 50%). Most of the industrially extracted pectins are initially in form of HM pectins, from which LM pectins are produced through treatments with dilute acid [30]. Amidation of pectin is carried out via ammonolysis to convert some methyl-ester groups to amide groups in the alkaline de-esterification process [31]. In addition, the C-2 and C-3 hydroxyl groups in the pectin galacturonan can be esterified by acetic acid. Pectin extracted from raw materials like sugar beet may contain a large amount of O-acetyl groups mainly at the C-3 with some at the C-2 position of galacturonan. Figure 1 shows the chemical structure of galacturonic acid, methyl esterified galacturonan, acetylated galacturonan and amidated galacturonan.

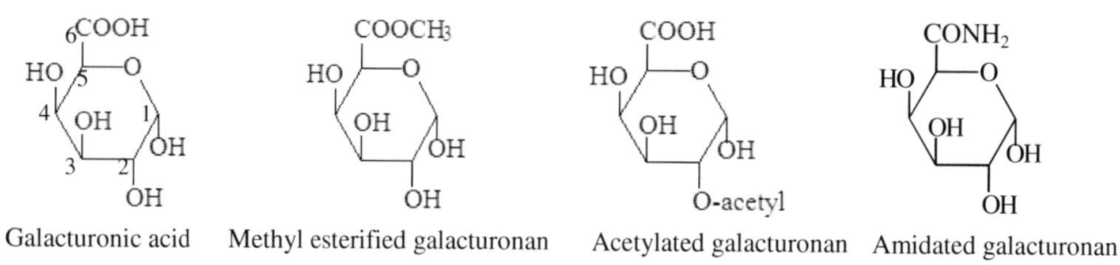

Figure 1. Structure of galacturonic acid, methyl esterified galacturonan, acetylated galacturonan and amidated galacturonan.

The structure and composition of a pectin preparation depend largely on its origin including plant growth conditions like climate, soils and maturity, as well as the extraction method used e.g., aqueous versus ethanolic method, absence or presence of a chelating agent like EDTA and CDTA, the use of an enzyme treatment, supercritical fluid extraction and microwave heating under pressure [8, 32-34; United States Patent 6143337 2007]. The raw materials for manufacturing pectins are mainly industrial by-products such as pomace from apple and sugar-beet and peels from citrus and tropical fruits [35]. Table 1 shows that fruit peel wastes are a good source of pectins. The plant origin of peel by-products (e.g. fruit type and cultivar), extraction method (e.g. extraction medium and temperature), as well as the analysis method (e.g. acid hydrolysis and detection techniques) all influence the determination of pectin content [36-38]. The structural arrangement of extracted pectins varies with the material origin, extraction method and pectin concentration i.e. structural arrangement as networks of strands with spherical nodes embedded in the network or at the ends of strands [39]. When the same extraction methods is applied to different cultivars of the same type of plant species or tissue, the pectin content may still differ. For example, the skin of three apple cultivars had 9.4, 10.4 and 9.6%w/w pectin (on dry basis, as GalA) for the red-, pink- and white-flesh fruit, respectively [40]. This suggests the determining role of intrinsic differences in traits of plant material, e.g., the significant difference in microstructure as a result of the variations in genotype (Figure 2).

Figure 2. Field Emission Gun scanning electron microscopy images of fiber preparations produced via an aqueous extraction method from white-, pink- and red-fleshed apples.

Modification of pectins after extraction is often performed to improve their solubility and impart functionality. Thus, the industrial pectin preparations may have different macromolecular structures containing the same basic monomer unit but different linking arrangements e.g., different branching patterns through linkages at two or more positions on the same residue. These pectins would exhibit different interactive behaviours with other nutrients or bioactives in food or dietary products, in the digesta and along the human digestive tract [41, 42]. Pectins after extraction with/without purification are widely used as "commercial hydrocolloids" that function as thickeners and structuring agents. The molecular weight of commercial pectins is affected by processing typically ranging between 80 and 400 kDa. An industrial product must be abundant in α-D-galacturonic acid i.e. at least 65 % to be claimed as pectin in the European Union [29]. A standardization step in which batches of pectin are combined and sometimes sucrose is added depending on the food or pharmaceutical applications, is often included to ensure the uniformity of pectin product that fulfil the target specification of the final product. The launch and development of pectin ingredients such as CM203 citrus pectin from Herbstreith & Fox KG of Switzerland, has been based on the processing functional properties created by innovative food technology and engineering, and the nutritional quality and organoleptic acceptability assessed by clinical trials and consumer studies.

Table 1. Pectin content of some fruit peel wastes

Peel waste	Citrus (extracted with dilute HCl & 95% acetone, analysed via titration with NaOH)			Green kiwifruit					Feijoa							
				Extracts hydrolysed with concentrated H_2SO_4, measured colorimetrically)												
				Extraction medium (ethanol %)					Extraction at 20°C, ethanol%				Extraction at 50°C, ethanol%			
	Sweet orange	Grape fruit	Lemon	0	30	50	75	96	0	30	50	80	0	30	50	80
Pectin (% dry basis, as GalA)	3.2	7.9	4.0	1.9	1.7	0.8	0.6	1.7	5.6	1.9	0.5	1.4	3.7	3.9	2.2	1.0

Note: GalA represents galacturonic acid.

PECTIN-RICH PLANT CELL WALL PREPARATIONS

Pectin is abundant in the plant kingdom. Plant cell walls (as primary DFs) contribute to the structure of a plant and form a physical and chemical barrier between the cell and its environment. Plant cell walls generally contain complex polysaccharides, proteins or glycoproteins, lignins, other inorganic components and water [8]. The types and relative proportions of polysaccharides in cell walls vary between species [8] and during developmental processes particularly fruit ripening [43]. Cell walls are classified as primary cell walls (usually unlignified, composed of cellulose and non-cellulosic polysaccharides like pectic polysaccharides, xyloglucans and galactoglucomannans) or secondary cell walls (thicker than primary cell walls, usually lignified, rich in cellulose and 4-O-methyl glucuronoxylans with some glucomannans). The middle lamella between the primary walls of adjacent plant cells is rich in pectic polysaccharides with a low degree of chain branching [8]. These pectins are readily solubilised by CDTA, whilst those in primary cell walls (more highly branched rhamnogalacturonans) are solubilised by dilute alkali.

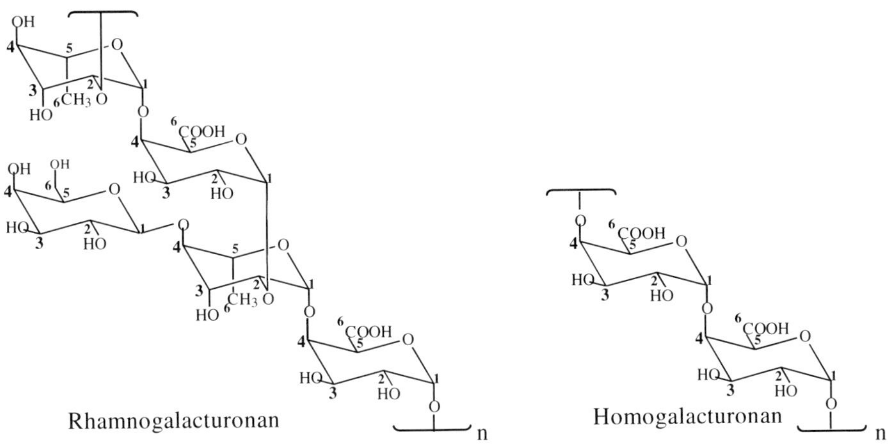

Figure 3. Structures of some galacturonans [8].

Pectins are complex polysaccharide polymers. Pectin's major constituents include homogalacturonans (helical homopolymers, also called 'smooth blocks') and rhamnogalacturonans (contorted rod-like heteropolymer, also called'hairy blocks' such as rhamnogalacturonan-I (RG-I) (Figure 3) and some xylogalacturonans [44]). The linear D-galacturonosyl chains of pectin (i.e. galacturonan) connected by α-(1-4) glycosidic linkage may contain side chains made up by neutral sugars such as rhamnose, galactose, xylose and arabinose. Pectins are embedded in the cellulose-xyloglucan three-dimensional (3-D) plant cell wall network of dicotyledons and some monocotyledons and mostly located in the middle lamella of plant cell walls with some also occurring in cell walls [8, 45]. Their structures vary between plant types, plant cultivars, tissues, location (even within the same cell wall), age and maturity.

The functionality of pectins depends on the polysaccharide composition and location and orientation of polysaccharides in the cell-wall networks [8]. Physical characteristics of pectins such as the 3-D network, chain length, surface morphology, hydrophilicity/hydrophobicity, charge and area, pore and particle size, are important attributes of plant cell wall preparations. Small particle size provides increased external surface area for interactions with other food components, reception of environmental stress (e.g. heat, shear and pressure), attack by microbial deterioration, and contact by digestive microorganisms and enzymes.

Apple is a good example of the dicotyledon family. Apple cell walls contain homogalacturonan, arabinan-rich rhamnogalacturonan and xylogalacturonan [46] (Figure 4). Arabinans are present either alone as simple sugar side chains or in the outer branch of arabinogalactan side chains. Some pectin fractions may contain galacturonan main chains and (1→3)-or (1→6)-linked galactans side chains [47]. Different cultivars differ in their monosaccharide composition (Table 2) [43, 48, 49].

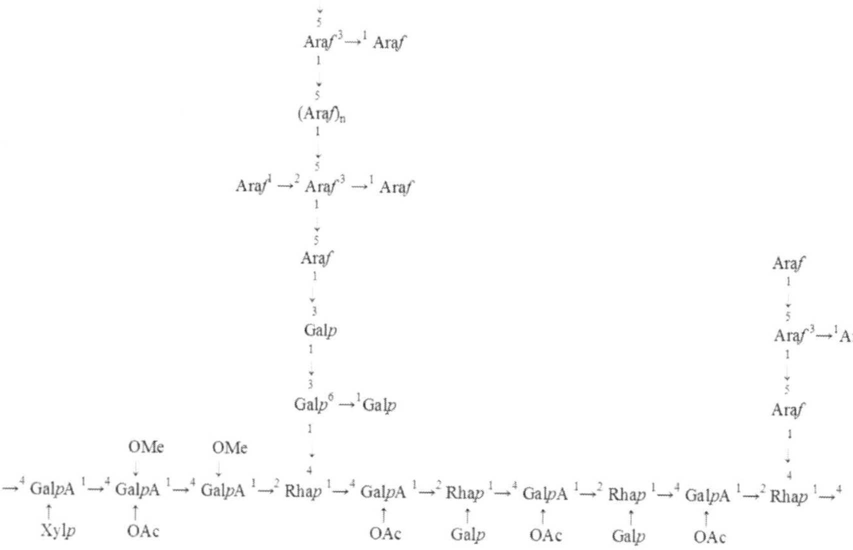

Figure 4. Modified hairy regions of rhamnogalacturonan (after [46]).

To understand the functionality (especially physical properties) of plant cell walls, examination on the microstructural characteristics should be performed using microscopy techniques such as bright field microscopy or cryo-scanning electron microscopy. These techniques allow direct visualization of the morphology and packing pattern of plant tissues and cell walls, degree of cell openness during cell wall isolation. Figures 5 and 6 demonstrate the raw tissues and derived isolated cell walls of apple (a dicotyledons) and onion (a monocotyledons), respectively.

Table 2. Neutral monosaccharide and uronic acid content of different apple cell walls

Raw Apple	Pacific Rose*	Braeburn**	Braeburn***	Royal Gala***
Total neutral monosaccharide content (mg/g dried cell walls)	539.5	429.1	327.2	311.5
Uronic acid (mg/g dried cell walls, as GalA)	239.0	250.6	146.1	212.6

Data are mean values. * values of the same column are derived from [48], ** from [43], and *** from [49], respectively.

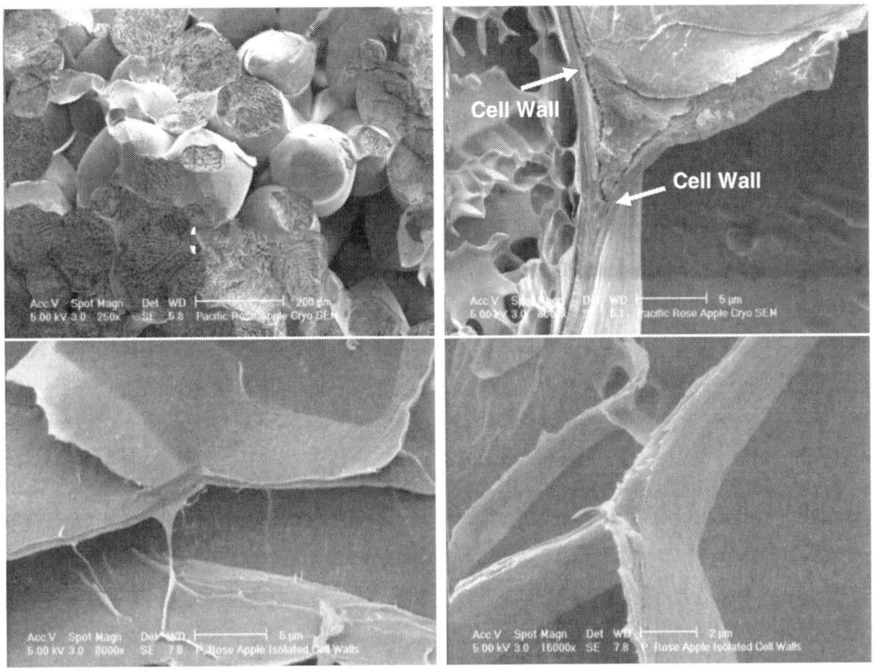

Figure 5. The parenchyma cells of raw apple (upper row) and derived cell walls isolated by an aqueous extraction method (lower row).

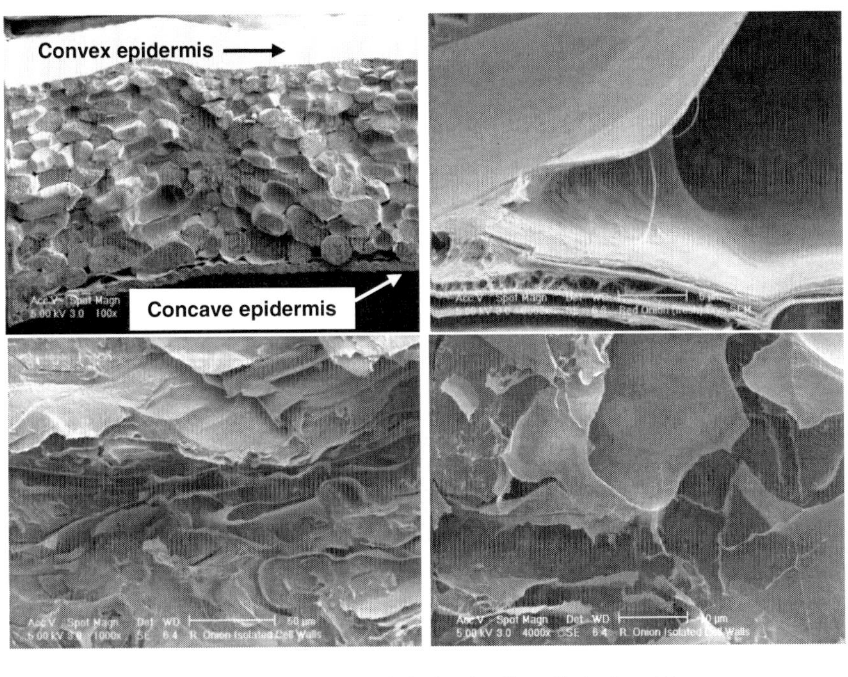

Figure 6. The cells of raw onion (upper row) and derived cell walls isolated by an aqueous extraction method (lower row).

Apple or onion cells differ in size, shape and morphology. Apple tissues contain more loosely packed cells, whilst cells in the onion tissues are more densely packed with few intercellular spaces. These structural differences affect the ease of DF extraction from apple and onion cell walls (e.g. extractabilities in different extraction media), and determine the composition and physical attributes of obtained cell wall polysaccharide preparations. As a result, the cell wall polysaccharides in these preparations may exhibit different interactions with other components in food or digesta matrices.

Fractionation of plant cell walls is performed to obtain various cell wall preparations that are dominated by specific classes of polysaccharides. The fractionation process (Figure 7) normally starts with the collection of water-soluble polysaccharides after cell wall isolation. Then the less branched pectic polysaccharides such as those in the middle lamella of cell walls and the primary cell walls are extracted using a chelating agent (e.g. CDTA) to extract the Ca^{2+} ions cross-linking polygalacturonic acids. Further treatments with mild alkali (e.g. Na_2CO_3) followed by strong alkali solutions of increasing concentration are performed to extract the highly branched rhamnogalacturonans and some hemicelluloses. The polysaccharides retained in the final residue of fractionation are mainly cellulose.

Table 3 shows the monosaccharide composition of the different apple cell wall fractions obtained by fractionation of the isolated whole cell walls. A relatively high amount of soluble pectins were extracted in both the CDTA and Na_2CO_3 fractions although the degree of pectin branching of these two fractions differed (as indicated by the ratio of uronic acid: arabinose + galactose). The detected rhamnose content and arabinose and galactose contents in these fractions indicate the occurrence of rhamnogalacturonans and arabinan and galactan side chains. A relatively low yield of neutral monosaccharides was obtained in the 1 M KOH step, followed by a high yield in both the 4 M KOH and residue wash fractions.

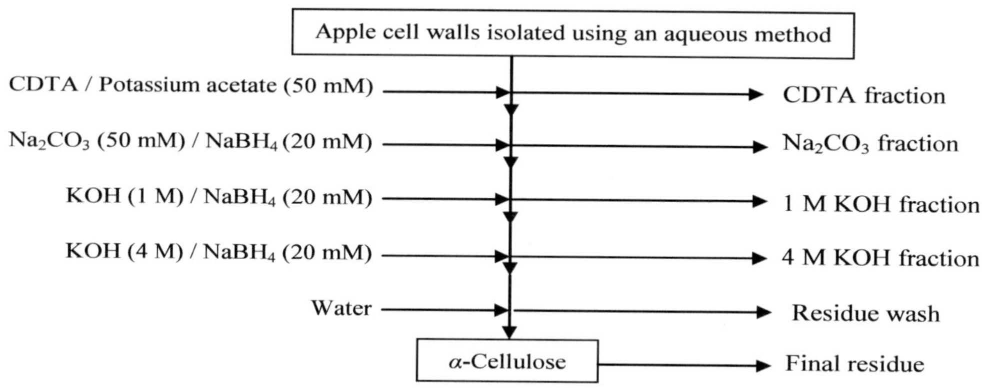

Figure 7. Scheme for fractionation of cell walls.

The presence of xylose indicates heteroxylans and xyloglucans. Galactose may be derived from xyloglucans, pectins and galactoglucomannans. The ratio of glucose to xylose in the KOH fractions can be used as an indicator of the proportion of different hemicelluloses.

The lower ratio of the 1 M KOH fraction (i.e. 0.68: 1) was possibly due to more heteroxylans (including arabinoxylans) and less xyloglucans, while the higher ratio in the 4 M KOH fraction (i.e. 0.99: 1) perhaps resulted from a high xyloglucan content rather than heteroxylans [50]. The 1 M KOH fraction still had a significant amount of pectins because of its uronic acid content.

The 4 M KOH fractions had the lowest uronic acid content but the highest neutral monosaccharide content indicates the least dominant pectin content.

Table 3. Neutral monosaccharide and uronic acid content of apple cell wall fractions

Monosaccharide	Yields in cell wall fraction (mg / g dried cell wall fraction)					
	CDTA	Na_2CO_3	1 M KOH	4 M KOH	Residue wash	Final residue
Rhamnose	5.2±0.5	1.2±0.2	5.5±1.5	3.8±0.2	19.0±1.5	9.5±0.2
Fucose	1.8±0.2	–	4.8±0.5	48.0±1.2	10.6±1.5	15.5±0.5
Arabinose	29.2±1.0	23.3±0.5	46.3±2.0	46.0±1.5	252.7±1.0	104.0±1.6
Xylose	54.7±1.8	8.1±0.4	32.0±2.0	189.7±1.4	82.0±1.7	99.8±1.7
Mannose	–	–	7.8±0.2	20.1±0.2	–	3.3±0.2
Galactose	61.0±1.9	60.4±1.6	25.8±1.3	91.0±1.3	97.0±1.9	200.2±2.8
Glucose	27.3±1.4	9.5±0.6	21.8±0.4	188.3±1.5	80.0±1.7	127.4±1.7
Total neutral monoaccharide content	179.2±1.7	102.5±1.1	144.0±1.8	586.9±2.1	541.3±2.2	559.7±2.4
Uronic acid (as GalA)	285.3±0.2	312.8±1.3	202.8±5.3	108.7±4.0	390.8±7.5	207.0±2.8

Data are expressed as a mean value ± standard variations. "–" Not detected. NM and UA refer to neutral monosaccharide and uronic acid. GalA, Ara and Gal represent galacturonic acid, arabinose and galactose.

Table 4. Changes in antioxidant activity after incubation of the apple cell-walls or derived fractions with *L*-ascorbic acid

Samples	Apple whole cell walls	CDTA fraction	Na_2CO_3 fraction	1 M KOH fraction	4 M KOH fraction	Residue wash fraction	Final residue fraction	Commercial apple pectin	Commercial PolyGalA
Change in antioxidant activity (% per mg dry basis)	1.6-1.9	2.8-3.0	1.6-2.1	0.7-1.0	0.7-1.0	1.1-1.2	1.3-1.7	7.1-8.5	12.5-14.7

[a]Data are expressed as a range of the calculated results after statistical analysis by the Q tests (at 90% confidence level) with n= 4 (number of different cell-wall samples prepared at different times) × 2 (each cell-wall sample was used to prepare duplicate incubation mixtures) × 3 (each incubation mixture was measured by the FRAP assay in triplicate). Incubation conditions: pH 6.5 and 37°C. FRAP and PolyGalA represent "Ferric Reducing Antioxidant Power" and "Polygalacturonic Acid".

The Residue Wash fraction (the wash after the KOH treatments) had the third highest neutral monosaccharide content along with the highest uronic acid content, suggesting the presence of highly branched pectins (although still less branched than those in the 4 M KOH and Final Residue fractions). The ratio of glucose to xylose in the Residue Wash fraction (i.e. 0.98: 1) suggests the presence of xyloglucans. It was possible that the 4 M KOH treatment loosened the remaining cell wall structure, causing a significant amount of xyloglucan to be solubilized in the water washing step. The Final Residue fraction contained mainly α-cellulose although some pectins, xyloglucans, fucogalactoxyloglucans, galactoglucomannans and glucomannans may also be present.

It is worth noting that the method for isolating whole cell walls prior to fractionation can considerably influence the composition of all the cell wall fractions. For example, an ethanolic isolation method might lead to less pectin in the Final Residue fraction (cellulose fraction) but more pectins in the earlier fractions (e.g. the first two fractions) [8, 51]. The studies described in Tables 3 and 4 employed an aqueous method for isolating whole plant cell walls, in an effort to simulate real consumer eating practices when consuming meals containing fruits, vegetables and cereals. The DFs (plant cell walls) prepared using the aqueous method present advantages not only in cost-effectiveness for industrial scale-up, but also in the intrinsically high pectin and bound polyphenol contents [32], significant protection of protein nutrients during snack bar making [52] and desired health-promoting effects in gut health (*in vitro* studies and *in vivo* rat trials) [53, 54].

The fractionation of whole plant cell walls into a series of different cell wall preparations, each of which contains different polysaccharide species and varied functional groups, imparts each fraction with unique physical and chemical properties. Table 4 shows the changes in antioxidant activity after incubation of each of the above apple cell wall fractions with ascorbic acid at 37°C for 2 hours. It can be found that all these apple cell wall fractions, as well as commercial available apple pectin and polygalacturonic acid, largely stabilized or enhanced the antioxidant activity of ascorbic acid. As shown in Table 3, the CDTA and Na_2CO_3 fractions had relatively high pectin contents (i.e. 285 and 313 mg uronic acid/g dried cell wall fraction, respectively). In contrast, the KOH fractions had relatively lower uronic acid contents (203 and 109 mg/g dried cell wall fraction for the 1 M KOH and 4 M KOH fractions, respectively).

Thus, a high uronic acid content in the incubation mixture was likely associated with a greater increase in antioxidant activity during incubation. The pectins especially galacturonans appear to be responsible for the observed increment increase of antioxidant activity. Furthermore, the Residue Wash fraction, having the highest uronic acid content (391 mg/g dried cell wall fraction) among all the fractions, only caused an intermediate increase in antioxidant activity. It was possible that its high neutral monosaccharide content (541 mg/g dried cell wall fraction) had a negative effect on the antioxidant activity of ascorbic acid. This assumption was further supported by considering the 4 M KOH fraction (i.e. the lowest uronic acid content but the highest neutral monosaccharide content) which showed the smallest increment in the antioxidant activity.

Other factors might contribute to the change of antioxidant activity including the type or proportion of other co-existing polysaccharides (e.g. xyloglucans and heteroxylans), their relative extractabilities and charge in different solutions, as well as the absence or presence of different amino acids and polyphenols. In summary, there are interactions between components of a meal, which may slow the degradation of some antioxidants like ascorbic acid.

There are a wide range of characterization techniques that can be used to examine the interactions between the plant cell wall-derived fiber fractions and bioactive substances. Cyclic voltammetry and solid-state NMR spectroscopy are two techniques for tracking the changes caused by the interactions between pectin-containing cell wall preparations and antioxidants. The antioxidant activities of the mixtures of antioxidant and pectin-containing cell walls in aqueous solutions (Figures 8 and 9) can be measured via cyclic voltammetry. The obtained results were in good correlation with the antioxidant activities analysed by the ferric reducing antioxidant power (FRAP) assays [48, 55]. The oxidation of quercetin and its degradation products (either in water solution or in aqueous ethanolic solution) was partially reversible at a carbon electrode, while ascorbic acid was irreversibly oxidized, as the oxidation product in this case cannot be reduced at a carbon electrode (Figure 8). During the forward sweep the antioxidant (in reduced form) is oxidized, generating a positive (anodic) current with characteristic peaks at different potentials, whilst in the reverse sweep the oxidized form near the electrode is reduced. The integrated area under the anodic current peak Q_{500} (with charge passed to 500 mV and after subtraction of background spectra) provides an excellent evaluation of total antioxidant capacity. Generally, the lower the oxidation potential under the same conditions, the more powerful the antioxidant as a reducing agent [56]. Overall, the reducing power of ascorbic acid was retained to different extents (apple 33%, onion 82%) if the incubation involved apple or onion cell wall materials.

Onion cell walls stabilized ascorbic acid very effectively, as did apple cell walls but to a lesser extent (ascorbic acid completely disappeared during incubation in the absence of apple or onion cell walls). Neither of these two cell walls had a significantly protective effect on quercetin degradation. Furthermore, cyclic voltammetry can also be applied to the ethanol-water systems (Figure 9). For the same concentrations of quercetin and pectin-containing plant cell wall, the presence of 18%v/v ethanol resulted in significantly different cyclic voltammograms. For the quercetin-containing systems, the difference in the electrochemical responses between the aqueous solution and the aqueous ethanolic solution of quercetin (although at the same concentration) was mainly due to the different degradation processes of quercetin in purely aqueous or aqueous ethanol (18% v/v). Two anodic peaks occurred in the cyclic voltammograms of aqueous quercetin — one attributed to the *ortho*-diphenol group on the B- ring, the other due to oxidation of the hydroxyl group at C-3 on the C-ring (Figure 8). Only one broad peak was present in the ethanol-water solution of quercetin (Figure 9). In the presence of pectin-containing cell wall, a greater change in the antioxidant activity of quercetin was found for the aqueous system, compared to the ethanolic system.

Solid-state ^{13}C NMR spectroscopy has been used widely to examine the structure and organization of polysaccharides and obtained from cell walls [57, 58]. This technique can be used to characterize systems containing both antioxidant and fiber polysaccharides [59]. In particular, this technique is advantageous in its ability to analyse cell walls or their derived polysaccharides with minimal sample preparation. It can also be used to study polysaccharide-antioxidant interaction.

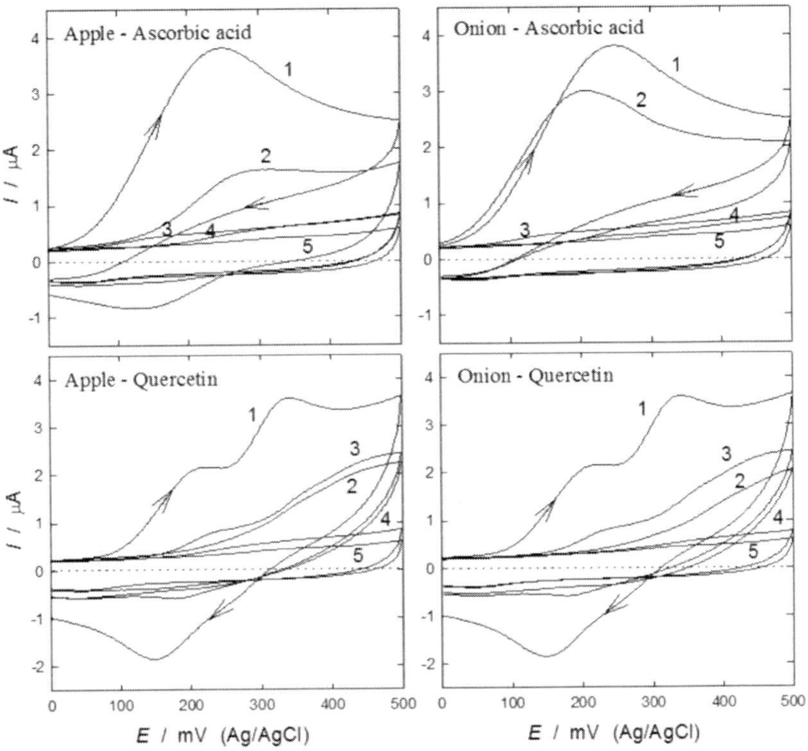

Figure 8. Cyclic voltammograms (aqueous systems, from top to bottom): 1, Antioxidant before incubation (pH 6.5, 37ºC); 2, Antioxidant+cell walls after incubation; 3, Antioxidant after incubation; 4, cell walls after incubation; 5, solvent background (HEPES 15 mM).

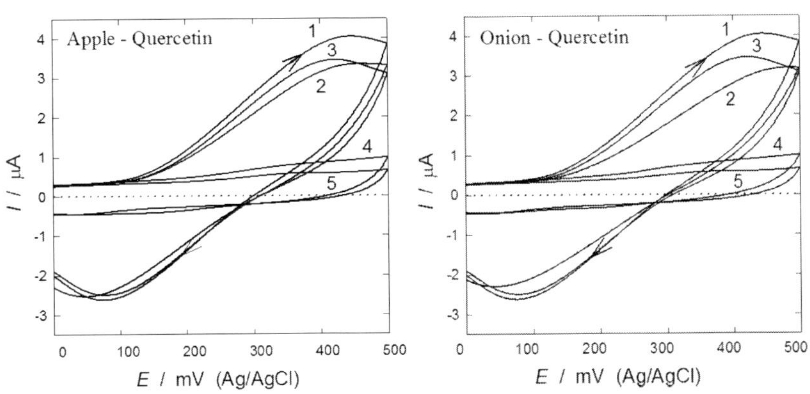

Figure 9. Cyclic voltammograms (aqueous systems in the presence of 18%v/v ethanol, from top to bottom): 1, Antioxidant before incubation (pH 6.5, 37ºC); 2, Antioxidant+cell walls after incubation; 3, Antioxidant after incubation; 4, cell walls after incubation; 5, solvent background (HEPES 15 mM + ethanol 18%v/v).

Figures 10 and 11 are the CP/MAS ^{13}C NMR spectra for incubation mixtures of Figures 8 and 9 (i.e. apple or onion cell walls with quercetin or ascorbic acid). Overall, the major signals present in the spectra were mostly derived from cellulose, pectic polysaccharides (e.g.

galacturonic acid, arabinose, galactose) and xyloglucans, consistent with the monosaccharide composition of cell wall polysaccharides from apple and onion. Signals in solid-state ^{13}C NMR spectra can be assigned as follows [60, 61]. Signals at 64-65, 72-75, 88-89 and 104-105 ppm were assigned to C-6, C-2,3,5, C-4 and C-1 of crystalline cellulose. Signals at 170-175, 100-101, 79-80, 68-69, 53-54 and 21 ppm were assigned to the C-6, C-1, C-4, C-2, methoxyl groups (CH_3O-) and acetyl groups (CH_3C=O) of galacturonic acid residues in pectins. The signal at 99-100 ppm was assigned to C-1 xylose of xyloglucans. The C-1 of glucose residues from xyloglucans are usually found at 103 ppm but are likely to be obscured here by the strong signal at 105 ppm from C-1 of cellulose. No signals arising from ascorbic acid or quercetin and their degradation products could be detected in spectra of the apple and onion cell walls. The spectra of ascorbic acid and qurecetin chemicals are as Figures 12 and 13. Taking the low concentrations of these compounds in the incubation mixtures, it can be concluded that either no or very little of these compounds effectively bound or adhered to the cell wall materials, or that binding to wall-bound compounds was too mobile to be detected. Alternatively, the concentrations of these compounds were simply too low for detection.

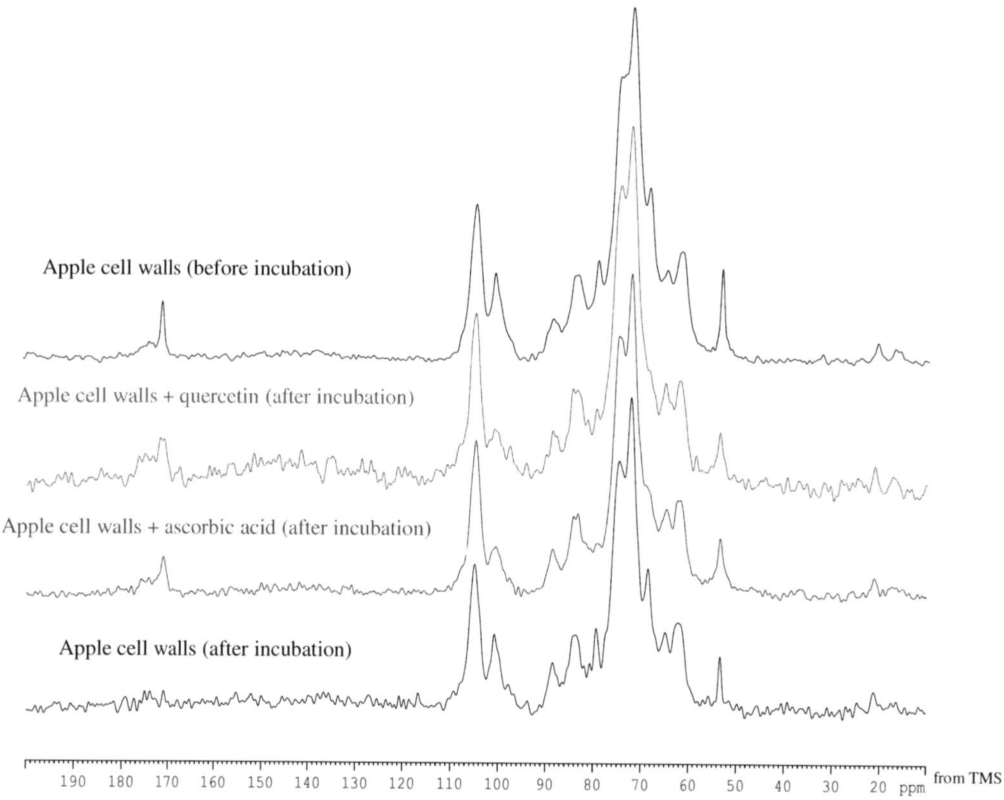

Figure 10. CP/MAS ^{13}C NMR spectra of apple cell walls before and after incubation at 37°C for 2 h, with HEPES buffer (pH 6.5, 15 mM) only (control sample), ascorbic acid (0.125 mM) in HEPES buffer, or quercetin (0.0625 mM) in HEPES buffer.

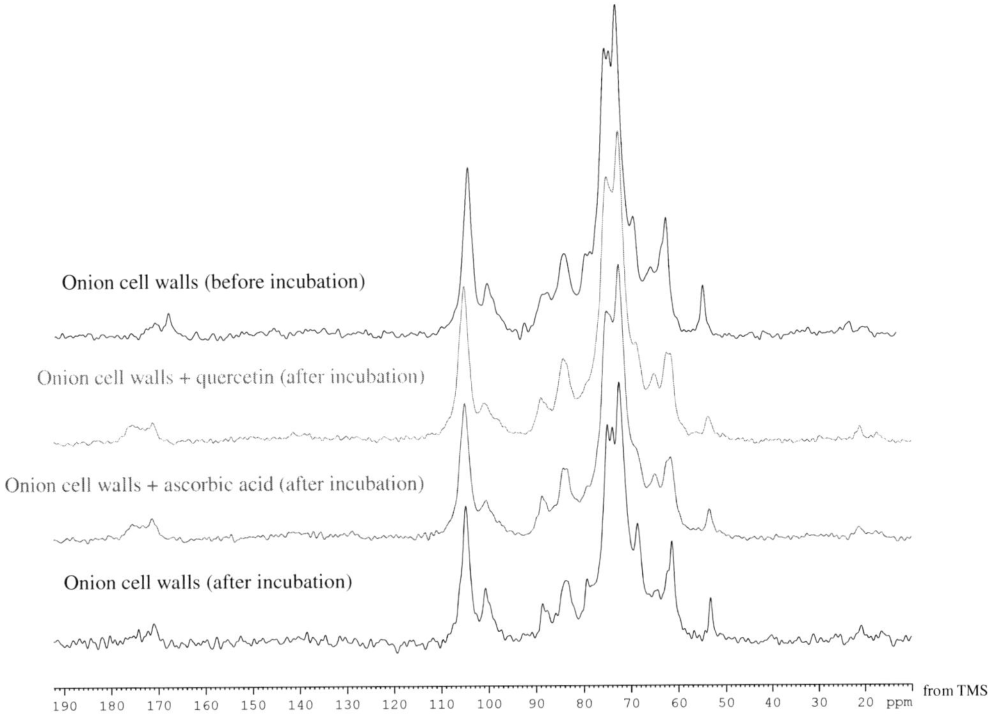

Figure 11. CP/MAS ^{13}C NMR spectra of onion cell walls before and after incubation at 37ºC for 2 h, with HEPES buffer (pH 6.5, 15 mM) only (control sample), ascorbic acid (0.125 mM) in HEPES buffer, or quercetin (0.0625 mM) in HEPES buffer.

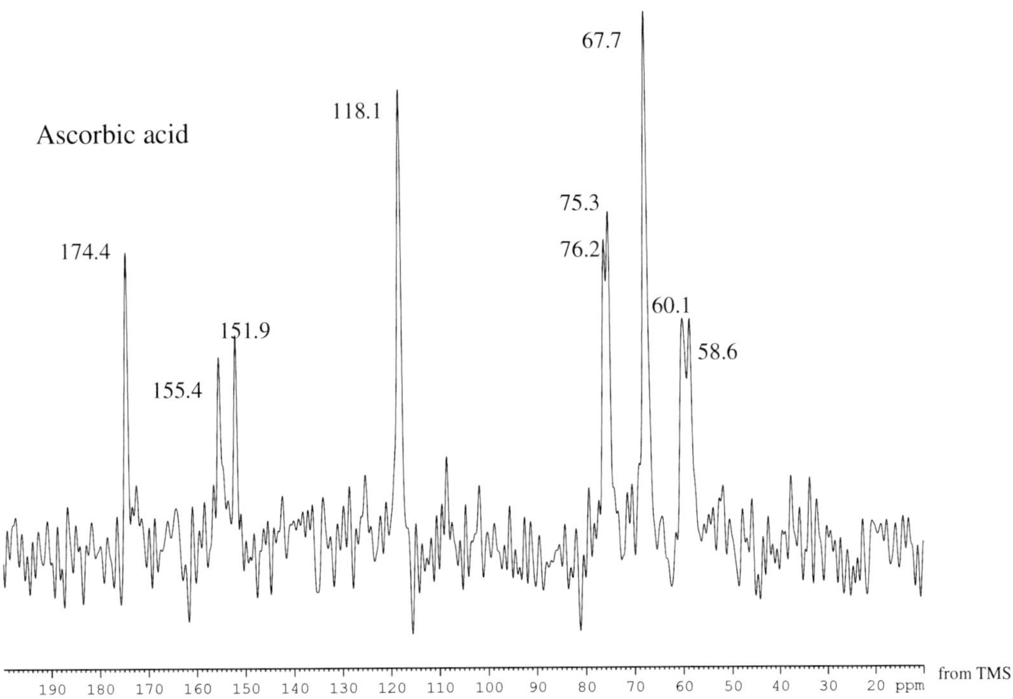

Figure 12. Solid-state CP/MAS ^{13}C NMR spectrum of *L*-ascorbic acid.

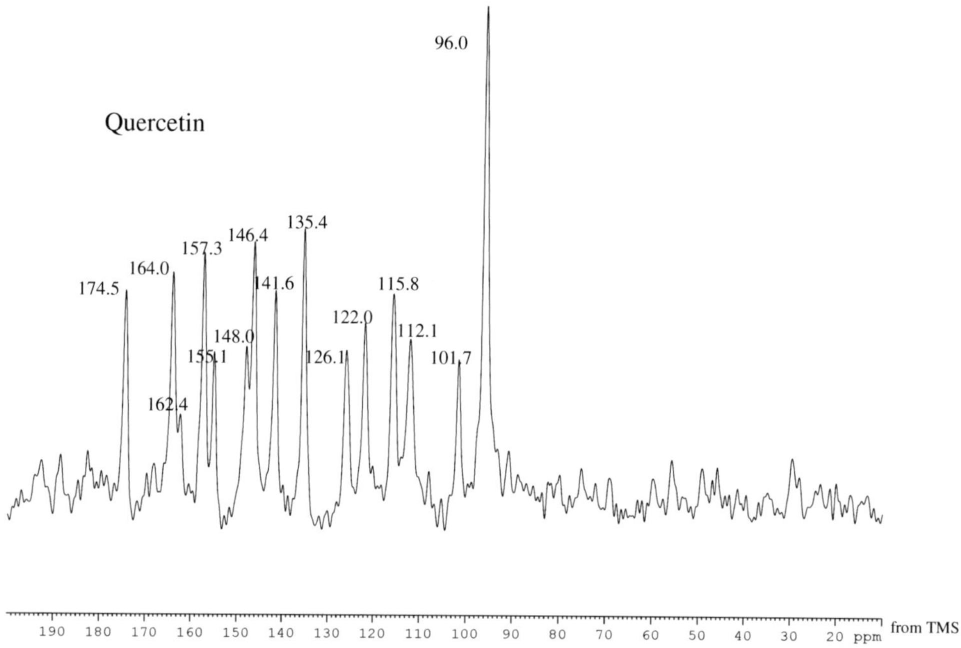

Figure 13. Solid-state CP/MAS ^{13}C NMR spectrum of quercetin.

Intermolecular associations involving pectins are affected by the configuration of hemi-acetal bonds, extent of branching and distribution of side chains. Any change in the surrounding chemical environment of the carbons of galacturonic acid would lead to variations in the signals [62], for example, the cleavage of portions of the (1→4)-linkages between the C-1 and C-4 of the neighboring free or esterified galacturonic acids, or the presence and absence of methylated (at C-6) in each galacturonic acid, possibly affect these signals to some extent. The major changes for the samples before and after incubation (Figures10 and 11) were the 170-175, 100-101, 79-80, 68-69, 53-54 ppm and 21 ppm signals derived from the C-1, C-4, C-2 and C-3, methoxyl groups (CH_3O-) and acetyl groups (CH_3C=O) of pectins (Figure 14). These differences suggest that incubation in the presence of ascorbic acid or quercetin antioxidant alters the mobility or extractability of pectins [63-66]. The peak at 174 ppm due to C-6, the carboxyl carbon, is shifted by approximately 3 – 4 ppm (to 171 ppm) when the free acid is methyl esterified [63]. The ratio of peak area at 171 and 174 ppm can provide an estimation of the extent of methylation [49]. The 2 h-incubation might induce some changes in the chemical environment of this carboxyl carbon, causing the loss of signal at 171–175 ppm. The 2 h-incubation may result in loss of polysaccharides from the cell wall, formation of new interactions among the remaining polysaccharides, and the carboxyl carbon might even become too mobile to be detected by NMR. Furthermore, at neutral pH, heat treatment may induce β-elimination on the polygalacturonic chain, hydrolysis of the glycoside bonds in neutral sugars and in polyuronide polymers, and the saponification of the methoxyl groups borne by the galacturonic acid moieties, to different extents [8, 67]. These reactions are associated with the dissolution and depolymerization of pectins, and the releasing of significant amounts of neutral sugars and some low molecular weight compounds into aqueous solution. It is possible that a small proportion of the loss in signals at 171 ppm resulted from the demethylation caused by residual pectic methylesterase.

The presence of antioxidants might affect the extractability and mobility of the cell wall polysaccharides, e.g., galacturonans. There might also be some interactions between antioxidants and plant cell wall components, via direct effects (e.g. functional group specific reactions or hydrogen bondings between antioxidants and cell wall monosaccharide residues on polysaccharides) or indirect effects (involving enzymes or other proteins). Since there was no evidence to show that ascorbic acid or quercetin bound to the cell walls, it was possible that the interactions between cell walls and these antioxidants occurred in a non-binding manner and/or in liquid state (e.g. redox reactions in solution).

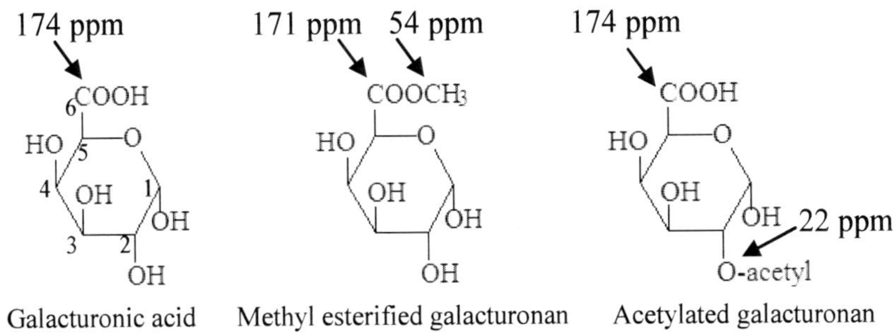

Figure 14. Solid-state NMR signals of galacturonic acid, methyl esterified galacturonan and acetylated galacturonan.

PECTINS AS ISOLATED INGREDIENTS

There is often insufficient fiber consumption in modern human diets, which is partly due to their absence in popular product categories such as convenience or snack foods. Thus, production of versatile functional fiber ingredients on an industrial scale that could then be added to popular foods, is attracting high interest. Pectins isolated through industrial processes from various raw materials are marketed as isolated ingredients or additives such as a gelling, thickening or stabilizing agent for food, pharmaceutical and cosmetic applications [29]. The intrinsic variability in the molecular chemical structure and polymer physical attributes of pectin ingredients confers different chemical and physical properties. As a result, favourable physiological effects on human health can be tailored via manipulating the interactions with other nutrients and digestive enzymes.

Pectin is a weak acid and as a result, deprotonation occurs with increasing pH. The pKa of pectin is approximately 2.9–4.5 but the precise value is governed by the dissociation degree, methylation degree, branching degree, distribution of the methylated groups, free carboxyl groups and neutral sugar side chains [68]. The pKa is about 3.3 or 5.4, respectively, when pectin is fully protonated or fully deprotonated, and is 4.1 for pectic acid (zero DE) and ~3.6 for HM pectin solution with 65% DE[69, 70]. Stability of pectins in aqueous solutions is affect by temperature and pH. Pectins are mostly stable at pH 3.5 to 4. Most pectins are branched and have multiple negative charges. Low pHs increase the percentage of unionized carboxyl groups, thus reducing electrostatic repulsion between adjacent pectin chains [71]. The more charges on pectin molecules, the greater the expansion of pectin polymer chains.

Pectins are generally soluble in water, although do not necessarily dissolve immediately upon exposure to water. Highly branched polysaccharides are typically more soluble than those less branched or linear polysaccharides. Branching reduces hydrogen bonding between polymers therefore favouring dissolution. Pectin does not dissolve in water if the conditions favor gelation. A gel is an intermediate phase between a solution and a solid. Pectins form cohesive gels via hydrogen bonds, Ca^{2+} bridges and other weak forces to creating junction zones. For either HM- or LM-pectin, molecular weight plays an important role in gelling, i.e. a higher molecular weight tends to cause higher breaking strength to gels. Irregularities in pectin molecules caused by the different distribution of methyl ester or O-acetyl groups, rhamnosyl residues in the backbone, and side chains give shape to the freely stretched macromolecules. Long and uncharged pectin chains can form junction zones at high concentrations, and charged galacturonans interact with calcium.

When the carboxyl groups are free, the pectin polymers are able to chelate divalent ions like calcium and gel at acidic pHs. For the partially methylated pectins, gelling in acidic solutions occurs at elevated solute concentrations e.g., 65% sucrose concentration.

The hydroxyl groups on the ring can enable the hemi-acetal bonding with other monosaccharide residues as well as hydrogen bonding between polysaccharides and with water. Acetyl groups could impart steric hindrance during pectin gelation.

The structural arrangement of poly-α-D-galacturonate sequence of pectin renders it negative charge characteristics. The bonds between each residue are parallel but offset from each other by the full width of the sugar ring generating geometries with large cavities for accommodating cations like Ca^{2+} (known as "egg-box" structure). The individual polymer chains are held together mostly by non-covalent bonds.

Pecins have a combination of liquid-like and solid-like characteristics. Pectin has relatively low viscosity in water, compared to other plant hydrocolloids.

The viscosity of pectin solutions increases with elevated concentration. Viscosity is the most important physical parameters for pectins in terms of food processing and physiological effects on human [72, 73]. While very diluted pectin solutions are still Newtonian liquids, pectin solutions mostly exhibit non-Newtonian behaviours. A large particle size generally leads to a longer dissolution time. Factors that influence the solubility and colloidal properties of pectin are the degree of methylation, charge, concentration, degree of branching, composition of side chain, extent of polymerization, molecular weight, polymer particle size, as well as environmental conditions like temperature, pH ion strength and electrolytes. Charged polysaccharides are usually more readily soluble than neutral chains.

The rheological attributes along with the water-holding and gelling capacity of these soluble fibers have impacts on the stomach and small intestine media or digesta of the upper gastrointestinal tract, e.g., modify the morphology, thickness and surface characteristics of the intestinal barrier layers and consequently causing changes in stomach emptying and digesta mobilization [74].

Pectin in water solutions is most stable at pH around 4 (i.e. HM pectin 3.7-3.8, LM pectin 4.2-4.3), otherwise may degrade via depolymerisation and/or deesterification at undesired temperatures and $pH > 4.5$ or $pH < 3$. Two kinds of degenerative reactions are possible for pectins in aqueous base solutions i.e. β-elimination that breaks the pectin chain and demethylation that decreases degree of esterification [75]. These reactions also take place at a neutral pH (e.g. in natural milk) over a prolonged time period in the absence or presence of heating [75]. Pectins are readily degraded by oxidants except for chlorite and chlorine dioxide.

Different extraction/manufacturing methods for pectin lead to products with varied rheological properties and hydration behaviours (which reflects the usefulness of pectin as a food additive) [76]. Figure 15 compares the strain-time plots of aqueous solutions of HM, LM and amidated pectins. The differences between HM and LM pectins lie in not only the degree of methylation, but also the molecular weight, polymer length and extent of branching or crosslinking [77].

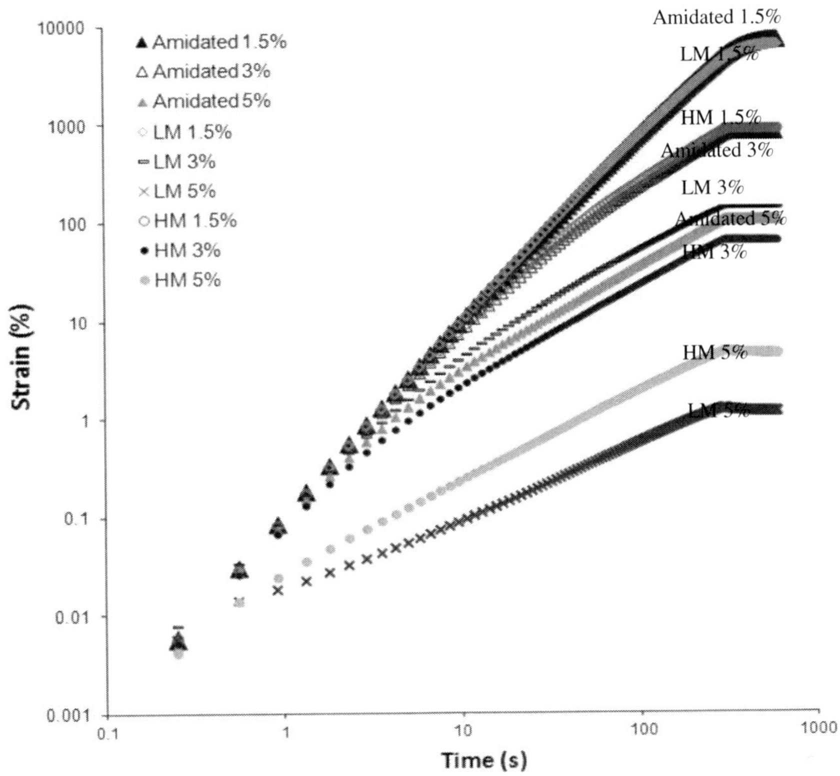

Figure 15. Strain-time plots for aqueous solutions of HM, LM and amidated pectins at concentrations of 1.5, 3 and 5%w/w.

HM pectin generally has a higher molecular weight, longer and greater entangled chains, compared to its derived LM pectin. The differences in chemistry between HM and LM pectins account for their varied physicochemical attributes and functional properties in food applications and during human digestion. Acetyl groups negatively affect the gelation of pectin, because of the hindrance caused by chain-chain association within the junction zones. But acetylated pectin can still impart viscosity and effectively stabilize oil-in-water emulsions partly due to the presence of the hydrophobic O-acetyl groups.

HM pectin requires a minimum soluble solid content, an acidic pH ≤ 3.0 and low water activity to form gels via hydrophobic interactions between methoxyl groups and hydrogen bonding between non-dissociated carboxyl and hydroxyl groups [29]. The resultant crosslinking occurs as aggregates rather than through junction zones. HM pectin requires high sucrose content and a low pH to gel [78]. HM pectin is ideal for rapidly set and thermo-irreversible gels with high solute/low pH such as jams, acid milk products and confectionery

[79]. The gelation of HM pectin involves various intermolecular interactions, with the junction zones being stabilised by both hydrogen bonds and hydrophobic interactions involving ester methyl groups [79, 80]. In pure HM pectin, the proportion of the ionized carboxyl groups present is smaller (< 50%) than the proportion of methylated carboxylate groups. The Ca^{2+} binding can also occur with the unmethoxylated chains (< 50%) in HM pectin, although > 50% methoxylated polymers in HM pectin do not gel via Ca^{2+} but instead rely on sugar or acid [71]. Thus, only a small number of hydrogen bonds are formed with strong interactions like ion-dipole. HM pectin gels exhibit temperature-dependent gel elastic modulus in different temperature ranges. With increasing temperature, the increase in entropy reduces the hydration of pectin chains, and hydrophobic interactions are strengthened and become the most important contribution to macromolecular interactions [81].

LM pectin requires a controlled amount of calcium or other divalent cations to form a gel. The LM pectin gels are thermo-reversible and bridged via divalent ions such as calcium and dimerization of polygalacturonate chains in a two-fold conformation to form 'egg-box' structures [82]. The ability of LM pectin to gel over a wide range of pH and soluble solid contents, especially under the conditions of no or low sugar but at a high pH, renders its common use for low or sugar-free products [83].

LM pectins are very responsive to calcium even at very low concentrations and can be precipitated by calcium ions [84]. LM pectin gels are relatively weak and their strength depends on the Ca^{2+} concentration and the pectin's molecular characteristics. LM pectin gels have a smaller difference between the setting and melting temperatures compared with HM pectin gels. LM pectin shows advantages over HM pectin in improving the consistency of fruit jams [29].

Amidated pectins are important pectin derivatives with good gelling properties at low-sugar conditions. Amidated pectins are generally thermoreversible whilst non-amidated LM pectins generate thermostable gels. The apparent viscosity of solutions of amidated pectin can be either similar or larger than solutions of LM and HM pectins. Further amidation of the carboxyl groups on LM pectins promotes gelling capacity. By amidating pectin, some methyl-ester groups may be converted to amide groups. Thus, amidated pectin also belongs to the LM pectin category but possess advantages in a higher degree of thermoreversibility, greater tolerance of calcium variation, firmer gels at lower calcium concentrations and reduction of precipitation risk at high calcium concentrations [83].

Amidated pectins are often used to replace LM pectins [85]. The gelation temperature generally increases with elevated amidation. But amidated pectins may not form gels at the same strength or as rapidly as LM pectin because of the positive charges introduced along the polymer chain. For amidated pectins, the formation of hydrogen bonds by the free carboxyl groups and the "egg box" structure with divalent cations (e.g. Ca^{2+}), as well as the clustering of pectin chains via hydrophobic interactions between the methylester groups, also occur though to a lesser extent.

The presence of amide groups allows additional hydrogen bonding (leading to stronger gels), although this process is slower than the complexation with Ca^{2+}.

The structural characteristics of various pectins determine the way in which they interact with themselves and with other food components including small molecules such as water, and macro-molecules such as polysaccharides, proteins and digestive enzymes like lipase even in simple interactive systems. Formulation of mixed hydrocolloids is a common practice in the food industries to generate products with specific properties and cost advantages.

Figure 16 demonstrates the compatibility of pectin with hydrocolloid alginate or locust bean gum.

For the alginate-containing systems, the presence of HM pectin led to a much greater viscosity than LM and amidated pectins (i.e. LM pectin exhibits a higher viscosity than amidated pectin). In contrast, the trend for the locust bean gum (LBG)-containing systems was different i.e. while the presence of amidated pectin caused the lowest viscosity, there existed a crossover point for the HM and LM pectin systems. If these mixed gels are prepared in milk, differences are found in their appearance (as shown in Figure 17) and in pH (i.e. pH 6.4, 6.3 and 6.5 for the LM, HM and amidated pectin system, respectively, compared to Anchor Super Trim milk pH 6.9).

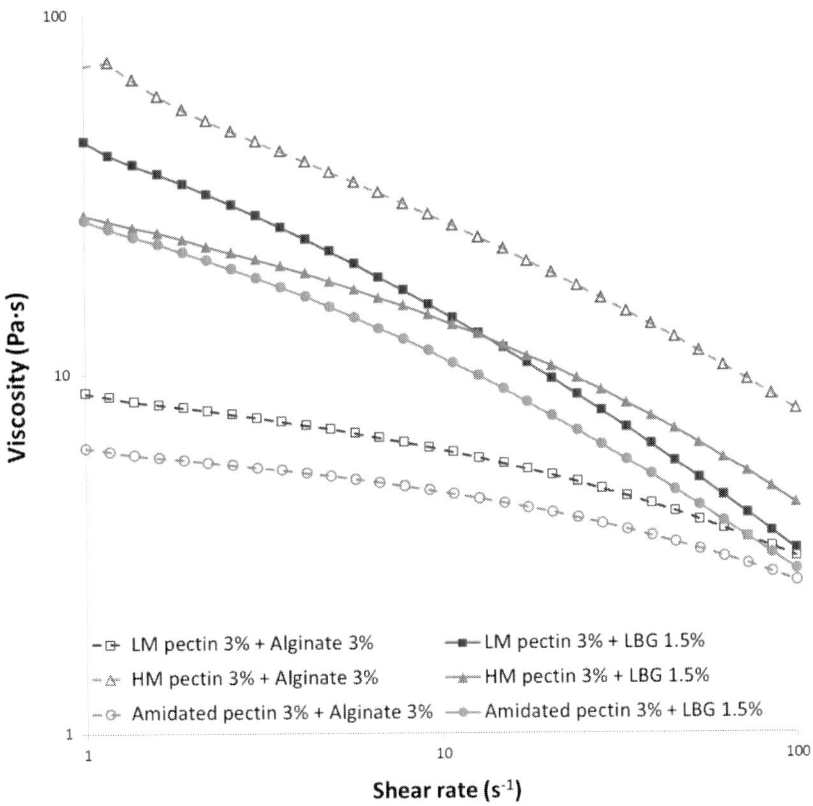

Figure 16. The viscosity of different types of pectin i.e. high or low methoxyl (HM/LM) pectin and amidated pectin, blended with another hydrocolloid i.e. alginate or locust bean gum (LBG).

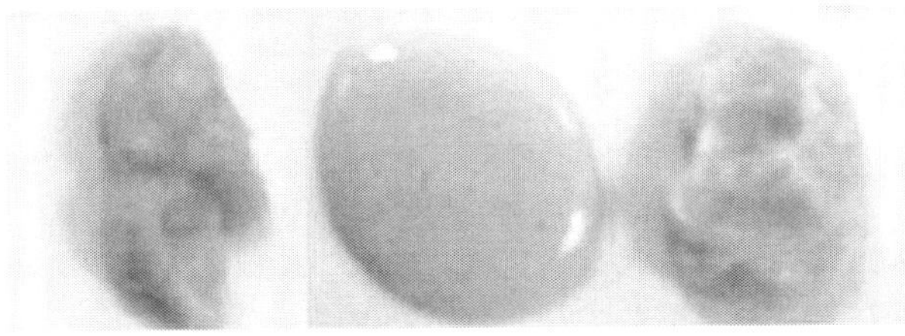

Figure 17. Blending milk with of 3% pectin-1.5% locust bean gum (LBG) gel (from left to right: LM pectin, HM pectin and amidated pectin). HM or LM pectin refers to high or low methoxyl pectin.

Figure 18 further shows the interaction between HM or LM pectin and another hydrocolloid, carboxy methyl cellulose (CMC) [86]. CMC is a typical anionic polysaccharide and is often used as a stabilizer in foods due to its ability to dissolve in both hot and cold water to impart increased viscosity. When mixing 1% CMC with 1.5% LM pectin, LM pectin seemed to play a dominant role in final viscosity. In comparison, HM pectin and CMC both contributed evenly to the final viscosity of a 1% CMC-1.5% HM pectin solution. It is important to consider the interplay between critical concentration and charge characteristics of the two different types of polymer when combining them in an aqueous solution.

For those having like charges, segregation into two aqueous phases with each phase dominated by one polymer is possible. In comparison, those having opposite charges tend to attract each other and may precipitate or gel. Changes in polymer network including polymer chain length and entanglement complexity, may lead to the requirement of different energies for separating or aggregating hydrocolloid chains [87].

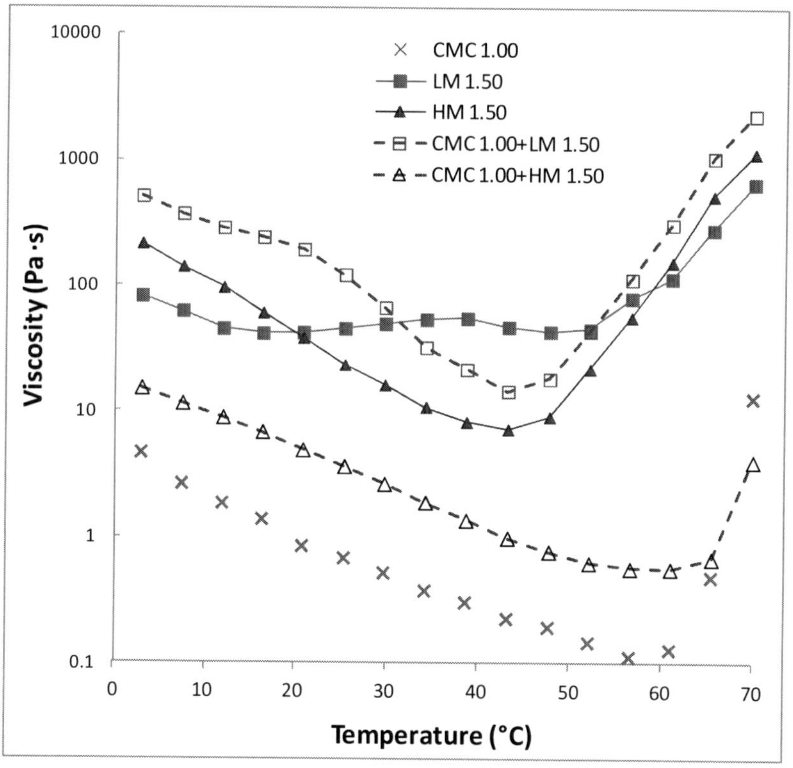

Figure 18. The viscosity of aqueous solutions of high or low methoxyl (HM/LM) pectin blended with carboxy methyl cellulose (CMC).

HM and LM pectins both interact with milk proteins but exhibit different behaviours. Figure 19 shows the change in the balance of bound and free calcium on pectin upon the addition of β-lactoglobulin (β-lg) at pH 4.5. As discussed earlier, Ca^{2+} binds to pectin and forms "egg box" structures. Binding of Ca^{2+} is greater for LM pectin than for HM pectin. At pH 4.5, β-lg has a net positive charge because of its PI value is ~5.1. Thus, β-lg can compete with Ca^{2+} for negatively charged pectin (especially non-methylated galacturonans) to form complexes at this pH. A greater binding to β-lg is expected for LM pectin than HM pectin, which is reflected in the data of Figure 19 where the amount of bound calcium on LM pectin decreased significantly after *β*-lg addition.

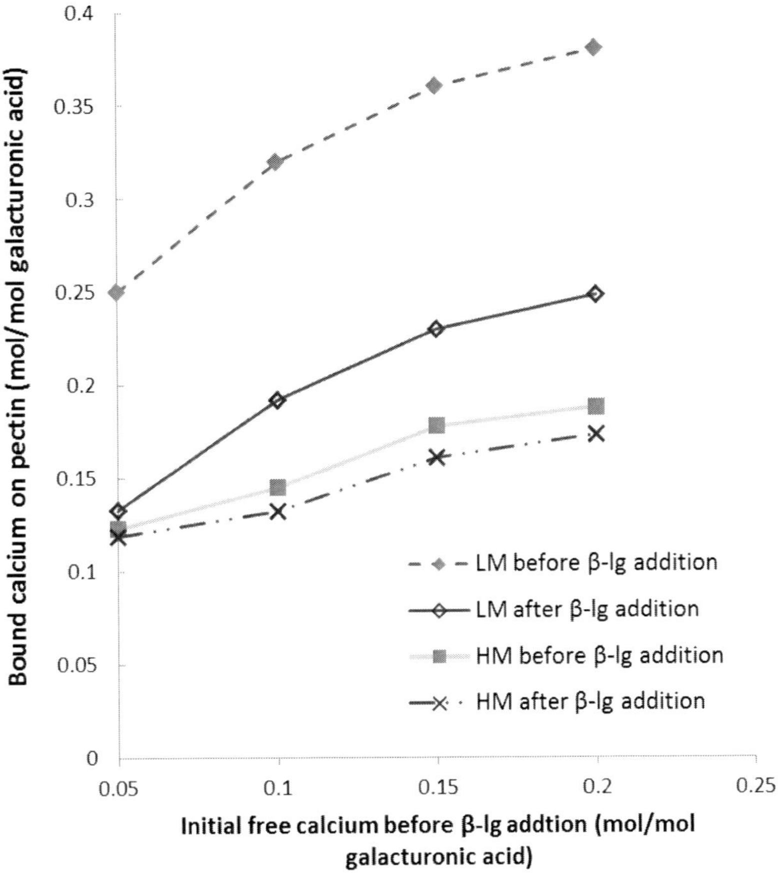

Figure 19. Effect of adding *β*-lactoglobulin (*β*-lg) on the bound calcium to pectins at pH 4.5.

The protein-HM pectin system did not change remarkably suggesting the stability of protein-pectin complexes in this system. The binding of HM pectin to protein via electrostatic repulsion and the gel network created by the pectin prevent protein from aggregation and minimize syneresis [88]. HM pectins are less sensitive to ionic changes, and their higher molecular weight and less compact structure enable extra steric effects on protein aggregation. HM and LM pectins differ significantly in their charge distribution at this pH, i.e. LM pectin has a greater charge density than HM pectin, thus likely form large complexes or even sediments by neutralizing protein aggregates. HM pectin is often used to stabilize protein-containing acidified beverages and yoghurts that have low ionic strength and pH near pI e.g., pH 4 (which facilitates the formation of soluble rather than insoluble protein–polyanion complexes).

Pectins can exert desired effects on the activity of pancreatic lipase and amylase *in vitro* and *in vivo* [89, 90]. Figure 20 compares the *in vitro* inhibitory effects of pectin and other soluble fibers at varying concentrations on calf pregastric lipase-catalyzed hydrolysis.

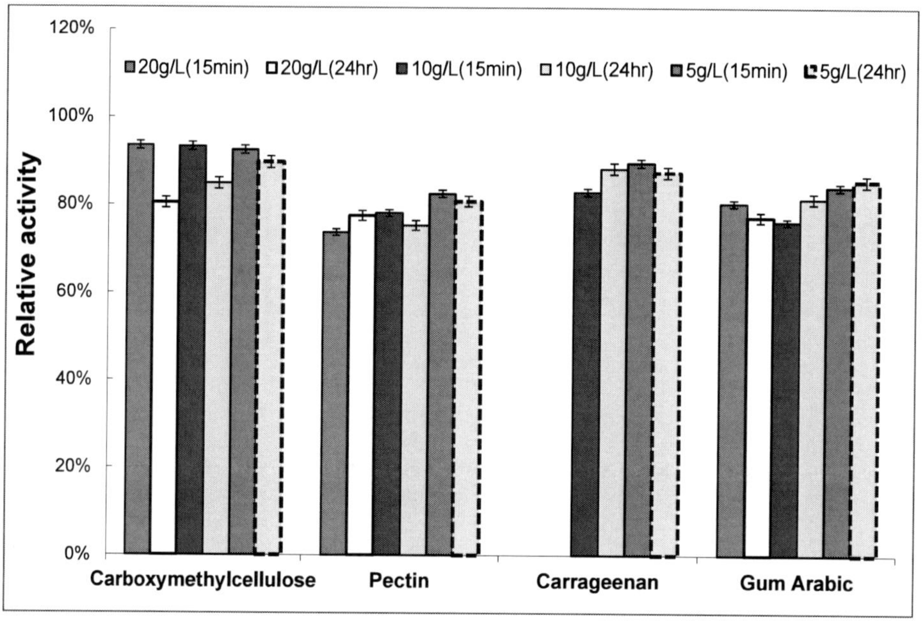

Figure 20. The inhibitory effects of carboxy methyl cellulose, pectin, carrageenan and gum arabic on the activity of calf pregastric lipase (50 μL, 20 mg/mL). Reaction conditions: tributyrin100 mg/50 mL, 37°C, pH 6.5, stirring speed 300 rpm, soluble fiber concentration and reaction time were 20 g/L, 15 min; 20 g/L, 24 h; 10 g/L, 15 min; 10 g/L, 24 h; 5 g/L, 15 min; 5 g/L, 24 h (left to right). No data were collected for carrageenan at 20 g/L (15 min or 2 h).

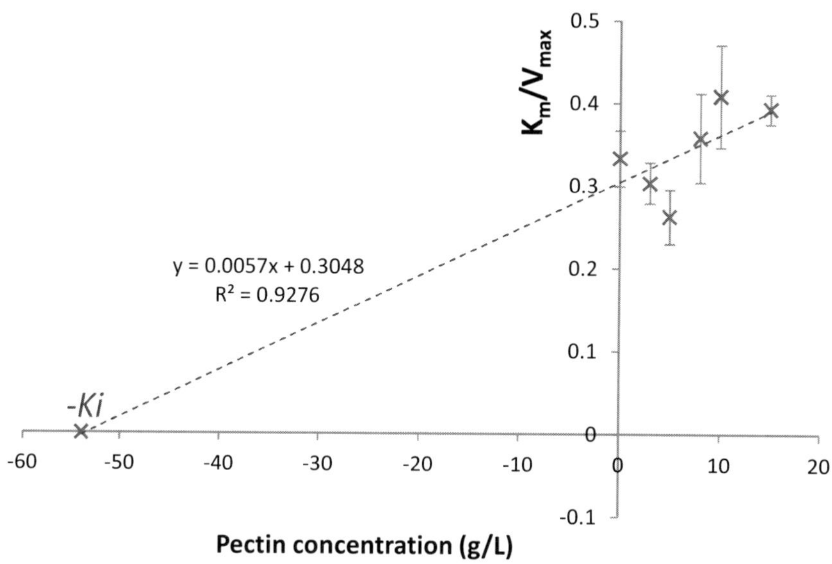

Figure 21. Determination of the enzyme-inhibitor dissociation constant for the calf pregastric lipase-catalyzed hydrolysis of tributyrin substrate in the presence of pectin inhibitor.

Pectin appeared to exert the greatest inhibition on the lipase activity. The interaction between soluble fibers and the lipase was partly physicochemical (depending on fiber composition), chemical (including ion-exchange capacity), and structural (polymer arrangements such as matrix, pore size, intercellular space).

Figure 21 shows the plot of the slopes of the Line weaver–Burk plots against the concentration of pectin inhibitor. From this plot, the enzyme-inhibitor dissociation constant, K_i, of pectin-lipase complex can be measured. This K_i (–54 g/L) is equivalent to a value ranging from 0.76 to 2.34 mM, indicating tight binding of pectin to lipase or entrapment of lipase in pectin network. In the initial Michaelis–Menten plots (not shown), the Km was found to be 0.21-0.35 mM, which suggests a strong binding between lipase and substrate tributyrin.

It can be concluded that there is an element of competitive inhibition between pectin fiber and tributyrin substrate for the active site on the lipase. The lipase might be partially bound or entrapped by pectin, allowing the significant part of lipase to remain unaffected.

PECTINS IN CONSUMER FOODS AND OTHER DIETARY PRODUCTS

The health benefits of pectins as DFs justifies increasing their content in the daily diet. The relatively low viscosity of pectin solutions makes this technically feasible for many foods and beverages. The versatile processing functionality of pectins offers opportunities for creating foods or other dietary products that possess high consumer acceptability. This section describes the interactions of pectins with various food and dietary systems. Selected case studies demonstrate the differences in product quality attributes, and chemical and physical properties caused by the pectin fortifying agent, as well as the insights offered by advanced characterization techniques.

Pectins are considered as a valuable ingredient for controlled colonic-specific drug or bioactive release, due to their low calorie supply, high nutritional value (e.g. soluble fiber attributes), distinct physicochemical properties (e.g. alterable viscosity and entrapment capacity), and favourable technological effects in food processing (e.g. emulsion stabilization) [77, 91]. For pharmaceutical applications, pectin is used either alone or in combination with other polymers to form microspheres using processes like emulsification and ionotropic gelation to deliver bioactive substances or drugs [92, 93]. Figure 22 shows the FT-IR spectra of the canola oil beads prepared via co-extrusion encapsulation using alginate or alginate-HM pectin as encapsulant (shell material). FT-IR spectroscopy is a powerful tool for chemical structure elucidation and qualitative compositional analysis. FT-IR was used successfully in this study to confirm the overall chemical stability of the encapsulated beads that were also evaluated chemically based on oil Totox values (listed in Table 5).

The addition of pectin as a co-encapsulant of oil led to a much greater bead size and an increased water activity. The difference in oil total oxidation (indicated by the Totox value) between the alginate alone-oil (A+O) beads and the alginate-HM pectin-oil (A+P+O) beads, was insignificant after a 30-day storage at either 20°C or 38°C. Both the alginate alone and the alginate-HM pectin shell formulations are acceptable and comparable for preserving canola oil. Considering the nutritional properties of pectin, addition of pectin fiber to the alginate-based shell may be beneficial for encapsulating unsaturated oils for food ingredient applications [92].

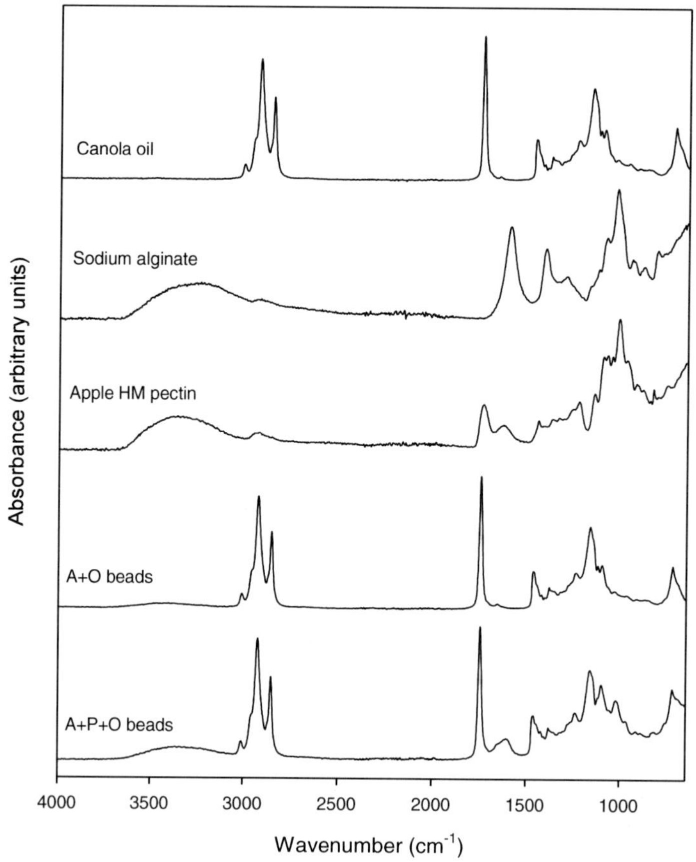

Figure 22. Normalised FT-IR absorbance spectra of ingredients used for preparing encapsulated beads i.e. canola oil, sodium alginate, apple high methoxyl (HM) pectin, and the resultant freeze dried encapsulated oil beads i.e. alginate alone (A+O) or a combination of alginate and HM pectin (A+P+O).

Table 5. Comparison between alginate encapsulated beads with/without added pectin

Encapsulated bead	Average diameter (μm)	Water activity	Totox value (30 days storage)	
			20°C	38°C
Alginate only	348±11	0.352±0.001	17.8±0.2	18.9±0.2
Alginate-pectin	533±16	0.460±0.001	17.8±1.1	19.1±0.6

Water–binding capacity and imparted gel texture are two important functional properties of pectin. Fruit-derived jellies are a consumer food application where these two functional properties are important. Proper setting to allow any air bubbles to escape, excellent water binding to minimize syneresis, and reduced osmosis to avoid colour migration are critical for jellies.

Figure 23 compares the total phenolic contents retained in the pectin jellies fortified with fruit juice or a fruit phenolic extract. LM pectin gel retained a greater amount of the phenolic antioxidants from fresh juices of pineapple, apple, green and gold kiwifruits, or phenolic extracts of kiwifruit or blackcurrant. This is encouraging as LM pectin is traditionally used for producing soft and partly thixotropic fruit gels across the entire solids and pH range, while HM pectin is normally limited to gel preparations containing soluble solids above 60% and pH below 3.5. A greater quantity is required for non-amidated LM pectin than for amidated LM pectin, although the former results in a greater degree of thixotropy [83].

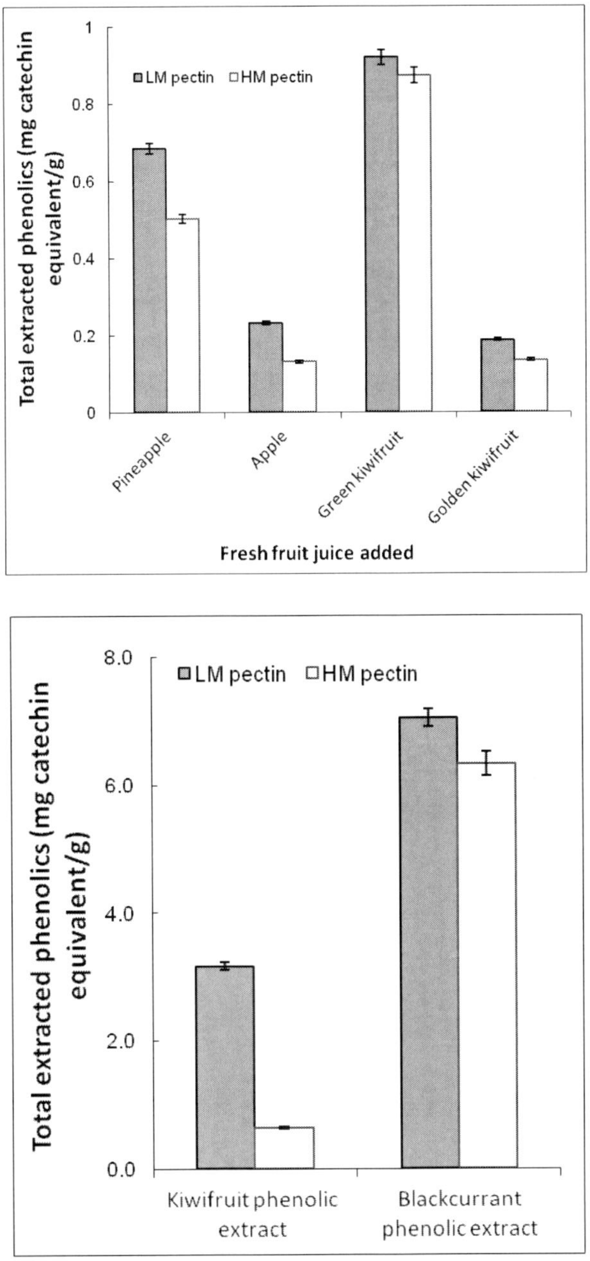

Figure 23. The effect of adding fresh fruit juice or fruit phenolic extract on the total extracted phenolics in the final pectin jelly products. HM or LM pectin refers to high or low methoxyl pectin.

The distinct water-holding or water-binding capacity of various pectins can lead to the competition for water between pectin and other co-existing macromolecules (such as proteins) or small molecules (such as phenolics). Figure 24 demonstrates the effect of HM or LM pectin addition on the water absorption and antioxidant activity of pastas (in the absence and presence of an elderberry juice concentrate). It is well known that denaturation of gluten proteins occur upon pasta cooking. During this process, pasta protein subunits aggregate and form a firm and viscoelastic network via hydrophobic interactions, disulfide bonding, disulfide interchange reactions and nonpolar group interactions.

Any substance such as phenolic antioxidants that can alter redox potential would influence these reactions as well as water-protein or water-pectin interactions [94].

LM and HM pectins have the ability to interact with the polar groups of pasta proteins (e.g. hydroxyl and carboxyl groups), and thus affect the protein-water interactions [71]. The different chemical structure (e.g. the number of hydroxyl groups and the degree of methylation in pectin), and different internal polymer arrangements of LM or HM pectins create variations of water-polymer interactions and subsequently water absorption [71]. Ultimately, all these effects caused the differences in protein secondary structure, pasta microstructure, water-holding property, and stability of entrapped phenolic bioactives.

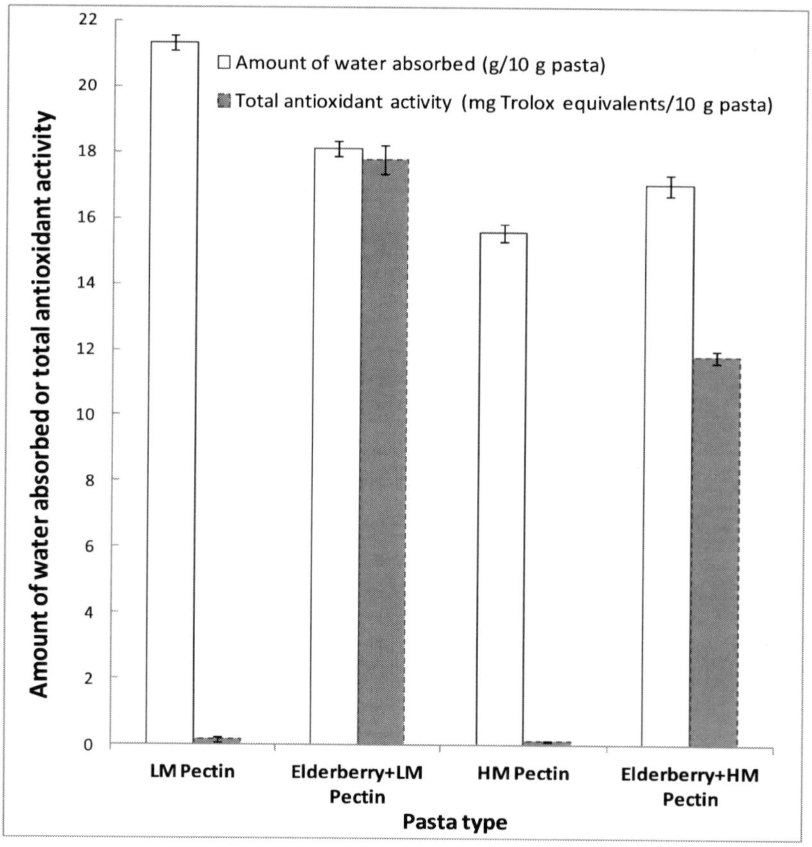

Figure 24. Effect of high and low methoxyl (HM and LM) pectins on the amount of absorbed water and total antioxidant activity of pastas enhanced with pectin in the absence and presence of elderberry juice concentrate.

Consumers increasingly demand dairy products that possess high nutritional quality and pleasant sensory attributes. Increasing the total solids via addition of pectin is an efficient approach to improve the texture of yoghurt [95]. Moreover, pectins as hydrocolloids can improve product consistency and mouth feel and prevent sedimentation [96]. Pectins exhibit advantages over hydrocolloids such as carrageenan, as pectins like the LM pectins unlikely co-precipitate with casein at reduced pH. Moreover, protein-containing acidified beverages and yoghurts can be quite unstable because protein may gradually lose water-binding capacity and aggregate at low pHs. Hydrocolloid stabilisers like pectins have long been used as stabilisers to maintain product body and viscosity because of their favourable interactions with milk proteins. When the pH of a pectin–milk system is greater than the PI of protein, the pectins and proteins repel each other and the pectin phase has a greater affinity for water than the protein phase, milk protein precipitation occurs. When the pH of a pectin–milk system is lower than the PI of protein, the pectins and proteins are attracted to each other and interact. As discussed in previous section, different types of pectin have different physicochemical properties, so they may interact with dairy proteins differently.

Figure 25 compares the interactions of HM or LM pectin with milk casein proteins, and shows that the size of casein micelles depends on HM and LM pectins.

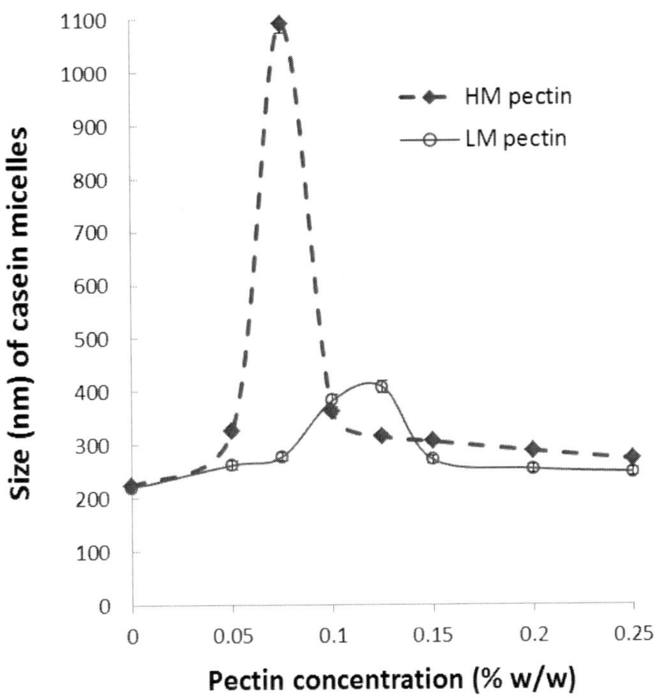

Figure 25. Effect of high and low methoxyl (HM or LM) pectins on the size of casein micelles.

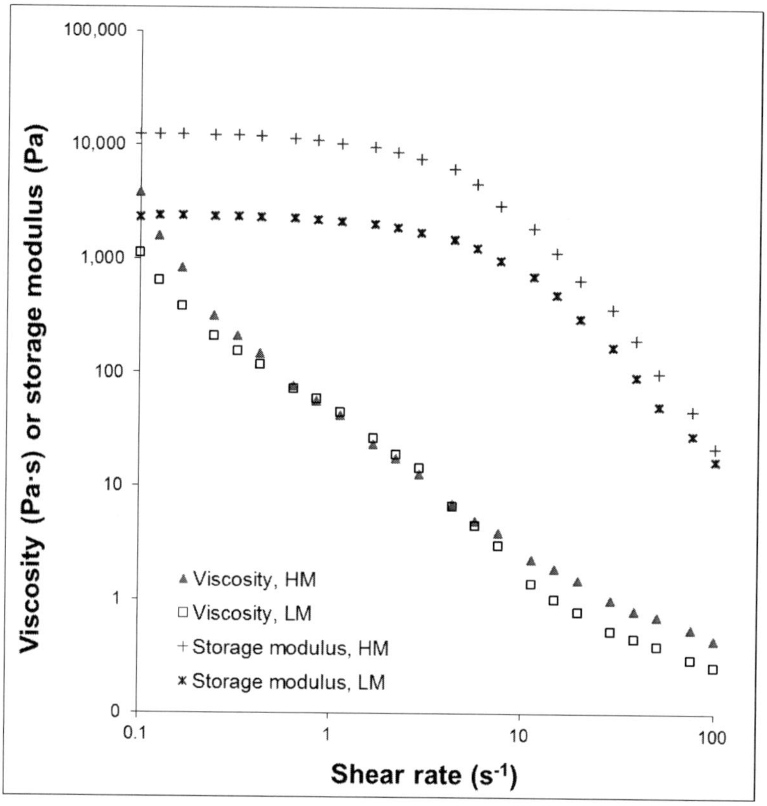

Figure 26. Effect of pectin on yoghurt viscosity and storage modulus.

In either case, a maximum is present over the testing range of pectin concentration, i.e. adding HM pectin at concentrations around 0.07-0.08% resulted in a remarkably large micelle size (i.e. a sharp and high peak) while only a very small and broad peak occurred in the LM pectin system (i.e. at a higher concentration of 0.12-0.13%). Figure 26 further indicates the HM and LM pectins affect the rheological properties of yoghurt differently. Much greater difference between the HM and LM pectin yoghurts was observed in the storage modulus (G´, the ratio of in-phase stress/applied strain) than in the viscosity (i.e. only different at shear rates $< 0.65\ s^{-1}$ and $> 8\ s^{-1}$).

The difference in the degree of methylation of pectin not only affects the interactions between pectin and milk proteins, but also influences the stability of the bioactive substances that co-exist in a yoghurt matrix e.g., probiotic starter cultures (Figure 27) and fortified polyphenol antioxidants (Figure 28, i.e. the anthocyanin content analysed by High Performance Liquid Chromatography-Mass Spectrometry, HPLC-MS) [97-99]. These may result from their differences in yoghurt's microstructure and rheological properties. LM pectin likely facilitates a higher viscosity while HM pectin tends to result in greater gel strength [97, 98, 100].

In summary, it is important to choose the most suitable type of pectin for the formulation of a particular food or dietary product. Aspects that should be considered are: 1) What are the specifications including storage attributes of final product? 2) Which function of pectin is of the most importance, viscosity, gelation, emulsification, stabilizing effect, binding/affinity for a co-existing component such as protein, enzyme or bioactives like polyphenols? 3) What processing technologies and equipment are used for product manufacturing. Correct choice of pectin ensures target sensory attributes and nutritional qualities of finished foods or pharmaceutical products are achieved.

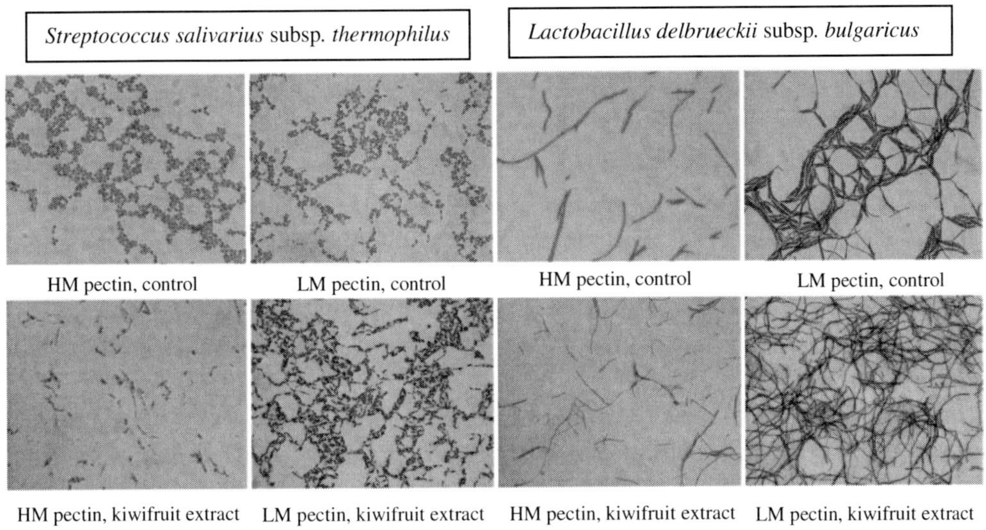

Figure 27. Effect of pectin and kiwifruit extract on the starter cultures in yoghurt.

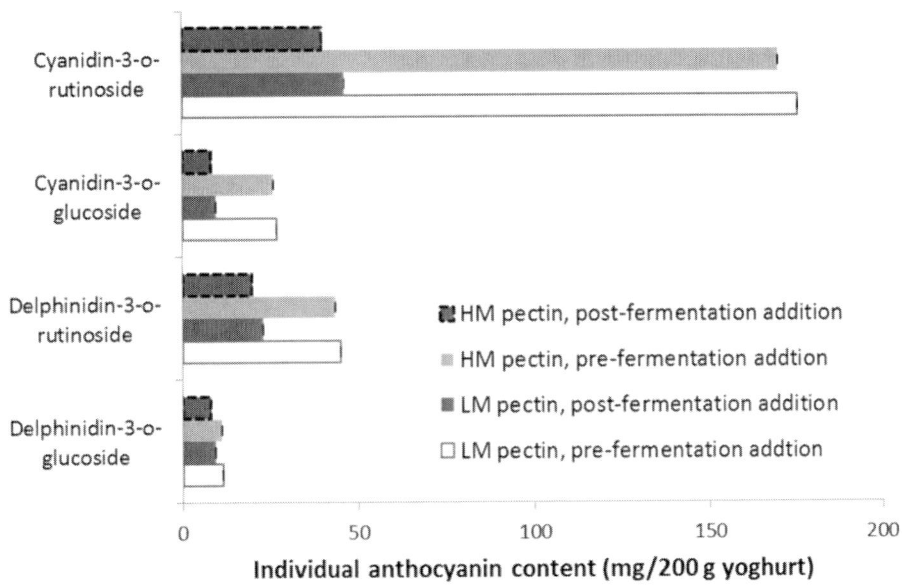

Figure 28. Individual anthocyanin content in blackcurrant extract-enhanced yoghurts. HM or LM pectin refers to high or low methoxyl pectin, and pre- or post-fermentation addition represents adding blackcurrant polyphenol extract before and after fermentation, respectively.

CONCLUSION AND FUTURE OUTLOOK

The rising consumer awareness of a healthy life-style drives industrial manufacturers to develop biomedical products and functional foods that deliver wellness to consumers. Pectins, as the major form of soluble fiber, an active constituent of plant cell walls, and a versatile food ingredient or additive, have considerable potential in these applications. Although the physicochemical properties and nutrition values of pectins especially within the context of dietary fibers have been well studied previously, there is a need to re-examine the roles of pectins in the new dietary matrices. This is especially important for manufactured foods that have produced to contain novel and desirable consumer traits (to increase consumer enjoyment, e.g., new composition and texture) as well as other bioactives (to confer specific health functionality e.g., phenolics and probiotics). Novel and desirable synergies between pectins and dietary matrices can be discovered and exploited in large scale production and during digestion. This chapter increases the understanding of pectins as ingredients for pectin-enriched functional foods with targeting health outcomes. It is very important to picture pectins as a food component in a polymeric matrix rather than an isolated ingredient.

During pectin ingredient extraction, modification, addition to manufactured foods, and ingestion and digestion by human, it is possible that the mobility of water and small molecules is modified or even restricted (including products resulting from hydrolysis and fermentation of food macro-components), and pectins are altered into forms different from initial ingredients via food component interactions.

Thus, it is challenging to deliver the desired nutritional properties of pectin in real consumer foods, as the intrinsic factors (e.g. monosaccharide composition, intra-/inter-molecular linkages, chain branching degree, and polymer structure and surface characteristics), and the external factors (e.g. food matrix, and processes of food manufacture and human digestion) all contribute to the ultimate functionality of pectin-enriched foods. Up-to-date characterization methodologies are essential to the success of elucidating the structure-function relationship and examining the chemical and physical interactions between pectin and other matrix components. Characterization techniques include GC-MS, HLPC-MS, cyclic voltammetry, FT-IR spectroscopy, solid-state NMR spectroscopy and rheometry provide useful information that facilitates the development of pectin-fortified healthy foods and pharmaceutical products.

REFERENCES

[1] ALINORM 09/32/26 (2009). Report of the 30th session of the Codex committee on nutrition and foods for special dietary uses, 3–7 November 2008 (p. 46, Appendix II).

[2] Schneeman, B. O. (1987). Soluble vs insoluble fiber-different physiological responses. *Food Technology*, 41(2), 81–82.

[3] Bingham, S. A., Day, N. E., Luben, R., Ferrari, P., Slimani, N., Norat, T., Clavel-Chapelon, F., Kesse, E., Nieters, A., Boeing, H., Tjønneland, A., Overvad, K., Martinez, C., Dorronsoro, M., Gonzalez, C. A., Key, T. J., Trichopoulou, A., Naska, A., Vineis, P., Tumino, R., Krogh, V., Bueno-de-Mesquita, H. B., Peeters, P. H., Berglund, G., Hallmans, G., Lund, E., Skeie, G., Kaaks, R., & Riboli, E. (2003). Dietary fibre in

food and protection against colorectal cancer in the European prospective investigation into cancer and nutrition (EPIC): an observational study. *The Lancet*, 361(5), 1496–1501.

[4] Block, G., Patterson, B., & Subar, A. (1992). Fruit, vegetables, and cancer prevention: A review of the epidemiological evidence. *Nutrition and Cancer*, 18, 1–29.

[5] Joshipura, K. J., Hu, F. B., Manson, J. E., Stampfer, M. J., Rimm, E. B., Speizer, F. E., Colditz, G., Ascherio, A., Rosner, B., Spiegelman, D., & Willett, W. C. (2001). The effect of fruit and vegetable intake on risk for coronary heart disease. *Annals of Internal Medicine*, 134, 1106–1114.

[6] Kushi, L. H., Doyle, C., McCullough, M., Rock, C. L., Demark-Wahnefried, W., Bandera, E. V., Gapstur, S., Patel, A. V., Andrews, K., & Gansler, T. (2012). American Cancer Society Guidelines on nutrition and physical activity for cancer prevention: reducing the risk of cancer with healthy food choices and physical activity. *CA: A Cancer Journal for Clinicians*, 62(1), 30-67.

[7] Pelucchi, C., Talamini, R., Galeone, C., Negri, E., Franceschi, S., Dal Maso, L., Montella, M., Conti,E., & La Vecchia,C. (2004). Fibre intake and prostate cancer risk. *International Journal of Cancer*, 109, 278–280.

[8] Fry, S. C. (1988). The Growing Plant Cell Wall: Chemical and Metabolic Analysis (pp. 1-333). New York: Longman and Scientific Technical.

[9] Prosky, L. (2000). What is fibre? Current controversies. *Trends in Food Science and Technology*, 10, 271–275.

[10] Burton-Freeman, B. (2000). Dietary fiber and energy regulation. *Journal of Nutrition*, 130(2S), 272s-275s.

[11] Slavin, J. L. (2005). Dietary fiber and body weight. *Nutrition*, 21, 411-418.

[12] Schulze, M. B., Liu, S., Rimm, E. B., Manson, J. A. E., Willett, W. C., & Hu, F. B. (2004). Glycemic index, glycemic load, and dietary fiber intake and incidence of type 2 diabetes in younger and middle-aged women. *American Journal of Clinical Nutrition*, 80, 348–356.

[13] Slavin, J., & Green, H. (2007). Dietary fibre and satiety. *Nutrition Bulletin*, 32, S32–S42.

[14] Lunn, J., & Buttriss, J. L. (2007). Carbohydrates and dietary fibre. *Nutrition Bulletin*, 32, 21–64.

[15] Scott, K. P., Duncan, S. H., & Flint, H. J. (2008). Dietary fibre and the gut microbiota. *Nutrition Bulletin*, 33, 201–211.

[16] Slavin, J. (2013). Fiber and Prebiotics: Mechanisms and Health Benefits. *Nutrients*, 5(4), 1417–1435.

[17] Shankar, S., & Lanza, E. (1991). Dietary fiber and cancer prevention. *Hematology/Oncology* Clinics of *North America*, 5, 25-41.

[18] Eliasson, K., Ryttig, K. R., Hylander, B., & Rossner, S. (1992). A dietary fiber supplement in the treatment of mild hypertension. A randomized double-blind placebo-controlled trial. *Journal of Hypertension,* 10, 195-199.

[19] Raben, A., Christensen, N. J., Madsen, J., Holst, J. J., & Astrup, A. (1994). Decreased postprandial thermogenesis and fat oxidation but increased fullness after a high-fiber meal compared with a low-fiber meal. *American Journal of Clinical Nutrition,* 59, 1386-1394.

[20] Sriamornsak, P. (2001). Pectin: The role in health. *Journal of Silpakorn University*, 21, 60-77.

[21] Van Laere, K. M. J., Hartemink, R., Bosveld, M., Schols, H. A., & Voragen, A. G. J. (2000). Fermentation of plant cell wall derived polysaccharides and their corresponding oligosaccharides by intestinal bacteria. *Journal of Agricultural and Food Chemistry*, 48(5), 1644-1652.

[22] Glinsky, V. V., & Raz, A. (2009). Modified citrus pectin anti-metastatic properties: One bullet, multiple targets. *Carbohydrate Research*, 344, 1788-1791.

[23] Jenkins, D. J., Newton, A. C., Leeds, A. R., & Cummings, J. H. (1975). Effect of pectin, guar gum and wheat fiber on serum cholesterol. *Lancet*, 1, 116-117.

[24] Noble, O., & Danone, L. D. M. (2010). Acidic beverage containing low-methoxyl amidated pectin and no added calcium for a satiety effect. WO 2010116210. Paris, France.

[25] Wanders, A. J., Jonathan, M. C., van den Borne, J. J., Mars, M., Schols, H. A., Feskens, E. J. M., & de Graaf, C. (2013). The effects of bulking, viscous and gel-forming dietary fibres on satiation. *British Journal of Nutrition*, 109, 1330-1337.

[26] Zhang, S., & Vardhanabhuti, B. 2014. Intragastric gelation of whey protein–pectin alters the digestibility of whey protein during in vitro pepsin digestion. *Food and Function*, 5, 102-110.

[27] Hansen, W. E., & Schulz, G. (1982). The effect of dietary fiber on pancreatic amylase activity in vitro. *Hepato-Gastroenterology*, 29, 157-160.

[28] Tsujita, T., Sumiyosh, M., Han, L. K., Fuijiwara, T., Tsujita, J., & Okuda, H. (2003). Inhibition of lipase activities by citrus pectin. *Journal of Nutritional Science and Vitamins*, 49(5), 340-345.

[29] Willats, W. G. T., Knox, J. P., & Mikkelsen, J. D. (2006). Pectin: new insights into an old polymer are starting to gel. *Trends in Food Science & Technology*, 17, 97-104.

[30] Joye, D. D., & Luzio, G. A. (2000). Process for selective extraction of pectins from plant material by differential pH. *Carbohydrate Polymers*, 43, 337-342.

[31] Reitsma, J. C. E., Thibault, J. F., & Pilnik, W. (1986). Properties of amidated pectins. I. Preparation and characterization of amidated pectins and amidated pectic acids. *Food Hydrocolloids*, 1(2), 121-127.

[32] Sun-Waterhouse, D., Farr, J., Wibisono, R., & Saleh, Z. (2008). Fruit-based functional foods I: production of novel food grade apple fibre ingredients. *International Journal of Food Science and Technology*, 43, 2113–2122.

[33] Nawirska, A., & Kwaniewska, M. (2004). Dietary fibre fractions from fruit and vegetable processing waste. *Food Chemistry*, 91, 221–225.

[34] Shi, J., Yeoh, S., & Langrish, T. A. G. (2008). Comparisons between different techniques for water-based extraction of pectin from orange peels. *Desalination*, 218(1), 229-237.

[35] Srivastava, P., & Malviya, R. (2011). Sources of pectin, extraction and its applications in pharmaceutical industry – An overview. *Indian Journal of Natural Products and Resources*, 2(1), 10-18.

[36] Sun-Waterhouse, D., Wang, W., Waterhouse, G. I. N., & Wadhwa, S. S. (2012). Utilisation potential of feijoa fruit wastes as ingredients for functional foods. *Food and Bioprocess Technology*, 6(12), 3441-3455.

[37] Sun-Waterhouse, D., Wen, I., Wibisono, R., Melton, L. D., & Wadhwa S. (2009). Evaluation of the extraction efficiency for polyphenol extracts from by-products of green kiwifruit juicing. *International Journal of Food Science and Technology*, 44, 2644-2652.

[38] Aina, V. O., Barau, M. M., Mamman, O.A., Zakari, A., Haruna, H., Umar, M. S. H., & Abba, Y. B. (2012). Extraction and characterization of pectin from peels of lemon (Citrus limon), grape fruit (Citrus paradisi) and sweet orange (Citrus sinensis). *British Journal of Pharmacology and Toxicology*, 3(6), 259-262.

[39] Crandall, P. G., Braddock, J. R., & Rouse, A. H. (1978). Determining the yield and quality of pectin from fresh peel and pectin pomace. *Proceedings of the Florida State Horticultural Society*, 91, 109-111.

[40] Sun-Waterhouse, D., Luberriaga, C., Jin, D., Wibisono, R., Wadhwa, S., Waterhouse, G. I. N. (2013). Juices, fibres and skin waste extracts from white, pink or red fleshed apple genotypes as potential food ingredients: A comparative study. *Food and Bioprocess Technology*, 6(2), 377–390.

[41] Eliaz, I., Hotchkiss, A. T., Fishman, M. L., & Rode, D. (2006). The effect of modified citrus pectin on urinary excretion of toxic elements. *Phytotherapy Research*, 20, 859-864.

[42] Zhao, Z. Y., Liang, L., Fan, X., Yu, Z., Hotchkiss, A. T., Wilk, B. J., & Eliaz, I. (2008). The role of modified citrus pectin as an effective chelator of lead in children hospitalized with toxic lead levels. *Alternative Therapies, Health and Medicine*, 14, 34-38.

[43] Percy, A. E., Melton, L. D., & Jameson, P. E. (1997). Xyloglucan and hemicelluloses in the cell wall during apple fruit development and ripening. *Plant Science*, 125, 31-39.

[44] Vincken, J. P., Schols, H. A., Oomen, R. J. F. J., McCann, M. C., Ulvskov, P., Voragen, A. G. J., & Visser, R. G. F. (2003). If homogalacturonan were a side chain of rhamnogalacturonan I. Implications for cell wall architecture. In: *Plant Physology* (Vol. 132). United States: American Society of Plant Biologists, Dartmouth Journal Services.

[45] Cosgrove, D. J. (2001). Wall structure and wall loosening. A look backwards and forwards. *Plant Physiology,* 125, 131-134.

[46] Schols, H. A., Posthumus, M. A., & Voragen, A. G. J. (1990). Structural features of hairy regions of pectins isolated from apple juice produced by the liquefaction process. *Carbohydrate Research*, 206, 117–130.

[47] De Vries, J. A., Den Uijl, C. H., Voragen, A. G. J., Rombouts, F. M., & Pilnik, W. (1983). Structural features of the neutral sugar side chains of apple pectic substances. *Carbohydrate Polymers*, 3, 193-205.

[48] Sun-Waterhouse, D., Melton, L. D., O'Connor, C. J., Kilmartin, P. A., & Smith, B. G. (2008). Effect of apple cell walls and their extracts on the activity of dietary antioxidants. *Journal of Agricultural and Food Chemistry*, 56(1), 289–295.

[49] O'Neill, M. B. (2001). Physicochemical Properties of Dehydrated Apple Pieces. Food Science PhD thesis (pp. 46-73). The University of Auckland, New Zealand.

[50] Hayashi, T. (1989). Xyloglucans in primary cell walls. *Annual Review of Plant Physiology,* 40, 139-168.

[51] Selvendran, R. R., & Ryden, P. (1990). Isolation and analysis of plant cell walls. In: P. M. Dey, J. B. Harbourne (Eds.) *Carbohydrates* (pp. 549-581). London: Academic Press.

[52] Sun-Waterhouse, D., Teoh, A., Massarotto, C., Wibisono, R., & Wadhwa, S. (2010). Comparative analysis of fruit-based functional snack bars. *Food Chemistry*, 119, 1369–1379.

[53] Parkar, S., Trower, T., Stevenson, D., Sun-Waterhouse, D., & Skinner, M. (2010). Food, fibre and satiety - how does fibre make you feel full. In:*The 2nd TNO Beneficial Microbes Conference*, Noordwijkerhout, March 2010, Netherlands.

[54] Butts, C., Paturi, G., de Guzman, C. E., Monro, J., Wibisono, R., Hedderley, D., Smith, H., Martell, S., Sun-Waterhouse, D., Lister, C., & Sutherland, J. (2009). The effect of dietary vegetable and fruit fibres on gut health in healthy rats. In: *Joint annual conference of the New Zealand Nutrition Society and the Australian Nutrition Society*. Newcastle, NSW.

[55] Sun-Waterhouse, D., Smith, B. G., O'Connor, C. J., & Melton, L. D. (2008). Effect of raw and cooked onion dietary fibre on the antioxidant activity of ascorbic acid and quercetin. *Food Chemistry*, 111, 580–585.

[56] Kilmartin, P. A., Zou, H., & Waterhouse, A. L. (2001). A cyclic voltammetry method suitable for characterizing antioxidant properties of wine and wine phenolics. *Journal of Agricultural and Food Chemistry*, 49, 1957-1965.

[57] Newman, R. H., Ha, M. A., & Melton, L. D. (1994). Solid-state 13C NMR investigation of molecular ordering in the cellulose of apple cell walls. *Journal of Agricultural and Food Chemistry*, 42, 1402-1406.

[58] Tang, H.-R., & Belton, P. S. (2002). Molecular dynamics of polycrystalline cellobiose studied by solid-state NMR. *Solid State NMR*, 21, 117-133.

[59] Sivam, A. S., Waterhouse, G. I. N., Zujovic, Z. D., Perera, C. O., & Sun-Waterhouse, D. (2013). Structure and dynamics of wheat starch in breads fortified with polyphenols and pectin: An ESEM and solid-state CP/MAS 13C NMR spectroscopic study. *Food and Bioprocess Technology*, 6(1), 110-123.

[60] Smith, B. G., Harris, P. J., Melton, L. D., & Newman, R. H. (1998). The range of mobility of the non-cellulosic polysaccharides is similar in primary cell walls with different polysaccharide compositions. *Physiologia Plantarum*, 103, 233-246.

[61] Smith, B. G., Harris, P. J., Melton, L. D. & Newman, R. H. (1998). Crystalline cellulose in hydrated primary cell walls of three monocotyledons and one dicotyledon. *Plant & Cell Physiology*, 39, 711-720.

[62] Harris, R. K. (1993). State of the art for solids. *Chemistry in Britain*, 29, 601-604.

[63] Keenan, M. H. J., Belton, P. S., Matthew, J. A., & Howson, S. J. (1985). Improvement of the gelation properties of sugar-beet pectin following treatment with an enzyme preparation derived from Aspergillus-Niger comparison with a chemical modification. *Carbohydrate Polymers*, 12, 295-306.

[64] Ha, M. A., Apperley, D. C., & Jarvis, M. C. (1997). Molecular rigidity in dry and hydrated onion cell walls. *Plant Physiology*, 115, 593-598.

[65] Hediger, S., Emsley, L., & Fischer, M. (1999). Solid-state NMR characterization of hydration effects on polymer mobility in onion cell-wall material. *Carbohydrate Polymers*, 322, 102-112.

[66] Tang, H.-R., Belton, P. S., Ng, A., & Ryden, P. (1999). ^{13}C MAS NMR studies of the effects of hydration on the cell walls of potatoes and Chinese water chestnuts. *Journal of Agricultural and Food Chemistry*, 47, 510-517.

[67] Greve, L. C., McArdle, R. N., Gohlke, J. R., & Labavitch, J. M. (1994). Impact of heating on carrot firmness: changes in cell wall components. *Journal of Agricultural and Food Chemistry*, 42, 2900-2906.
[68] Panchev, I. N., Kirtchev, N. A., Kratchanov, C. G., & Proichev, T. (1988). On the molecular weight of pectic substances and its relation to their gel strengths. *Carbohydrate Polymers*, 8, 257-269.
[69] Rudan-Tasic, D., & Klofutar, C. (1996). Potentiometric titration of poly-(α-D)-galacturonic acid. In: J. Visser, A.G. J. Voragen (Eds.) *Pectins and Pectinases*. United Kingdom: Elservier Science Ltd.
[70] Axelos, M. A. V., & Thibault, J. F. (1991). The chemistry of low-methoxyl pectin gelation. In: R. H. Walter (Ed.) *The Chemistry and Technology of Pectin*. New York: Academic Press.
[71] Whistler, R. L., & BeMiller, J. N. (1997). Carbohydrate chemistry for food scientists (pp. 203-210). St. Paul: The American Association of Cereal Chemists.
[72] Voragen, A. G. J., Pilnik, W., Thibault, J. F., Axelos, M. A.V., & Renard, C. M. G. C. (2010). Pectins. In: A. M. Stephen, G. O. Phillips (Eds.) *Food Polysaccharides and their Applications* (pp. 287-339). Boca Raton, FL: Taylor and Francis.
[73] Gallaher, D., Hassel, C. A., Lee, K. J., & Gallaher, C. M. (1993). Viscosity and fermentability as attributes of dietary fiber responsible for the hypocholesterolemic effect in hamsters. *Journal of Nutrition*, 123, 244-252.
[74] Hellendoorn, E. W. (1983). Fermentation as the principal cause of the physiological activity of indigestible food residue. In: G. A. Spiller (Ed.) Topics in dietary fiber research (pp. 127-168). New York: Plenum Press.
[75] Renard, C. M. G. C., & Thibault, J. F. (1996). Pectins in mild alkaline conditions: β-elimination and kinetics of demethylation. In: J. Visser, A. G. J. Voragen (Eds.) *Pectins and Pectinases*. United Kingdom: Elsevier Ltd.
[76] Fishman, M. L., Chau, H. K., Hoagland P. D., & Hotchkiss, A. T. (2006). Microwave-assisted extraction of lime pectin. *Food Hydrocolloids*, 20, 1170–1177.
[77] Nguyen, S., Alund, S. J., Hiorth, M., Kjøniksen, A.-L., & Smistad, G. (2011). Studies on pectin coating of liposomes for drug delivery. *Colloids and Surfaces B: Biointerfaces*, 88, 664– 673.
[78] Tibbits, C. W., MacDougall, A. J., & Ring, S. G. (1998). Calcium binding and swelling behaviour of a high methoxyl pectin gel, In: *Carbohydrate Research* (Vol. 310). United Kingdom: Elsevier Ltd.
[79] Oakenfull, D. G., & Scott, A. (1984). Hydrophobic interaction in the gelation of high methoxyl pectins. *Journal of Food Science*, 49(4), 1093-1098.
[80] Walkinshaw, M. D., & Arnott, S. (1981). Conformations and interactions of pectins. II. Models for junction zones in pectinic acid and calcium pectate gels. *Journal of Molecular Biology*, 153, 1075-1085.
[81] Oakenfull, D. G., & Fenwick, D. E. (1977). Thermodynamics and mechanism of hydrophobic interaction. *Australian Journal of Chemistry*, 30, 741-752.
[82] Williams, P. A., & Phillips, G. O. (2000). Introduction to food hydrocolloids. In: G. O. Phillips, P. A. Williams (Eds.) *Handbook of hydrocolloids* (pp. 1-19). Cambridge, UK: Wood head Publishing Ltd.

[83] Matia-Merino, L., Lau, K., & Dickinson, E.(2004). Effects of low-methoxyl amidated pectin and ionic calcium on rheology and microstructure of acid-induced sodium caseinate gels. *Food Hydrocolloids*, 18(2), 271-281.

[84] Thibault, J. F., & Rinaudo, M. (1985). Interactions of mono- and divalent counterions with alkaline- and enzyme-deesterified pectins. *Biopolymers*, 24, 2131-2144.

[85] Lutz, R., Aserin, A., Wicker, L., & Garti, N. (2009) Structure and physical properties of pectins with block-wise distribution of carboxylic acid groups. *Food Hydrocolloids*, 23, 786-794.

[86] Sun-Waterhouse, D., & Wadhwa, S. S. (2012). Synergistic effects on the rheological behaviour and cloudy-point of aqueous hydrocolloid dispersions. In: the *2012 NZIFST conference*, 26-28 June 2012, Hamilton, New Zealand.

[87] Walkenström, P., Kidman, S., Hermansson, A.-M., Rasmussen, P. B., & Hoegh, L. (2003). Microstructure and rheological behaviour of alginate/pectin mixed gels. *Food Hydrocolloids*, 17, 593-603.

[88] Sejersen, M. T., Salomonsen, T., Ipsen, R., Clark, R., Rolin, C., & Engelsen, S. B. (2007). Zeta potential of pectin-stabilised casein aggregates in acidified milk drinks. *International Dairy Journal*, 17, 302–307.

[89] O'Connor, C. J., Sun, D., Smith, B. G., & Melton, L. D. (2003). Effect of soluble dietary fibres on lipase-catalysed hydrolysis of tributyrin. *Journal of Food Science*, 68(3), 1093-1099.

[90] Hillman, L. C., Peters, S. G., Fisher, C. A., & Pomare, E. W. (1986). Effects of fibre components pectin, cellulose, and lignin on bile salt metabolism and biliary lipid composition in man. *Gut*, 27, 29-36.

[91] Das, S., & Ng, K. Y. (2010). Resveratrol-loaded calcium pectinate beads: Effects of formulation parameters on drug release and bead characteristics. *Journal of Pharmaceutical Sciences*, 99, 840-860.

[92] Wang, W., Waterhouse, G. I. N., & Sun-Waterhouse, D. (2013). Co-extrusion encapsulation of quercetin-fortified canola oil: Effect of pectin addition to alginate shell. *Food Research International*, 54(1), 837-851.

[93] Sriamornsak, P. (2003). Chemistry of pectin and its pharmaceutical uses: A review. *Silpakorn University International Journal*, 3, 206-228.

[94] Sun-Waterhouse, D., Jin, D., & Waterhouse, G. I. N. (2013). Quality attributes, polyphenol contents and antioxidant activity of pastas enhanced with elderberry juice concentrate and fibre polysaccharides. *Food Research International*, 54(1), 781-789.

[95] Robinson, R. K., Lucey, J. A., & Tamime, Y. A. (2006). Manufacture of Yoghurt. In: Y. A. Tamime (Ed.) *Fermented Milks*. UK: Blackwell Science Ltd.

[96] Yanes, M., Durán, L., & Costell, E. (2002). Effect of hydrocolloid type and concentration on flow behaviour and sensory properties of milk beverages model systems. *Food Hydrocolloids*, 16, 605-611.

[97] Sun-Waterhouse, D., Zhou, J., & Wadhwa, S. S. (2013). Drinking yoghurts with berry polyphenols added before and after fermentation. *Food Control*, 32(2), 450–460.

[98] Sun-Waterhouse, D., Zhou, J., & Wadhwa, S. S. (2010). Effects of adding apple polyphenols before and after fermentation on the properties of drinking yoghurt. *Food and Bioprocess Technology*, 5(7), 2674–2686.

[99] Benitez, K., & Sun-Waterhouse, D. (2008). Properties of yoghurt enhanced with fruit fibre and phytochemicals. In: *the Functional Foods & Edible Oils - The Future*, 12-13 November 2008, Auckland, New Zealand.

[100] Aportela-Palacios, A., Sosa-Morales, M. E., & Velez-Ruiz, J. F. (2005). Rheological and physicochemical behaviour of fortified yogurt with fibre and calcium. *Journal of Texture Studies*, 36(3), 333–349.

In: Pectin: Chemical Properties, Uses and Health Benefits
Editor: Phillip L. Bush
ISBN: 978-1-63321-438-5

Chapter 8

PECTINS APPLIED TO THE DEVELOPMENT OF ANTIOXIDANT EDIBLE FILMS: INFLUENCE OF THE MACROMOLECULAR STRUCTURE IN THE L-(+)-ASCORBIC ACID STABILIZATION

Carolina D. Pérez[1,2,4,*], María D. De'Nobili[1,5], Eliana N. Fissore[1,4], María F. Basanta[1,5], Lía N. Gerschenson[1,4], Randall G. Cameron[3,†] and Ana M. Rojas[1,4,‡]

[1]Industry Department, School of Natural and Exact Sciences (FCEN), Buenos Aires University (UBA). Ciudad Universitaria, Ciudad Autónoma de Buenos Aires, Argentina
[2]Current affiliation: Institute of Food Technology (ITA), Instituto Nacional de Tecnología Agropecuaria (INTA), Morón, Argentina
[3]Citrus and Subtropical Products Unit, U.S. Horticultural Laboratory, Agricultural Research Service, United States Department of Agriculture (USDA), Pierce, US
[4]Member and [5]Fellow of the National Scientific and Technical Research Council of Argentina (CONICET)

ABSTRACT

Pectins of different nanostructure were assayed in their ability to develop film networks able to stabilize L-(+)-ascorbic acid (AA) to hydrolysis in view of antioxidant protection at interfaces, nutritional supplementation or therapy. Compartmentalization into edible films can permit not only to increase the AA stability but also to achieve localized antioxidant activity and controlled release. The AA hydrolysis was specifically studied in the present work. Hence, films were stored at controlled relative humidity (RH) in the absence of air. Films were made with each one of the enzymatically tailored (50, 70 and 80% DM) pectins (Cameron et al., 2008) or commercial high methoxyl pectin (HMP; 72% DM). A random distribution of demethylated blocks is expected to

* Phone and Fax numbers: +54 11 4576 3366. Phone number: +54 11 4576 3397.
† Tel.: +1 772 462 5856; fax: +1 772 462 5986. E-mail address: randall.cameron@ars.usda.gov.
‡ Corresponding author: Ana M. Rojas. Email: arojas@di.fcen.uba.ar. Phone and Fax numbers: +54 11 4576 3366.

characterize commercial pectins whereas ordered patterns are obtained by enzymatic action. Calcium ions are necessary for crosslinking of low methoxyl pectins. Hence, the ability of Ca-mediated junction zones to stabilize AA into the edible films made with commercial pectins of low (LMP; 40%) or high (HMP; 72%) DM, at the same Ca^{2+} concentration (film systems called Ca-LMP and Ca-HMP, respectively), was also evaluated. Glycerol was used for plasticization. Kinetics of AA loss and subsequent browning development were determined by film storage at constant 57.7% RH and 25°C, in the dark. Since AA stability was dependent on water availability in the film network, determined by ^{1}H-NMR, it was observed that the pectin nanostructure affected the AA kinetics. Higher AA retention and lower browning rates were achieved in HMP films than in enzymatically tailored pectin films, and the immobilization of water and consequent AA stability increased with the proportion of calcium-crosslinked junction zones present in the film network. As determined through tensile assays, the presence of Ca^{2+} in the film network produced significant decrease in elongation at fracture. This assay also revealed some sensibility of the HMP (commercial 72% DM pectin) to calcium ions. The glass transition temperature values of pectin films decreased ($T_g \approx -88$ to -102°C) with the moisture content increase, indicating the contribution of water to the network plasticization by glycerol. However, water was mostly confined in the Ca-LMP network ($T_g \approx -93.99$°C) followed by Ca-HMP ($T_g \approx -88.56$°C), as inferred from the water availability determined by the ^{1}H-NMR. This was attributed to the water interaction at the Ca^{2+}-junction zones. Random distribution of demethylated blocks in the HG chains in addition to the presence of some disordered (amorphous) regions of RG-I may produce better immobilization of water than more rigid networks like those developed from the tailored pectin macromolecules.

Keywords: Tailored pectins, pectin nanostructure, low and high methylated pectins, antioxidant edible films, ascorbic acid hydrolysis, water

1. Introduction

1.1. Antioxidant Edible Films: Pectin Networks to Carry L-(+)-Ascorbic Acid

In recent years, edible films and coatings have received increasing attention from researchers and industry as an interesting alternative for food packaging (Khwaldia et al., 2004). They constitute an application of the active food packaging (Han, 2005). Although edible coatings and films may not provide a good water vapor barrier, they can act as sacrificing agents retarding moisture loss from food products (Bourtoom, 2008). Actually, at relatively low relative humidity (RH) values of storage such as 57%, edible films made with high methoxyl pectin and carrying L-(+)-ascorbic acid were *per se* very good barriers to oxygen and produced additional antioxidant protection, preserving a hydrophobic functional food interface such as walnut oil from tocopherol and lipid oxidation (Pérez et al., 2013). Food preservation can then clearly profit from the edible films and coatings as they can be used to support active compounds such as food antimicrobials and natural antioxidants. Biodegradable edible films can constitute a technological hurdle for food preservation because they can act as selective gas-barriers (e.g. oxygen, aroma) while their microstructure is applied to carry, localize the activity and to control-release of food additives at interfaces (De'Nobili et al., 2011).

Also known as 'vitamin C', L-(+)-ascorbic acid (AA) is a water soluble reducing agent and a natural antioxidant which can be used for pharmaceutical and food preservation as well as for nutritional supplementation (Durschlag et al., 2007). It interferes with oxidative-reductive and other metabolic processes in an organism. It is important for the activity of enzymes keeping the balance among some enzymatic groups, and has great importance for physiological permeability of capillaries (Deng et al., 2013). The stability of AA is affected by processing and storage conditions because it depends on a large number of factors such as temperature, equilibrium RH, oxygen partial pressure, light (Kitts, 1997). AA is oxidized by oxygen or other oxidative compounds to produce L-dehydroascorbic acid (DHA) that also has vitamin C activity *in vivo* because of the vital cellular function of the dehydroascorbate reductase enzyme, which regenerates the AA in a glutathione dependent reaction (Zhou et al., 2012). However, the biological activity is irreversibly lost during food processing and storage, when DHA is hydrolyzed in a subsequent reaction. Furthermore, anaerobic degradation of AA through hydrolysis also occurs simultaneously to AA oxidation when oxygen is present, producing 2-keto-L-gulonic acid (KGA) (Figure 1) (Kurata and Sakurai, 1967). On the other hand, non enzymatic browning also proceeds with the AA concentration decay since the products of the reactions that follow the first step of AA destruction are also part of the browning reaction chain (León and Rojas, 2007). Compartmentalization of AA into a film network for antioxidant preservation at food interfaces could help achieve the AA stabilization because it can preclude its interaction with oxygen, with other food preservatives and food components, as well as with water (De'Nobili et al., 2011). Also, these films can be applied to or be used as pharmaceutical products for controlled release (De'Nobili et al., 2013).

Pectin is a filmogenic polysaccharide due to its ability to develop intermacromolecular physical bonds (hydrogen bonding, electrostatic and hydrophobic interactions) which are strengthened by dehydration. Hence, it can be applied to edible active film development. Pectin is a structural component of the cell walls of all land plants and in a normal western diet around 4-5 g of pectin are consumed each day like part of the dietary fiber (Willats et al., 2006). Their chemical composition, macromolecular structure, molecular weight and molecular weight distribution, as well as degree of methyl esterification (DM), distribution of unmethylated blocks and degree of acetylation (presence of *O*-acetyl groups) determine the pectin functionality as thickener and gelling agent as well as the physiological properties as dietary fiber. These characteristics are dependent on the origin and conditions of extraction. Pectins consists primarily of partially methyl esterified poly α-D-(1,4) linked galacturonic acid chains (called homogalacturonan –HG- domains), which are the 'smooth' ordered regions of pectins (Vincken et al., 2003; Morris et al., 2010). In the HG chains, the α-D-galacturonic acid (GalA) residues can be methyl esterified in their carboxyl groups, constituting pectins with high DM and pectins with low DM. Pectin macromolecule also shows some kinks of (1→2)-linked α-L-rhamnose (Rha) residues that alternate with GalA monomers.

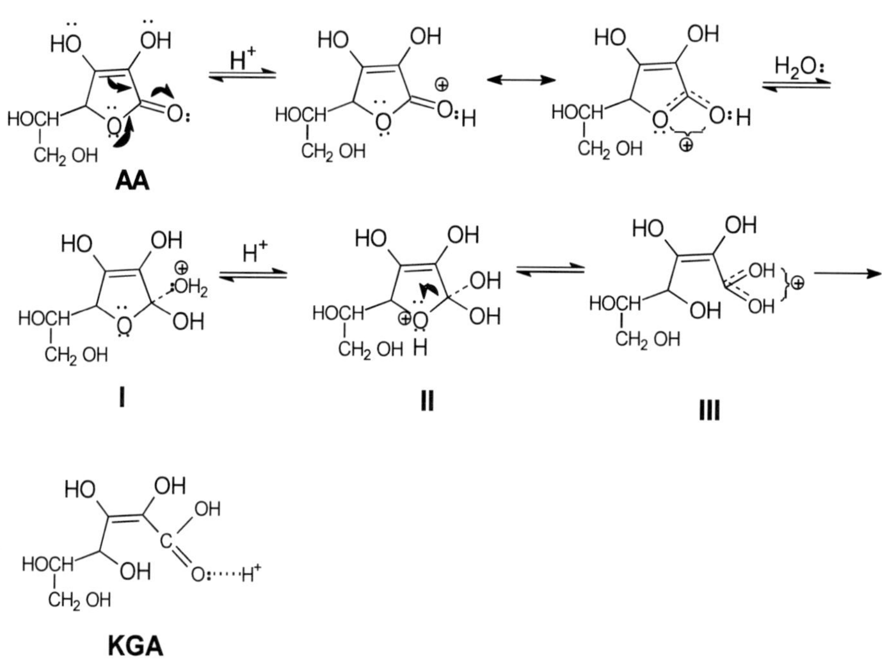

Figure 1. The acid catalyzed reaction of L-(+)-ascorbic acid hydrolysis (anaerobic condition) where the bimolecular nucleophilic substitution mechanism (S_N2) of attack by a water molecule (nucleophile) on the L-(+)-ascorbic acid molecule is shown, as proposed by León and Rojas (2007) in order to justify the influence of water concentration in the kinetics of L-(+)-ascorbic acid destruction under anaerobic conditions.

These alternating rhamnose residues present lateral substitution by a single galactosyl (Gal) residue [β-D-Gal*p*-(1→4)] but also polymeric chains such as arabinogalactan I (1,4-linked β-D-Gal*p* backbone) and arabinans (1,5-linked α-L-Ara*f* backbone, laterally substituted by other arabinan side chains). This rhamnose-complex constitutes the rhamnogalacturonan I (RG-I) domain, which is the amorphous 'hairy' region of pectins (Vincken et al., 2003; Morris et al., 2010). Depending on the origin, a RG-II domain can be additionally found in some extracted pectins isolated from primary cell walls of higher plants by endo-α-1,4-polygalacturonases. These enzymes releases this kind of low molecular weight (5-10 kDa) structurally complex pectic "mega-oligosaccharide" called RG-II, where some short segments of GalA residues in the HG chains are laterally substituted by different and rare monosaccharides such as apiose, KDO, methyl-xylose, acetyl methyl fucose and aceric acid, among others (Mazeau and Pérez, 1998; Pérez et al., 2003). The chain lengths of the HG and RG-I domains can vary considerably. It was first accepted that the RG-I and HG domains are interspersed, constituting the "backbone" of pectin polymers (Carpita and Gibeaut, 1993; Pérez et al., 2003). However, an alterative structure has also been proposed in which the HGs are long side chains of the RG-I cores (Vincken et al., 2003). One thing that is not disputed is that pectins are an extremely complex and structurally diverse group of polymers. The fine structures of pectins can be extremely heterogeneous between plants, between tissues, and even within a single cell wall. Additionally, the original pectin structure is affected by the extractive procedure used (Fissore et al., 2007; Fissore et al., 2010; Fissore et al., 2011). Pectin is a high value functional pharmaceutical and food ingredient widely used as a thickener, gelling agent and stabilizer (Fissore et al., 2012). New applications are constantly developing and their use as emulsifiers is one of the latest new-comes (Siew and Williams, 2008; Fissore et al., 2013). In high methylesterified or methoxylated pectins (DM >50%), HMP, gelation is promoted by high sugar concentration and low pH (<3.5), whereas in low methylesterified pectins (DM $\leq$ 50%), LMP, gelation occurs in presence of divalent ions. Therefore, LMP can be used in low-calorie jams and jellies (Hotchkiss et al., 2002; Willats et al., 2006). As above mentioned, the functionality of pectins is deeply affected by their macromolecular structure. Most of the pectin used by the food industry originates from citrus or apple peel from which it is extracted at low pH and high temperature, and it is primarily a homogalacturonan (Voragen et al., 2009). Functional properties of the HG region are dependent on the amount and distribution of the methyl and acetyl esters. Methyl esters are the dominant form in most species but acetyl esters are abundant in sugar beets. Commercial sugar beet pectins show a high degree of *O*-acetylation at the *O*-2 and/or *O*-3 into the HG chains, which is responsible for a characteristic functionality (Ralet et al., 2005). Lateral substitution by acetyl groups produces the expansion of these pectin macromolecules in water, increasing the solution viscosity. On the other hand, Willats et al. (2001) demonstrated that for water holding capacity of gels made from pectins with differing methyl ester distribution patterns in their HG region, the distribution pattern was more important than the DM. Spatial distribution patterns of methyl esters within HG regions are demonstrably related to the mechanism used for demethylation. Demethylation occurs either by enzymatic action or alkaline (chemical) hydrolysis, respectively resulting in either ordered or random patterns. Enzymatic tools potentially can be used both to modify the amount and pattern of methylesterification and to characterize chemically the polymer chain. Plant pectinmethylesterases (PMEs), with a basic isoelectric point (pI), have been shown to produce ordered distributions of demethylated stretches within the HG regions, whereas fungal PMEs, with an acidic pI, as well as base (chemical) catalyzed demethylations produce random distributions (Cameron et al., 2008). Using endopolygalacturonase (EPG) digested pectins, and based on the amounts of monomer, dimer and trimer of GalA liberated from pectic substrates, Daas et al. (1999) introduced the term ''Degree of Blockiness'' (percentage of non-esterified GalA residues released during the EPG digest) to differentiate pectins with numerous small unesterified sequential GalA residues from those with larger unesterified blocks. Cameron et al. (2011) determined the masses for each oligomer ($GalA_n$, where n is the number of demethylesterified GalA residues in the oligomer) of different degree of polymerization (DP) by high-pressure anion-exchange chromatography (HPAEC) coupled to an evaporative light-scattering detector (ELSD). Mass concentrations for each $GalA_n$ were used to calculate $GalA_n$ molar concentration per milliliter (C_n), which was used with the molar concentration of pectin per milliliter (C_p) to estimate the average number of demethylated blocks released from a molecule of length n ($\bar{B}_n$) for each oligomer of $GalA_n$:

$$\bar{B}_n = \frac{C_n}{C_P}$$

The average number of blocks per molecule ($\bar{B}$) is the sum of the average number of demethylated blocks of size n:

$$B = \sum_{n=3}^{z} B_n$$

The size of that average number of demethylated blocks (called $\overline{BS}$) that was enzymatically released, was estimated according to:

$$\overline{BS} = \frac{\sum_{n=3}^{z} nB_n}{B}$$

The number of blocks of the average block size ($\overline{BS}$) per molecule ($\overline{BN}$) was then calculated as:

$$\overline{BN} = \frac{B}{\overline{BS}}$$

The frequency distribution of $\overline{BS}$ can be controlled and the frequency distribution of $\overline{BN}$ can also be manipulated. Hence, pectin nanostructure is amenable to enzymatic engineering. It can be tailored according to a desired functionality for especial uses in pharmaceutical and food formulation as well as to determine the macromolecular structure of pectins (Cameron et al., 2008; Luzio and Cameron, 2008; Cameron et al., 2011).

Rombouts and Thibault (1986) demonstrated the presence of ester-linked ferulic acid in pectin fractions isolated from certain plants. Ferulic acids were attached to arabinosyl and galactosyl residues in the side-chains of RG-I, crosslinking them covalently as diferulate ester bridges, anchoring pectins to the cell wall. According to Waldron et al. (1997), red beet (*Beta vulgaris* L. var. conditiva) tissue contains ferulic acid, being responsible for the thermally resistant cell-cell adhesion in tissue containing diferuloyl bridges (Parker & Waldron, 1995; Ng and Waldron, 1997).

In the present work, pectin was selected to develop edible films by considering that the presence of some alternating 'disorder' (hairy) regions of the remaining RG-I together with ordered HG regions, could sufficiently immobilize water to achieve higher AA stabilization to hydrolysis into the film network and, consequently, higher antioxidant effect at interfaces. This can also contribute to develop lower browning of films. Therefore, the *hydrolytic* stability of AA into pectin films, plasticized with glycerol, was specifically studied by film storage *under vacuum* (anaerobic condition), at controlled RH (57.7%) and temperature (25.0ºC).

Table 1. Molecular weight, average length of the longest demethylated block (DP), average demethylated block size ($\overline{BS}$) and number of blocks ($\overline{B}_n$) per molecule of the average block size for each series of tailored pectins used are reported[1]

Pectin series (DM [2]; %)	Molecular weight (Da)	Oligomer DP	$\overline{BS}$	$\overline{B}_n$
50	43,280	52	8.6	3.9
70	43,280	42	10.2	2.0
80	43,280	14	5.3	1.7

[1] Cameron et al. (2008). [2] DM: degree of methyl esterification.

The influence of the pectin (macromolecule) nanostructure on the development of film networks which can be able to stabilize the AA molecule to its hydrolysis was analyzed. Hence, films were developed with each one of the enzymatically tailored pectins (50%, 70% and 80% DM) (Cameron et al., 2008) and commercial high methoxyl pectin (HMP; 72% DM). Enzymatically tailored pectins presented known patterns of demethylated HG blocks (Table 1). The commercial HMP showed higher proportion of RG-I than the enzymatically tailored pectin with 70%-DM, as well as a random distribution of demethylated HG blocks.

Since calcium ions are necessary for crosslinking of low methoxyl pectins, the ability of Ca-mediated junction zones to stabilize AA into the edible films made with commercial pectins of low (LMP; 40%) or high (HMP; 72%) DM, at the same Ca^{2+} concentration, was also evaluated and analyzed, after film storage at the same conditions previously indicated.

2. Study of the Influence of the Macromolecular Structure of Pectins in the L-(+)-Ascorbic Acid Stabilization into Edible Films

2.1. Hydrolytic Stability of L-(+)-Ascorbic Acid: Use of Enzymatically Tailored Pectins and Commercial High Methoxyl Pectin for Film Developments

2.1.1. Macromolecular Characteristics of the Pectins Used

Edible films were developed with one of the pectins tailored by Cameron et al. (2008) through demethylation of a model HG (94–97% of GalA, DM = 94%; only 3-6% of galactose -Gal-) with a monocomponent citrus pectinmethyl esterase (PME) at pH 4.5 (to produce a pectin with a DM = 50%) or 7.5 (for pectins with DM of 70% and 80%). The nanostructures of the macromolecules obtained were characterized by the average demethylated block size ($\overline{BS}$) and the average number of demethylated blocks ($\overline{B}_n$) of size n ($\overline{B}_n$) per molecule (Cameron et al., 2008), which are summarized in Table 1. The parent HG presented a molecular weight ≈ 43,280 Da which means a HG stretch with a $GalA_n$ chain of $n = 246$.

Also, films were made with a commercial high methoxyl pectin, herein called “HMP” (GENU™ pectin type B rapid set-Z, CP Kelco, USA), with a molecular weight of ≈ 457,000 Da. Its chemical composition and molecular weight were determined by Pérez et al. (2009)

directly or after pectin dialysis against deionized water, and the results are summarized in Table 2.

Table 2. Molecular characteristics of the commercial high methoxyl pectin (HMP)[1] and dialyzed low methoxyl pectin (LMP)[2] used for film developments

	HMP	LMP
Molecular weight (kDa)[3]	457 (684 – 204)[7]	734.2
Total carbohydrate content (%)[3,4,5]	98 ± 1	99 ± 3
Galacturonic acid (%)[3,4]	82 ± 3	85 ± 2
Degree of methyl esterification[3,4,6]	73 ± 2	40 ± 3
Degree of acetyl esterification[3,4,6]	14 ± 2	6.0 ± 0.6
Neutral sugar composition[3] (mol/100 mol)		
Rhamnose	13.2	39.3
Fucose	0	0
Arabinose	36.6	5.9
Xylose	1.9	0
Manose	0	0
Galactose	44.1	46.9
Glucose	4.2	7.9
Cation content[8] (mol/100 g)		
Na	1.870×10^{-2}	2.41×10^{-4}
K	9.49×10^{-3}	1.33×10^{-5}
Ca	3.75×10^{-3}	6.15×10^{-5}
Mg	1.47×10^{-3}	7.66×10^{-6}

[1] Almost the same chemical composition was showed by the commercial high methoxyl pectin before and after its dialysis through a membrane with a molecular weight cutoff of 12,000 (Sigma, St. Louis, USA) (Pérez et al., 2009). The glucose proportion added for standardization was < 8%, being in part observed as a significant proportion of glucose moles. [2] Composition showed by the low methoxyl pectin after dialysis through a membrane with a molecular weight cutoff of 12,000 (Sigma, St. Louis, USA) (De'Nobili et al., 2011). [3] Chemical assays and molecular weight analysis were performed according to Fissore et al. (2009). [4] Mean and standard deviation (n = 3) are shown. [5] Total carbohydrates are expressed as *g* of galacturonic acid (standard for calibration curve) per 100 *g* of product (Fissore et al., 2010). [6] Degree of methyl esterification was calculated as the molar ratio between the methanol and galacturonic acid contents (Pérez et al., 2009; De'Nobili et al., 2011), whereas the degree of acetylation was calculated as the molar ratio between the acetyl and the total carbohydrate contents. [7] The molecular weight range is also shown. [8] Cation content was determined through atomic absorption spectrometry (Pérez et al., 2009; De'Nobili et al., 2011).

The proportion of glucose added for pectin standardization was low (< 8%) and, hence, the commercial form was directly used for film development. The HMP was characterized by a DM of 72% and a degree of acetylation of 14%. The latter can contribute to hinder the hydrogen bonding and hydrophobic interactions at the junction zones, which lead to the film network development, modifying or modulating the macromolecular aggregation (Oosterveld et al., 2000). As observed in Table 2, the commercial HMP was mainly constituted by GalA (82% per 100 g of pectin), with a 2.3%w/w of Rha, 6.0%w/w of arabinose (Ara) and 8.6% w/w of galactose (Gal). This composition gives a GalA/Rha molar ratio of 32.6 as well as an

(Ara+Gal)/Rha, Ara/Gal and a Gal/Rha molar ratios of 6.0, 0.8 and 3.4, respectively (Pérez et al., 2009). Hence, HMP showed significant content of Gal in the hairy regions of RG-I. From these ratios, it can be inferred that the commercial HMP macromolecule with a molecular weight of 457,000 Da could be structurally constituted by ≈ 49 repeating units of:

(a) at least 28 residues of GalA (≈ 22 of them, methyl esterified), that is, the HG chain,
(b) followed by other two residues of free GalA alternating with 2 residues of L-rhamnose (RG-I backbone), the latter two being laterally substituted by two arabinogalactan I side chains of four Gal and 2 to 3 Ara residues.

It was then estimated that disordered (hairy) regions of RG-I contributed in an 18% to the total carbohydrate content of this HMP, whereas the rest was constituted by GalA residues organized in ordered HG "smooth" regions.

2.2. Film Making Procedure

Films were developed by casting (dehydration in a convection oven for 2.5 h, at 60ºC) from each film making aqueous solution distributed, in a given weight (0.0001 g precision), among horizontally leveled polystyrene plates (55-mm diameter). The compositions are reported in Table 3. For each film making solution, pectin swelling and dissolution in deionized water was achieved by stirring under controlled high speed (1,400 rpm-constant) shear, followed then by heating up to 90ºC at a constant rate of 5.0 ºC/min, after which the other components were added while stirring at 85ºC constant. Glycerol was used as plasticizer at the proportions shown in Table 3.

Table 3. Composition of film making solutions as well as initial L-(+)-ascorbic acid concentration, pH and thickness of films equilibrated at 57.7% of relative humidity and 25.0°C are shown

	50%-DM[1]	70%-DM[1]	80%-DM[1]	HMP	Ca-HMP	Ca-LMP (dialyzed)	Ca-LMP
Pectin (g)	6.00	6.00	6.00	6.00	6.00	5.25	8.00
Glycerol (% w/w) [glycerol·100/(pectin + glycerol)]	36.80	36.80	36.80	36.80	36.80	46.8	46.8
L-(+)-ascorbic acid (g)	0.3000	0.3000	0.3000	0.3000	0.3000	0.3000	0.3000
Potassium sorbate (g)	0.0900	0.0900	0.0900	0.0900	0.0900	0.0900	0.0900
$CaCl_2 \cdot 2H_2O$ (g) (0.026 g Ca/g pectin)						0.4750	0.4750
Enough amount of deionized water added in order to achieve a total weight (g) of	300.00	300.00	300.00	300.00	300.00	300.00	300.00

Finally, a total weight of 300.00 g was obtained by adding the rest of the deionized water to the total solution, followed by homogenization at the condition above mentioned.

Casted films were peeled from the polystyrene plates and stored in darkness into desiccators over a saturated solution of known water activity (a_W°), in order to maintain a constant RH of 57.7% for film equilibration (Greenspan, 1977):

$$a_W = \left.\frac{RH\%}{100}\right|_{25°C} \tag{1}$$

The salt used was NaBr (a_W° = 0.577) at 25.0 °C. Equilibration was assessed by the measurement of the a_W of film samples every day, at 25.0°C, in a Decagon AquaLab (series 3 water activity meter, USA) until attaining the same value of RH%/100, corresponding to the a_W° of the saturated solution used (Pérez et al., 2012). Afterwards, the film pH and thickness were measured as reported by Pérez et al. (2012). Three batches of films (replicates) were prepared as indicated. The film samples obtained from each batch were identified and distributed among the desiccators for storage at the RH (57.7%) studied, in order to involve the influence of the film making in the following determinations.

Storage was carried out *under vacuum* (P = 132 Pa) in order to ensure that AA degradation initiated through the irreversible hydrolysis of its lactone ring as the first (and limiting) reaction step (Figure 1) (León and Rojas, 2007). Hence, the specific influence of water in the AA stability can be followed. The following analyses were performed on each film sample collected from the three batches at each corresponding time of storage, in order to determine the kinetics of AA concentration decay and of browning development. The AA concentration remaining at every time was determined by a spectrofotometric method (Pérez et al., 2012; De'Nobili et al., 2013), and browning was measured as the Yellow Index (YI) (ASTM E-1925, 1988 and 1995):

$$YI(\%) = \frac{100}{Y}\left(1.2769 \cdot X - 1.0592 \cdot Z\right) \tag{2}$$

wherein X, Y and Z are the tristimulus values of the sample measured in a Minolta colorimeter (Minolta CM-508d, Tokyo, Japan) under the D-65 (sodium) illuminant and a 2° angle of observer, using an aperture of 1.5 cm of diameter. A positive value of YI describes the presence and magnitude of a yellow component, whereas a negative value indicates a blue component in the sample. Also, L, a, and b (HunterLab) color parameters were measured, which ranged from L = 0 (black) to L = 100 (white or maximum) for lightness (L); $-a$ (greenness) to $+a$ (redness), and $-b$ (blueness) to $+b$ (yellowness). Standard values considered were those of the white background (Pérez et al., 2012).

The moisture content, reported on dry basis, was determined in six cut films of each formulation after their equilibration at 57.7% RH, by weighing with a precision of 0.0001 g before and after dehydration until constant weight (≈ 22-30 days) into a vacuum oven at 70°C.

The glass transition temperature (T_g), at the onset, was determined in triplicate using a modulated differential scanning calorimetry (MDSC, TA Instruments, USA) from the second scan performed on an equilibrated film sample (10–15 mg) placed into an hermetically sealed

40 μL-aluminium medium pressure pan. An empty pan served as reference. The temperature was brought down to −140ºC (20ºC/min) followed by a 5 min-isotherm at −140ºC. A ± 0.5ºC every 40 s modulation was applied. A ramp was then performed up to 40ºC (10ºC/min), followed by a second decrease in temperature to −140ºC (20ºC/min), and a 5 min-isotherm at −140ºC. Afterwards, a second ramp was performed up to 200ºC (10ºC/min), from which the T_g value was determined and reported from the onset value of the transition (Miao and Roos, 2004). MDSC was periodically calibrated with a sapphire disk, in the full temperature range at which the equipment is usually employed.

Proton nuclear magnetic resonance (^{1}H-NMR) assays to determine water mobility in the film network were performed on the equilibrated film samples using a Bruker Avance II spectrometer operating at 300 MHz for ^{1}H. The probe was a Bruker high power CP-MAS and was used under static conditions. The rotor sizes were 18-mm long with a 4-mm outer diameter. All the experiments were conducted on resonance at room temperature. The Carr Purcell Meiboom Gill (CPMG) pulse sequence was used to measure the transverse magnetization decay (T_2). The free induction decay (FID) was exported to WIN-NMR (Bruker) software where it was Fourier transformed, phase and baseline corrected. Overlapping spectra were deconvoluted by analysis. Peak intensity (area) and linewidth of the deconvoluted peaks were recorded. The exponential curve fitting was performed according to the Maxwell's model (eq. 3) at 50 and 10 μs using a nonlinear fitting program (OriginPro 7.5 SRO, Origin Lab Corporation, Northampton, MA, USA):

$$A(t) = \sum_{i=1}^{n} A_i \cdot \exp\left(-\frac{t}{T_{2_i}}\right) \qquad (3)$$

wherein T_{2i} is the spin-spin relaxation time of the "i" element, component of the relaxation process; $A(t)$ is the peak height at time t of relaxation; A_i is the equilibrium magnetization of the "i" element. An average value of replicates taken from the three batches of films, and the corresponding standard deviation (n = 3) were reported (Pérez et al., 2012).

Tensile assays were performed on rectangular specimens of 6 mm width and 60 mm-total length cut from each film in order to be sure that only uniaxial tension was accomplished along testing, as it was determined through previous assays carried out with specimens of different width. A gage length of 20 mm was used and elongation is reported as the relative deformation to this initial gage length. The force-elongation curves were then recorded at 5 mm/min-constant crosshead speed in an Instron Testing Machine (model 3345, Norwood, MA, USA), equipped with a load cell of 100 N and pneumatic grips with flat rubber coated faces. At least ten specimens with non significant difference in thickness ($p<0.05$) of a given formulation were used, each of them cut from a different film sample. The tensile strength (N/m) was calculated as the ratio between the tensile force (N) and the corresponding elongation or deformation (m) at film failure (Pérez et al., 2012).

2.3. Characteristics of the Pectin Films Developed

Films obtained from pectins tailored by demethylation or from commercial HMP showed similar appearance. They were homogeneous and flexible, transparent and slightly yellow (b = +9.0; YI = 19% ± 2), and with high initial lightness (L = 83.5% ± 0.8).

Table 4. The rate constants[1] of L-(+)-ascorbic acid (AA) hydrolysis (k'_{AA}) and browning development[2] (k_{YI}) determined by film storage at 57.7% constant relative humidity and temperature (25.0 °C) are summarized. The initial AA concentration and yellowness index (YI) values, as well as the mean lightness and pH recorded during the complete storage period are also reported

Pectin film system	Lightness [3] (L; %)	Initial YI (%)	pH [3]	$k'_{AA} \times 10^6$ (min^{-1})	$k_{YI} \times 10^4$ (YI%.min^{-1})
50% DM tailored	80 ± 3	20.0 ± 0.7	3.8 ± 0.4	8.3 ± 0.4	2.2 ± 0.1
70% DM tailored	82 ± 2	18.2 ± 0.9	4.0 ± 0.3	8.1 ± 0.3	2.01 ± 0.08
80% DM tailored	82 ± 2	20 ± 1	4.1 ± 0.4	16.3 ± 0.9	4.0 ± 0.2
HMP	79 ± 3	16 ± 3	3.9 ± 0.4	5.1 ± 0.6	4.6 ± 0.2
Ca-HMP	83 ± 2	17 ± 2	3.6 ± 0.4	0.87 ± 0.08	0.49 ± 0.07
Ca-LMP (dialyzed)	81 ± 1	22 ± 2	3.38 ± 0.03	0.63 ± 0.05	0.30 ± 0.02
Ca-LMP	81 ± 2	22 ± 2	3.4 ± 0.6	0.66 ± 0.08	0.33 ± 0.02

[1] The arithmetic mean and the corresponding standard deviation are shown (n = 3). [2] It was measured as yellowness index (YI) increase against time of storage. [3] Values reported are the mean and standard deviations corresponding to all values recorded during the total storage period.

Table 5. Moisture content and glass transition temperature (T_g) of the pectin films equilibrated at 57.7% of constant relative humidity and 25.0 °C, stored under vacuum

Pectin used for film	Moisture content [1,2] (g water/ 100 g dry mass)	T_g [2] (C)
50%-pectin	29.0 ± 0.5^{A}	-102.43^{A}
70%-pectin	23.4 ± 0.4^{B}	-99.31^{B}
80%-pectin	23.8 ± 0.2^{B}	-100.62^{B}
HMP	24.3 ± 0.4^{B}	-94.20^{C}
Ca-HMP	32.4 ± 0.8^{C}	-88.56^{D}
Ca-LMP	23.6 ± 0.4^{B}	-93.99^{C}

[1] The mean and standard deviation are shown (n = 6). [2] The same capital letter means non significant difference (p<0.05) among values shown in the same column.

The AA concentration initially determined was ≈ 3.00 g AA/100 g of film, which means that a 100% of AA recovery was obtained after casting. Film samples equilibrated rapidly during the first 24 h of storage when a water activity (a_W^o) of 0.577 (eq. 1) was determined on film samples. The average thickness, measured after equilibration at six different locations in each of the ten specimens through a digital micrometer (Mitutoyo, Kawasaki, Japan), was 0.134±0.030 mm. The film pH recorded through a flat surface electrode (Phoenix, AZ, USA) during film storage varied between 3.5 and 4.1 (Table 4) (Pérez et al., 2012). The moisture contents, reported as g water/100 g of film's dry mass, are summarized in Table 5. They were lower than 35% (Kou et al., 2000).

2.4. Stability of L-(+)-Ascorbic Acid to Chemical Hydrolysis in Pectin Films

When stored under vacuum, the AA compartmentalized into each kind of pectin film showed different stability to hydrolysis. Just from zero time of film storage, that is, before equilibration, the AA concentration [$C_{AA}(t)$] decreased significantly ($p<0.05$) with time (t) according to a pseudo-first order kinetic (Figure 2A) (Pérez et al., 2012). This concentration was the ratio between the mass of AA remaining at time t and the initial mass of AA charged into each identified film:

$$C_{AA}(t) = \frac{mass_{AA}(t)}{mass_{AA}{}^{o}{}_{film}} \tag{4}$$

wherein the mass is expressed in grams. At each time t, the AA concentration was measured in three film samples, each one taken from the three batches made, as above mentioned.

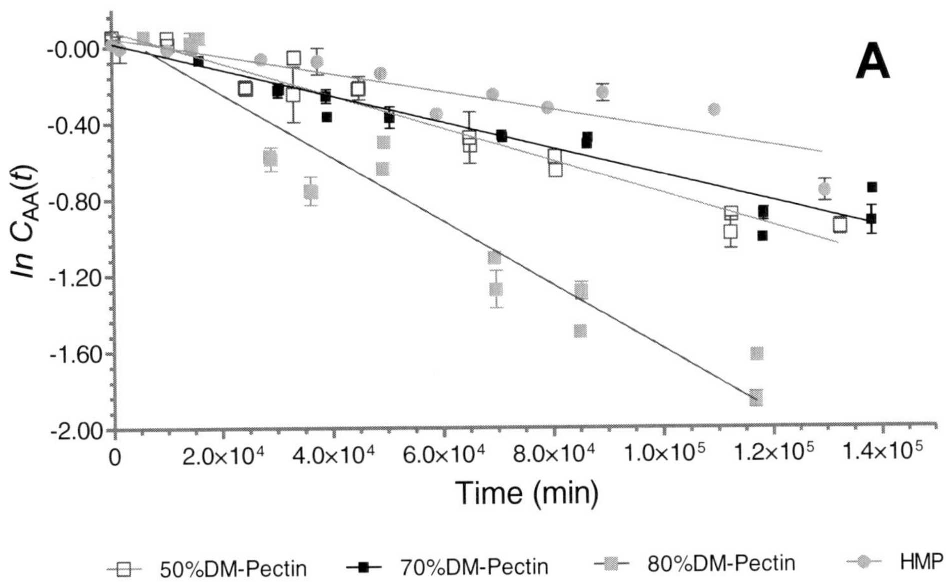

Figure 2. (Continued)

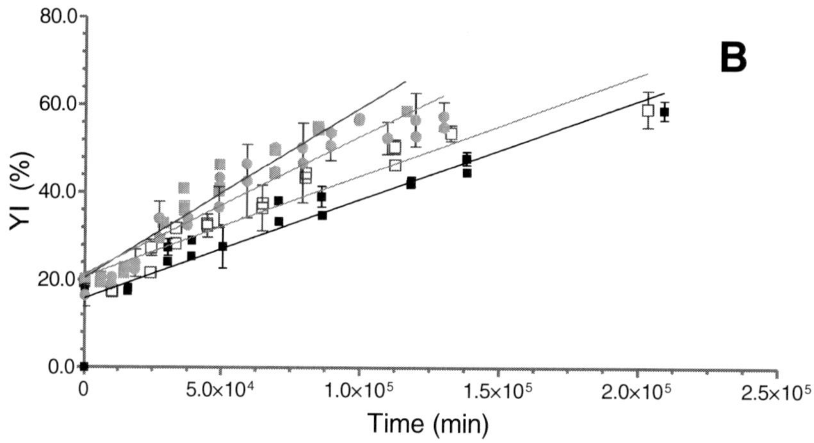

Figure 2. Kinetic curves recorded from the L-(+)-ascorbic acid concentration decay (A) and from the increasing in yellowness index (YI) or browning development (B) determined during storage under vacuum (57.7% constant RH, 25.0ºC) in films made either with 50%, 70% or 80% DM tailored pectin or with commercial high methoxyl pectin (HMP). The linear fitting is shown as a continuous line.

The rate constants (k_{AA}') calculated from the slope obtained by linear regression fitting of the natural logarithm of the $C_{AA}(t)$ values against time t (Figure 2A) are reported in Table 4. As observed in Figure 2A, the network microstructure developed by commercial HMP (DM = 72%) was the most able to stabilize the AA to hydrolysis. Considering the films made with tailored pectins, the highest AA stability was also observed in 70% as well as in 50% methyl esterified pectin networks, whereas the AA showed the lowest stability when supported in 80% methylated pectin films (Figure 2A). Considering the results obtained from tailored pectin films of different DM, the length and distribution of unesterified HG-blocks (Table 1) affected the stability of AA to hydrolysis. As inferred from comparing the AA retention in 70%-methylated tailored pectin film with that in commercial HMP (DM = 72%) film, the network nanostructure developed by these pectins affected the AA stability. It has to be remarked that these pectins have different molecular weight, being that of commercial pectin 10 times higher (Table 2) than the one of the tailored pectin (Table 1). Even with similar DM, it has been reported that different patterns of distribution of the unesterified GalA blocks in pectin macromolecules produce different gel characteristics (Willats et al., 2006; Cameron et al., 2011) and calcium sensitivity (Willats et al., 2006; Cameron et al., 2011; Tanhatan-Nasseri et al., 2011). Figure 3A shows that the pectin films developed from 50%, 70% or 80%-methylated tailored pectins with the block distribution summarized in Table 1, were in general characterized by similar tensile strength values at break with respect to those showed by commercial HMP films (Figure 3A). However, the tailored pectin films presented the lowest relative elongation at failure (Figure 3B). Although with the same proportion of glycerol and equilibrated at the same RH, all tailored pectin films were then more brittle than the films made with commercial HMP, as previously observed by Willats et al. (2006) with respect to gels assayed under compression. The citrus PME used for tailoring of pectins usually have a blockwise action pattern, producing long stretches (blocks) of unesterified HG with different lengths as well as characteristic distribution patterns into the macromolecules, like those reported in Table 1 (Willats et al., 2006; Cameron et al., 2008). Instead, chemical treatment with acids or a base produces a random distribution of methyl-esters. Since commercial pectins are generally obtained by acidic treatment at high temperature (Voragen et al., 2009), they are randomly methyl esterified like occurred with the commercial HMP (72% DM) herein used for film development (Willats et al., 2006). Consequently, unesterified HG-blocks are randomly distributed and, also, they are of random length.

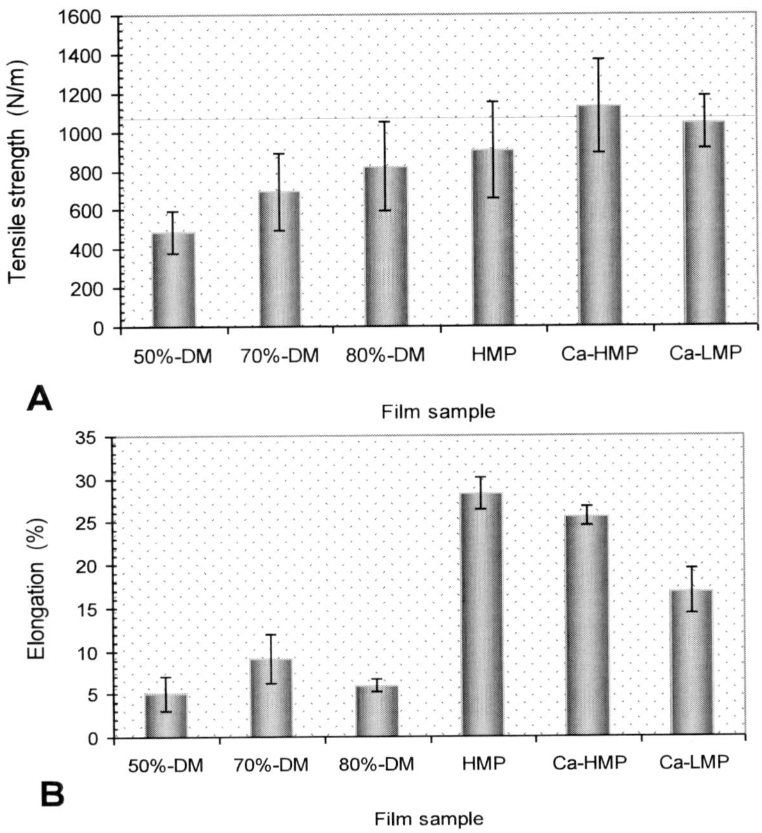

Figure 3. Tensile strength (A) and relative elongation (B) at break obtained from the force-elongation curves recorded from films developed from tailored 50% DM, 70% DM and 80% DM pectin or from commercial high methoxyl pectin films (HMP). The same results are reported for films developed in the presence of calcium ions either with HMP (Ca-HMP) or with low methoxyl pectin (Ca-LMP). Films were stored at 57.7% relative humidity and 25.0 °C under vacuum. Error bars correspond to the corresponding standard deviation (n = 10). Films used for tensile assays were those selected with film thickness values in the 0.110 ± 0.008 mm range.

The critical effect of the methylester distribution pattern was then evidenced by the contrasting values of tensile elongation at break showed by tailored pectin films and commercial HMP film (Figure 3B). These patterns are also associated with different degrees of water retention (Zsivanovits et al., 2004). These considerations allowed thinking that the polymeric networks developed by each kind of these pectins were essentially different with respect to water immobilization.

The water *available* in the film network is the nucleophile responsible for the irreversible opening of the lactone ring of the AA molecule (Figure 1). This irreversible opening constitutes the first reaction and limiting step of the AA hydrolysis in the absence of air (Kurata and Sakurai, 1967). León and Rojas (2007) found a dependence of the pseudo-first order rate constant of AA hydrolysis (k_{AA}') on the RH of film storage (33.3%; 57.7% and 75.2% RH) at 25.0°C. In the intrinsic kinetics, where physical processes (diffusion, convection) are not coupled (Smith, 1986), the rate constant of a reaction is only dependent on temperature (Levine, 2004). As the storage temperature of films was constant, it was then proposed that a water molecule may be involved through a bimolecular nucleophylic substitution (S_N2) mechanism at the first step of the chemical destruction of AA under anaerobic storage (Figure 1). This reaction leads to hydrolysis of the AA-lactone ring to irreversibly render KGA through an acid catalyzed reaction (Kurata and Sakurai, 1967). The attack of the AA (lactone) ring from behind by a water molecule (S_N2 mechanism) was also suggested to be slow enough in order to determine the total reaction rate and, hence, the kinetic equation. This proposed S_N2 mechanism explains a second order kinetics (Morrison and Boyd, 1990) for AA hydrolysis and, hence, a second order rate constant (k) of AA destruction can be written. Hence, the pseudo-first order rate constant (k_{AA}') herein determined from the only measurement of the AA concentration into each film [$C_{AA}(t)$] (eq. 4) against time (Table 2) included the water concentration ($k_{AA}' = k \cdot C_{water}$) in the film network, so the water *available* for reactions:

$$r_{AA} = -\frac{1}{\nu_{AA}}\frac{dC_{AA}}{dt} = k \cdot C_{water} \cdot C_{AA}(t) = k'_{AA} \cdot C_{AA}(t) \qquad (5)$$

wherein ν_{AA} is the stoichiometric coefficient for AA hydrolytic degradation reaction (here $\nu_{AA} = 1$) (Figure 1); r_{AA} is the AA-reaction rate/unit volume at a constant temperature; $C_{AA}(t)$ is the concentration of AA at the time t of storage (eq. 4); C_{water} is the concentration of water *available* in the polymeric network; k is the rate constant of the second order kinetics for the AA hydrolytic reaction (Figure 1); k'_{AA} is the rate constant of the pseudo first order kinetics for the AA hydrolytic reaction.

2.5. Water Dynamic in the Film Networks

The dynamic aspects of water interactions in the pectin films were hence studied in order to determine differences in the ability of the polymeric matrices to retain or immobilize water. For low-moisture biopolymer systems (water content < 35%), the slowing of water motion has been reported to be associated with bound water (i.e., immobile water) arising from hydrogen bonding (Kou et al., 2000). Moisture levels lower than 35% were determined for all film systems assayed in the present work (Table 5). Short-range motions or molecular (water) relaxations investigated by ^{1}H-NMR (Vittadini and Chinachoti, 2003) led to observe that the magnetization decay in the xy-plane fitted with two spin-spin time constants (T_{2a} and T_{2b}) (eq. 3), indicating the existence of multi-relaxation rate behavior. The T_{2a} and T_{2b} parameters may be associated with two fractions of water, which have different relaxation rates (T_{2i}^{-1}) or mobility degrees (Chen et al., 1997; Kerr and Wicker, 2000). One of the water fractions showed transverse relaxation values of T_{2a} in the order of magnitude of 10^{-4} s, ten times lower than the T_{2b} relaxation times, corresponding to the other water population. Pure water has only a T_2 characteristic time and with a value of ≈ 1–2 s (Chen et al., 1997). Figure 4A shows the increase of the pseudo-first order rate constants of AA hydrolysis (k_{AA}') with the water mobility (T_{2b} as an example) in the film network. It can be suggested that differences in the structure of pectin macromolecules produced different retention of water molecules by adsorption in the film network developed and, hence, different stability of the compartmentalized AA. Different degrees of water retention were previously determined in pectins by Zsivanovits et al. (2004). Tailored pectins were of similar molar mass. The 80%-methylated pectin produced the lowest water retention. Probably, concentration of methyl esterified carboxyl groups in blocks of the HG chains, as thought from considering the proportions of demethylated block size ($\overline{BS}$) and the average number of demethylated blocks ($\overline{B}_n$) present (Table 1), led to the prevalence of hydrophobic interactions between water and pectins over hydrogen bonds in most regions of the 80%-methylated pectin. Oakenfull and Scott (1984) indicated that mainly hydrogen bonds though necessarily in conjunction with hydrophobic interactions, contribute to the free energy of gelation or formation of junction zones in high methoxyl pectins. On the other hand, tailored 70%-methylated pectin whose $\overline{BS}$ and $\overline{B}_n$ values were not very different from those of 50%-methylated pectin (Table 1), stabilized AA similarly to the 50%-methylated pectin film. As observed in Figure 4, their macromolecular structures produced similar water mobility in their respective film networks.

The commercial HMP is characterized by almost the same DM as tailored 70%-methylated pectin but with higher molar mass (Table 2 against Table 1). Water presented the lowest mobility in the film network developed by the commercial pectin (HMP) with respect to tailored (70% DM) pectin films (Figure 4A). This may be ascribed to the fact that it is randomly demethylated (Willats et al., 2006) and presents some RG-I cores (amorphous, disordered, regions). Hence, the HMP network formed by alternating hydrophobic and hydrophilic channels embedded among the pectin helices provides an interesting motif in the formation of junction zones in this kind of pectin, being the hydrophilic channels filled with water molecules hydrogen bonded to the surrounding polymer chains (Chandrasekaran, 1998; Pérez et al., 2009).

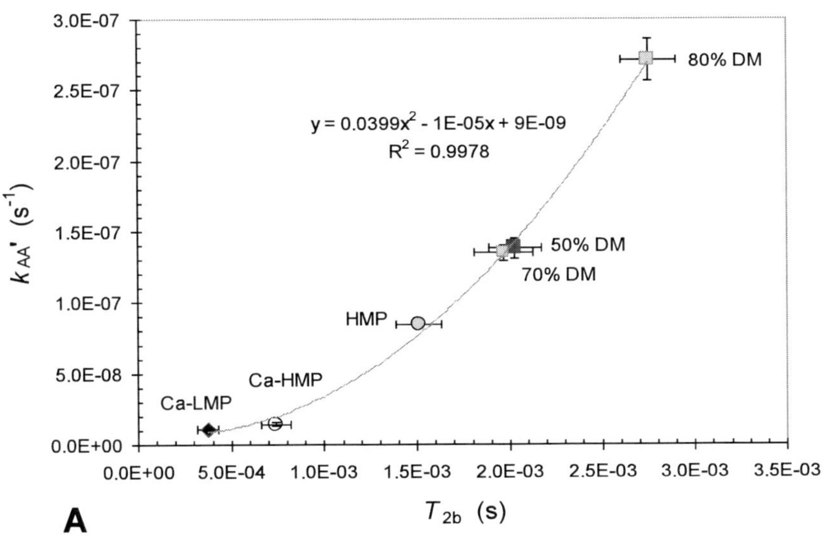

Figure 4. (Continued)

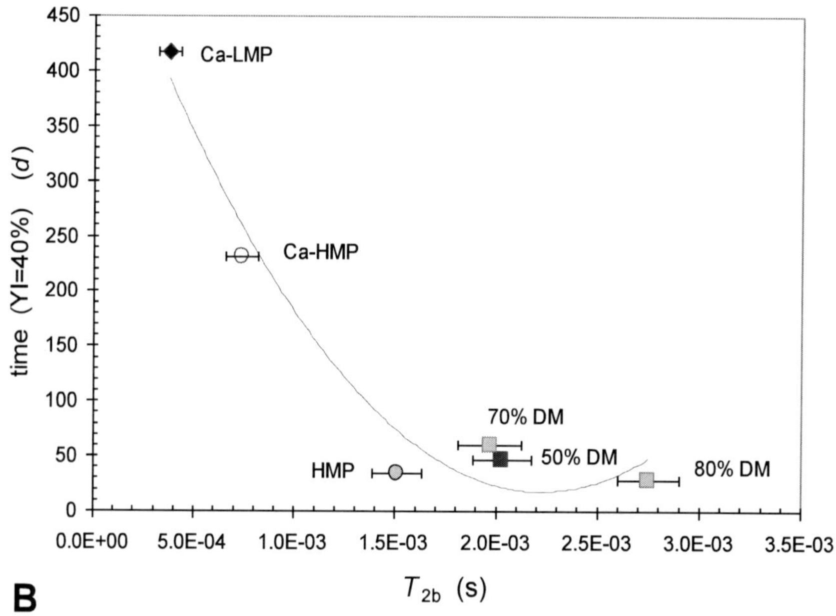

Figure 4. Rate constants for L-(+)-ascorbic acid hydrolysis (k_{AA}') (A) and the time necessary to reach a yellowness index (YI) value of 40% (B) are plotted against the spin-spin relaxation time (T_{2b}) for films made with tailored 50%DM, 70%DM and 80%DM pectin as well as for film made with commercial high methoxyl (72% DM) pectin (HMP). Films developed in the presence of calcium ions with HMP (Ca-HMP) or with low methoxyl pectin (Ca-LMP) are also shown. Films were stored at 57.7% relative humidity and 25.0 °C under vacuum. Error bars correspond to the standard deviations (n = 3).

2.6. Macromolecular Mobility

The macromolecular relaxation in the film networks was also analyzed by determination of the T_g film values through calorymetry. As observed in Table 5, the glass transition temperatures were well below the 25.0 ºC-storage temperature, indicating an amorphous-rubbery state as a consequence of the film plasticization by glycerol. As observed in previous works (León and Rojas, 2007; Pérez et al., 2009), water also contributed to plasticization since T_g values decreased in about 10 ºC as RH of film storage increased from 33.3 to 57.7% and 75.2%. Pectin films did not essentially differ from one another in their macromolecular mobility (Table 5), though some higher plasticizing effect was observed on tailored pectin films as they showed some lower T_g values (≈ −100ºC) than that observed in HMP film (≈ −94.20ºC). However, this observation was not accompanied by higher elongation at tailored pectin film failure (Figure 3B) as expected for more plasticized films. It can be suggested that the lowest T_g values of tailored pectin films may be ascribed to some higher macromolecular relaxation or mobility, so some higher plasticizing effect produced by the same level of glycerol also utilized in commercial HMP films. Water coming from the storage environment at constant 57.7% RH can also associate by hydrogen bonding with the glycerol interacting with pectin macromolecules, contributing to a down shift of T_g. It was previously established by FTIR analysis that water can only plasticize the polysaccharide networks when glycerol is formerly present in the HMP film network (Pérez et al., 2009). As observed from the T_g values (Table 5), the level of plasticization (macromolecular relaxation) in tailored pectin films was not very different to that observed in commercial HMP film. However, water mobility was higher in tailored pectin film networks and it correlated with the faster AA hydrolysis. Beyond the similar level of plasticization (T_g) observed for all of the films developed, the brittleness of tailored pectin films (Figure 3B), ascribed to the non random pattern of unesterified HG block distribution, was not overcome by the plasticization level.

Beyond the T_g, the DSC scans obtained from all film networks studied (36.8% w/w of glycerol content) did not show any other phenomena. Endotherm peaks at 0ºC and/or −38ºC, which respectively correspond to freezable water and freezable-bound water (Hatakeyama and Hatakeyama, 1998) were not observed in the scans performed at 10ºC/min between −140 ºC and +120 ºC. Hence, it can be mentioned that water could be sufficiently adsorbed and retained by the polymeric networks developed.

2.7. Browning

Irreversible hydrolysis of the lactone ring that constitutes the AA molecule leads to KGA through the acid catalyzed reaction (Kurata and Sakurai, 1967), which was suggested to occur through an S_N2 mechanism (Figure 1) (León and Rojas, 2007). Once KGA develops, this molecule constituted by α,β-unsaturated carbonyl and β-hydroxyl carboxyl reactive groups suffers successive transformations which involve dehydrations and decarboxylations producing different browning active compounds (Kurata and Sakurai, 1967). These chemical events are favored in *solid-like* systems like films. In these kind of systems, water is a limiting reactive because it is not available as solvent. Therefore, nucleophiles are highly reactive because they are not solvated by the water molecules like occurs in solution (León and Rojas, 2007). Browning development in the pectin films measured through the percentage of yellowness index (YI %) (eq. 2) showed pseudo-zero order kinetics (Figure 2B) and the calculated rate constants are reported in Table 4. As observed in Figure 2B, some higher rate of browning development is shown by 80% DM tailored pectin film.

On the other hand, Pérez et al. (2009) determined that the rate constants of browning development also increased with the RH (33.3-75.2%) of film storage. The time needed to reach a YI value of 40% (time $_{YI=\ 40\%}$) in the films was then calculated from the kinetic equations and plotted versus T_{2b} (Figure 4B). The values of time needed to reach a YI value of 40%, determined from tailored pectin films and commercial HMP film, were not sensibly dependent on the water mobility presented as T_{2b}, neither on macromolecular relaxation (T_g values; Table 5).

3. Hydrolytic Stability of L-(+)-Ascorbic Acid: Effect of Calcium Crosslinking in Commercial High Methoxyl and Low Methoxyl Pectin Films

3.1. Macromolecular Characteristics of the Pectins Used

As ionic polysaccharides, pectins show increasing ability to form gels in presence of divalent ions (e.g. Ca^{2+}) when their DM decreases. This is associated to their biological functions and it is useful for technological applications (Braccini and Pérez, 2001). Pectin macromolecules characterized by the regular occurrence of blocks constituted by at least 10 to 14 R-D-galacturonate residues generate ordered templates for polymer chain associations involved in physical gels (Braccini and Pérez, 2001; Fissore et al., 2013). On the other hand, calcium-sensitive pectins also have been described in which HMP can gel in the presence of calcium without the addition of sucrose as long as blocks of deesterified pectin are present. They retain more water than LMP (Joye et al., 2000; Willats et al., 2001; Hotchkiss et al., 2002). The effect of Ca^{2+}-crosslinking on the kinetics of AA degradation and subsequent browning was then studied in the edible film developed from the commercial HMP (DM = 72%) previously analyzed (now called Ca-HMP system) as well as in *dialyzed* commercial LMP (DM = 40%) films (Ca-LMP system), at the same calcium level. Commercial LMP is standardized with sucrose up to a solid content between 30 and 50% in order to modify the gelation temperature as well as specific gel properties like elasticity and syneresis. Chemical characteristics of the dialyzed LMP are reported in Table 2. Also, films were directly developed from the commercial LMP without previous dialysis. As herein determined, the sucrose content of the commercial or food grade form was 32.0% w/w. The molecular weight mean value (≈ 734 kDa) of the LMP macromolecules was higher than the one of the HMP (≈ 457 kDa), though both were of the same order of magnitude. Lower degree of acetylation was showed by the LMP. A GalA to Rha [GalA/Rha] molar ratio of 13 was calculated from the composition summarized in Table 2 for the LMP used for film development, which corresponded to ≈ 7.5% (molar ratio) of disorder (amorphous) regions of interspersed RG-I domains (Pérez et al., 2003), which was lower than the 18% proportion of disordered regions calculated for the commercial HMP. A galactose (Gal) to Rha [Gal/Rha] molar ratio of 1.2 and a Gal+Ara to Rha [(Gal+Ara)/Rha] molar ratio of 1.35 were also calculated from Table 2. They indicated that the chemical extraction of this commercial pectin produced hydrolysis or peeling of the hairy (RG-I) regions (Fry, 1986; Fissore et al., 2010) of the original pectin macromolecules. The dialyzed LMP used carried ions like Mg^{2+}, Na^{+}, K^{+} and 6.15×10^{-5} mol Ca^{2+}/g of pectin (Table 2). A Ca^{2+} amount of ≈ 450 moles of Ca^{2+} per mol of pectin polymer was used for LMP and HMP. By considering the pectin composition and molecular weight reported in Table 2, as well as the molar ratios between neutral sugars and GalA indicated above, it can then be calculated that ≈ 23% of the carboxylate groups of a LMP macromolecule (DM= 40%) can be electrostatically cross-linked by Ca^{2+} with other facing antiparallel chain.

3.2. Film Making Procedure

The commercial LMP used (GENU™ pectin type LM-12 CG, CPKelco, USA) was exhaustively dialyzed against deionized water for 48 h and finally freeze-dried. The film making solutions were made as previously indicated (section 1.2).

An aliquot of 5.25 g of the dialyzed pectin or 8.00 g of the commercial LMP was slowly poured in deionized water while stirring was being performed at controlled high speed (1,400 rpm-constant). Afterwards, heating up to 90ºC at a constant rate of 5.0 ºC/min was carried out. The glycerol proportions used to constitute Ca-HMP and Ca-LMP films were 36.80 and 46.80%w/w [glycerol·100/(pectin+glycerol)], respectively, which were the lowest one needed to obtain films with adequate and similar flexibility. When the rest of the components were added, calcium was aggregated as $CaCl_2.2H_2O$ pre-dissolved in 10 mL of deionized water, while the solution was maintained at ≈ 85°C. Finally, the total weight of the solution was led to 300.00 g with enough water and homogenized, placed under vacuum for some seconds to eliminate air bubbles, after which it was distributed among the horizontally leveled plates as previously described. After casting and peeling from plates, films of both systems made with commercial HMP and Ca^{2+} (called Ca-HMP) or with LMP and Ca^{2+} (called Ca-LMP) were also stored in darkness and under vacuum at the same RH (57.7%) and temperature (25.0°C) as the film systems previously studied.

3.3. Characteristics of the Pectin Films Developed

Homogeneous Ca-HMP and Ca-LMP films were obtained by casting, being transparent and slightly yellow (b = +8.4 to +10.0; YI = 17-22% ± 1%), with high initial lightness (L = 83%±1). A 100% of AA recovery was also obtained in Ca-HMP and Ca-LMP films after the casting process. Film samples reached the equilibrium with the environment (57.7% RH) at the fourth day of storage. The average thickness was then determined, being of 0.094±0.023 mm for Ca-HMP and of 0.104±0.004 mm for Ca-LMP, whereas the pH changed into a value of 3.6±0.4 and 3.4±0.6, respectively, during the complete storage period (Table 4). The moisture content of films is summarized in Table 5. They were lower than 35% (Kou et al., 2000).

2.4. Stability of L-(+)-Ascorbic Acid to Chemical Hydrolysis in Pectin Films

The AA concentration [$C_{AA}(t)$] (Eq. 4) decreased significantly (p<0.05) with time (t) according to a pseudo-first order kinetic (Eq. 5) from zero time of film storage for both calcium added systems (Figure 5A). Higher stabilization to hydrolysis of the compartmentalized AA was observed in Ca-HMP films than in the HMP system above studied.

The highest AA stabilization was observed in both Ca-LMP film systems, made either with dialyzed or commercial LMP. Therefore, only the AA kinetic in the film made with the dialyzed LMP is reported in Figure 5A. The rate constants (k_{AA}') calculated from each slope obtained by fitting of experimental data (Figure 5A) are reported in Table 4.

3.5. Water Dynamic in the Film Networks

Short-range motions or molecular (water) relaxations were investigated by ^{1}H NMR (Vittadini and Chinachotti, 2003). It was also observed that the magnetization decay in the xy-plane fitted exponentially through two spin-spin time constants (T_{2a} and T_{2b}) (Eq. 3), indicating the existence of multirelaxation rate behavior. As previously indicated, these parameters may be associated with two fractions of water with different relaxation rates or mobility (Kerr and Wicker, 2000). LMP film crosslinked by calcium ions was the best network to stabilize AA to hydrolysis (showed the lowest k_{AA}'). This coincided with the highest water immobilization, that is, the lowest T_{2b} value (Figure 4A). On the other hand, the presence of calcium ions in HMP films (Ca-HMP system) also stabilized AA (Figure 5A) and immobilized water (Figure 4A) in the polymeric matrix better that the other film systems made without Ca^{2+} (HMP and tailored pectin films; Figure 2A; Figure 4A).

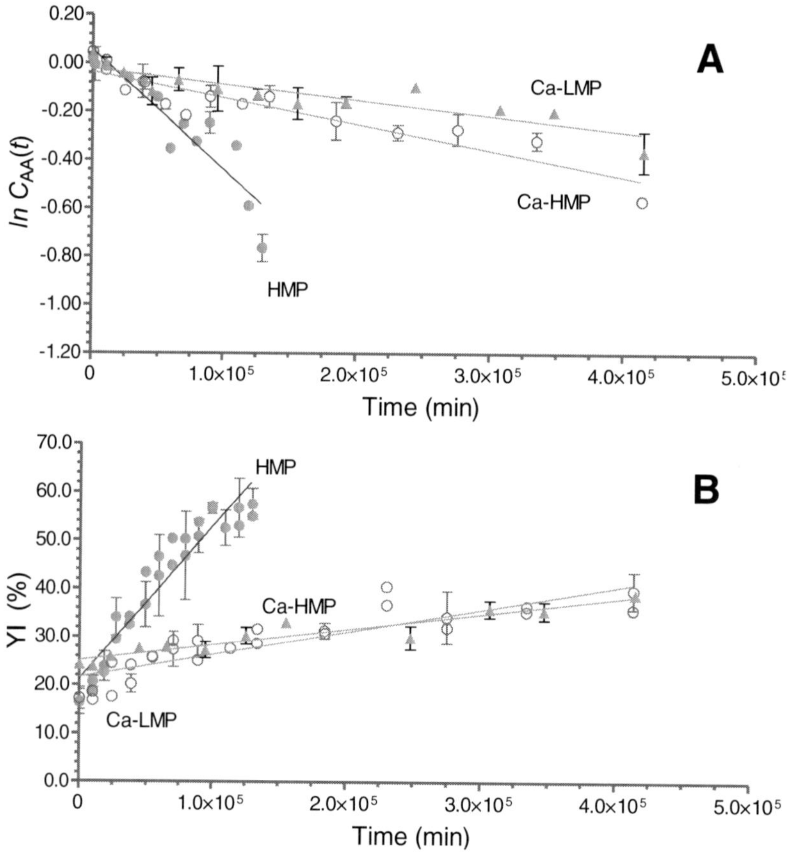

Figure 5. Kinetic curves recorded from the L-(+)-ascorbic acid concentration decay (A) and from the increasing in the yellowness index (YI) or browning development (B) determined during storage at 57.7% constant relative humidity and 25.0ºC under vacuum of films made in the presence of calcium ion either with commercial high methoxyl pectin (Ca-HMP) or with dialyzed low methoxyl pectin (Ca-LMP). The kinetics observed in films made with commercial pectin without calcium (HMP system) is also shown for comparison.

Calcium ions could contribute to crosslink HMP neighboring chains through electrostatic interactions only if blocks constituted by at least 10-14 unesterified GalA residues were present. This fact seemed to be verified in the Ca-HMP films through tensile assays: though only a trend to some higher tensile strength at failure was shown by Ca-HMP films than by HMP-films (Figure 3A), a significantly lower relative elongation (Figure 3B) was determined in Ca-HMP films.

Moreover, Ca-HMP films showed the highest T_g value (−88.56ºC) (Table 3) which is directly associated with some lower free volume and macromolecular mobility than in the other film networks. Hence, the commercial (randomly demethylated) HMP used for film constitution showed some calcium sensitivity (Willats et al., 2006). On the other hand, low methoxyl pectins have enough blocks of unesterified GalA residues distributed along the macromolecule chains in order to form calcium crosslinked junction zones, which produced gelling of the film forming solution before casting. Ca-LMP films made with the dialyzed pectin showed a non significant trend to higher tensile strength at fracture than in the case of Ca-HMP and HMP films (Figure 3A).

However, a significantly lower relative elongation characterized to Ca-LMP films, in comparison to Ca-HMP and HMP films (Figure 3B), even though the Ca-LMP films are plasticized by a higher proportion of glycerol (46.80% w/w). This behavior evidenced the magnitude of the calcium-crosslinking that developed the Ca-LMP network. Since antiparallel LMP chains, shifted in 1.7 Å, are crosslinked by calcium ions, the junction zones formed are not true “egg boxes” as in the case of alginates (Braccini and Pérez, 2001).

In calcium pectate gels, the junction zones involve three components: uronate chains, calcium ions and water (Braccini and Pérez, 2001), which corresponds to nonfreezable bound water (Ping et al., 2001). Water oxygen atoms may complete the coordination sphere of Ca^{2+} at the pectate junction zone (Braccini and Pérez, 2001). Adequate distribution patterns of demethylated carboxyl groups are also necessary to constitute Ca-junctions (Willats et al., 2001).

These patterns are also associated with different degrees of water retention (Zsivanovits et al., 2004). Hence, presence of calcium ions in the Ca-HMP and mainly in Ca-LMP film systems involved lower availability of water into the film networks (Figure 4A), which precluded AA from hydrolysis (Figure 5A) at a high enough level to also sensibly decrease the subsequent browning development (Figure 5B).

3.6. Macromolecular Mobility

As observed in Table 5, the T_g of Ca-HMP and Ca-LMP films were well below 25.0ºC, which was the storage temperature used. This indicated an amorphous-rubbery state for these systems due to their plasticization by glycerol. As occurred for Ca-HMP film, De’Nobili et al. (2011) also determined the contribution of water to the polymer plasticization: T_g values of this film system decreased in about 10 ºC as the RH of film storage increased from 33.3 to 57.7% and 75.2%. Ca-HMP film showed the lowest macromolecular mobility (the highest T_g), followed by the T_g values of Ca-LMP film. The latter showed increasing polymer mobility (Table 5) as a consequence of its higher glycerol proportion (46.80%w/w, [glycerol·100/(pectin+glycerol)]). Water coming from the storage environment at constant 57.7% RH can also associate by hydrogen bonding with the glycerol interacting with pectin macromolecules, contributing to decrease the T_g.

Other phenomena like endotherms at 0ºC and/or −38ºC, which respectively correspond to freezable water and freezable-bound water (Hatakeyama and Hatakeyama, 1998) were not observed in the DSC scans performed at 10ºC/min. Hence, it can be mentioned that water seems to be sufficiently adsorbed by the polymeric networks developed.

3.7. Browning

Browning development in the Ca-LMP and Ca-HMP films measured through the percentage of yellowness index (YI%) (eq. 2) also fitted through a pseudo-zero order kinetic (Figure 5B) and the rate constants calculated from the respective slope are reported in Table 4. As observed in Figure 5B, the rate of browning development is slowed down by the calcium presence in the LMP and HMP film networks. The same behavior was observed with respect to the AA stability to hydrolysis (Figure 5A). Similar browning rates were observed in Ca-LMP films stored at 57.7% RH, made either with dialyzed and non-dialyzed pectin. De'Nobili et al. (2011) found a significant (p<0.001) lower browning rate in Ca-LMP made with the dialyzed pectin when films were stored at 75.2% RH. The lowest rates of browning development were found for films made with dialyzed LMP and Ca^{2+}, whereas the presence of sucrose in films made with commercial LMP produced higher browning rate than in the former system (Table 4).

The time needed to reach a YI value of 40% ($t_{40\%=YI}$) in the films was also plotted versus T_{2b} for Ca-LMP and Ca-HMP films (Figure 4B). Contrary to that observed for the rest of the systems above studied, the level of browning (40% of YI) reached in the films crosslinked by calcium ions was sensibly dependent on the water mobility (T_{2b}) into the network. However, it seemed not to be dependent on the macromolecular relaxation (T_g values; Table 5).

Conclusion

AA was better stabilized in the film networks made with the randomly demethylated (commercial) HMP than in the tailored 70%DM-pectin films stored under vacuum. It can be suggested that a random distribution of demethylated blocks in the HG semiflexible chains in addition to the presence of some disordered (amorphous) regions of RG-I may produce better immobilization of water by physical adsorption in the pectin macromolecules than more rigid networks like those developed from the enzymatically tailored pectin macromolecules. LMP and HMP film networks reinforced by calcium junction zones led to lower rates of AA degradation and browning development as a consequence of the highest water immobilization. It can be concluded that the ability of the polymeric network to immobilize water seem to be an important factor to consider in order to succeed in retaining AA into film networks in view of preserving vitamin C under the AA form as well as for potential antioxidant activity localized at pharmaceutical and food interfaces.

Acknowledgments

The isolation and characterization of the PME and tailoring of demethylated pectins used in this work were supported by the U.S. Department of Agriculture, Agricultural Research Service, Project Numbers: 6618-41000-016-00, 6618-41000-015-00 and National Institute of Food and Agriculture, National Research Initiative, Award Number 2009-35503-05205. The rest of the work was funded by the University of Buenos Aires (UBA), National Scientific and Technical Research Council of Argentina (CONICET) and National Agency for the Promotion of Science and Technology (ANPCyT) of Argentina.

REFERENCES

ASTM. (1995). Annual book for ASTM standards. *Designation D1925, standard test method for yellowness index of plastics*. Philadelphia: America Society for Testing and Materials.

Bourtoom, T. (2008). Edible films and coatings: characteristics and properties. *International Food Research Journal,* 15(3), 237–248.

Braccini, I., & Pérez, S. (2001). Molecular basis of Ca^{2+}-induced gelation in alginates and pectins: the egg box model revisited. *Biomacromolecules*, 2, 1089–1096.

Cameron, R. G.; Luzio, G. A.; Goodner, K., & Williams, M. A. K. (2008). Demethylation of a model homogalacturonan with a salt-independent pectin methylesterase from citrus: I. Effect of pH on demethylated block size, block number and enzyme mode of action. *Carbohydrate Polymers,* 71(2), 287-299.

Cameron, R. G.; Luzio, G. A.; Vasu, P.; Savary, B. J., & Williams, M. A. K. (2011). Enzymatic modification of a model homogalacturonan with the thermally tolerant pectin methylesterase from citrus: 1. Nanostructural characterization, enzyme mode of action, and effect of pH. *Journal of Agricultural and Food Chemistry,* 59, 2717–2724.

Carpita, N. C., & Gibeaut, D. M. (1993). Structural models of primary cell walls in flowering plants: consistency of molecular structure with the physical properties of the walls during growth. *Plant Journal,* 3, 1–30.

Chandrasekaran, R. (1998). X-ray diffraction of food polysaccharides. *Advance Food Nutrition Research*, 42, 131-210 (pp. 131-137; 167-172; 185-191).

Chen, P. L.; Long, Z.; Ruan, R., & Labuza, T. P. (1997). *Nuclear resonance studies of water mobility in bread during storage*. Lebensm.-Wiss. und Technol. (LWT), 30, 178–183.

Daas, P. J. H.; Meyer-Hansen, K.; Schols, H. A.; De Ruiter, G. A., & Voragen, A. G. J. (1999). Investigation of the non-esterified galacturonic acid distribution in pectin with endopolygalacturonase. *Carbohydrate Research,* 318, 135–145.

De'Nobili, M. D.; Pérez, C. D.; Navarro, D. A.; Stortz, C. A., & Rojas, A. M. (2011). Hydrolytic stability of L-(+)-ascorbic acid in low methoxyl pectin films with potential antioxidant activity at food interfaces. *Food and Bioprocess Technology,* 6, 186–197.

De'Nobili, M. D.; Curto, L. M.; Delfino, J. M.; Soria, M.; Fissore, E. N., & Rojas, A. M. (2013). Performance of alginate films for retention of L-(+)-ascorbic acid. *International Journal of Pharmaceutics*, 450, 95–103.

Deng, P.; Xu, Z., & Li, J. (2013). Simultaneous determination of ascorbic acid and rutin in pharmaceutical preparations with electrochemical method based on multi-walled carbon nanotubes–chitosan composite film modified electrode. *Journal of Pharmaceutical and Biomedical Analysis,* 76, 234–242.

Durschlag, M. E.; Charlotte, N. C.; Kehoe, G. S., & Glendale, A. Z. (2007). *Edible film for transmucosal delivery of nutritional supplements*. United States Patent Application Publication, Pub. No.: US 2007/0087036 A1, Pub. date: April 19.

Fissore, E. N.; Ponce, N. M. A.; Stortz, C. A.; Rojas, A. M., & Gerschenson, L. N. (2007). Characterization of fiber obtained from pumpkin (*Cucumis moschata*, Duch.) mesocarp through enzymatic treatment. *Food Science and Technology International,* 16(1), 1–7.

Fissore, E. N.; Ponce, N. M. A.; Wider, E. A.; Stortz, C. A.; Gerschenson, L. N., & Rojas, A. M. (2009). Commercial cell wall hydrolytic enzymes for producing pectin-enriched products from butternut (*Cucurbita moschata*, Duchesne ex Poiret). *Journal of Food Engineering*, 93 (3), 293–301.

Fissore, E. N.; Ponce, N. M. A.; de Escalada Pla, M. F.; Stortz, C. A.; Rojas, A. M., & Gerschenson, L. N. (2010). Characterization of acid-extracted pectin-enriched products obtained from red beet (*Beta vulgaris* L. var. conditiva) and butternut (*Cucurbita moschata* Duch ex Poiret). *Journal of Agricultural and Food Chemistry,* 58, 3793–3800.

Fissore, E. N.; Ponce, N. M. A.; Matkovic, L.; Stortz, C. A.; Rojas, A. M., & Gerschenson, L. N. (2011). Isolation of pectin enriched products from red beet (*Beta vulgaris* L. var. conditiva) wastes: composition and functional properties. *Food Science and Technology International*, 17(6), 517–527.

Fissore, E. N.; Rojas, A. M., & Gerschenson, L. N. (2012). Rheological performance of pectin-enriched products isolated from red beet (*Beta vulgaris* L. var. conditiva) through alkaline and enzymatic treatments. *Food Hydrocolloids*, 26, 249–260.

Fissore, E. N.; Rojas, A. M.; Gerschenson, L. N., & Williams, P. A. (2013). Butternut and beetroot pectins: characterization and functional properties. *Food Hydrocolloids*, 31, 172–182.

Greenspan, L. (1977). Humidity fixed points of binary saturated aqueous solutions. *Journal of Research of the National Bureau of Standards Section A-Physics and Chemistry*, 81 A(1), 89–96.

Han, J. H. (2005). New technologies in food packaging: overview. In J. H. Han (Ed.), *Innovations in Food Packaging* (Chapter 1, pp. 9). Elsevier Academic Press, Oxford, U.K.

Hatakeyama, H., & Hatakeyama, T. (1998). Interaction between water and hydrophilic polymers. *Thermochimica Acta*, 308, 3–22.

Hotchkiss, A. T.; Savary, B. J.; Cameron, R. G.; Chau, H. K.; Brouillette, J.; Luzio, G. A., & Fishman, M. L. (2002). Enzymatic modification of pectin to increase its calcium sensitivity while preserving its molecular weight. *Journal of Agricultural and Food Chemistry*, 50, 2931–2937.

Joel, A.; Indictor, N.; Hanlan, J. F., & Baer, N. S. (1972). The measurement and significance of pH in paper conservation, *Bulletin of the American Group. Int. Inst. Conserv. Hist. Art. Works,* 12(2), 119–125.

Kerr, W. L., & Wicker, L. (2000). NMR proton relaxation measurements of water associated with high methoxy and low methoxy pectins. *Carbohydrate Polymers*, 42(2), 133–141.

Khwaldia, K.; Perez, C.; Banon, S.; Desobry, S., & Hardy, J. (2004). Milk proteins for edible films and coatings. *Journal of Critical Reviews in Food Science and Nutrition,* 44(4), 239–251.

Kitts, D. D. (1997). An evaluation of the multiple effects of the antioxidant vitamins. *Trends in Food Science and Technology*, 8, 198–203.

Kou, Y.; Dickinson, L. C., & Chinachoti, P. (2000). Mobility characterization of waxy corn starch using wide-line 1H nuclear magnetic resonance. *Journal of Agricultural and Food Chemistry*, 48(11), 5489–5495.

Kurata, T., & Sakurai, Y. (1967). Degradation of L-ascorbic acid and mechanism of non-enzymic browning reaction. Part II. *Agricultural and Biological Chemistry,* 31, 170–176.

León, P. G., & Rojas, A. M. (2007). Gellan gum films as carriers of L-(+)-ascorbic acid. *Food Research International*, 40, 565–575.

Levine, I. N. (2004). Physical Chemistry. Sixth edition. McGraw Hill, Thomas Timp Publisher, Higher Education, pp. 541-546.

Luzio, G. A., & Cameron, R. G. (2008). Demethylation of a model homogalacturonan with the salt-independent pectin methylesterase from citrus: Part II. Structure-function analysis. *Carbohydrate Polymers*, 71(2), 300–309.

Mazeau, K., & Pérez, S. (1998). The preferred conformations of the four oligomeric fragments of rhamnogalacturonan II. *Carbohydrate Research,* 311, 203–217.

Miao, S., & Roos, Y. H. (2004). Nonenzymatic browning kinetics of a carbohydrate-based low-moisture food system at temperatures applicable to spray drying. *Journal of Agricultural and Food Chemistry*, 52, 5250–5257.

Morris, G. A.; Ralet, M. C.; Bonnin, E.; Thibault, J. F., & Harding, S. E. (2010). Physical characterisation of the rhamnogalacturonan and homogalacturonan fractions of sugar beet (Beta vulgaris) pectin. *Carbohydrate Polymers*, 82(4), 1161–1167.

Morrison, R. T., & Boyd, R. N. (1990). *Química Orgánica*, Addison-Wesley Iberoamericana, S. A., Wilmington, Delaware, USA, pp. 190, 211, 233.

Ng, A., & Waldron, K. W (1997). Effect of cooking and pre-cooking on cell-wall chemistry in relation to firmness of carrot tissues. *Journal of the Science of Food and Agriculture*, 73, 503–512.

Oakenfull, D., & Scott, A. (1984). Hydrofobic interaction in the gelation of high methoxyl pectins. *Journal of Food Science*, 49, 1093–1098.

Oosterveld, A.; Beldman, G.; Schols, H. A., & Voragen, A. G. J. (2000). Characterization of arabinose and ferulic acid rich pectic polysaccharides and hemicelluloses from sugar beet pulp. *Carbohydrate Research*, 328, 185–197.

Parker, M. L., & Waldron, K. W. (1995). Texture of chinese water chestnut: involvement of cell wall phenolics. *Journal of the Science of Food and Agriculture,* 68(3), 337–346.

Pérez, S., Rodríguez-Carvajal, M. A., & Doco, T. (2003). A complex plant cell wall polysaccharide: rhamnogalacturonan II. A structure in quest of a function. *Biochimie,* 85, 109–121.

Pérez, C. D.; Flores, S. K.; Marangoni, A. G.; Gerschenson, L. N., & Rojas, A. M. (2009). Development of a high methoxyl-pectin edible film for retention of L-(+)-ascorbic acid. *Journal of Agricultural and Food Chemistry,* 57(15), 6844–6855.

Pérez, C. D.; Fissore, E. N.; Gerschenson, L. N.; Cameron, R. G., & Rojas, A. M. (2012). Hydrolytic and oxidative stability of L-(+)-ascorbic acid supported in pectin films: influence of the macromolecular structure and calcium presence. *Journal of Agricultural and Food Chemistry*, 60(21), 5414–5422.

Pérez, C. D.; De'Nobili, M. D.; Rizzo, S. A.; Gerschenson, L. N.; Descalzo, A. M.; Rojas, A. M. (2013). High methoxyl pectin-methyl cellulose films with antioxidant activity at a functional food interface. *Journal of Food Engineering,* 116(1), 162–169.

Ping, Z. H.; Nguyen, Q. T.; Chen, S. M.; Zhou, J. Q., & Ding, Y. D. (2001). States of water in different hydrophilic polymers-DSC and FTIR studies. *Polymer,* 42, 8461–8467.

Ralet, M. C.; Cabrera, J. C.; Bonnin, E.; Quéméner, B.; Hellin, P., & Thibault, J. F. (2005). Mapping sugar beet pectin acetylation pattern. *Phytochemistry*, 66, 1832–1843.

Rombouts, F. M., & Thibault, J. F. (1986). Sugar beet pectins: chemical structure and gelation through oxidative coupling. In M. L. Fishman, & J. J. Jen (Eds.), *Chemistry and functions of pectins* (pp. 49–59; Chapter 5). Washington DC, USA: American Chemical Society.

Siew, C. K., & Williams, P. A. (2008). Role of protein and ferulic acid in the emulsification properties of sugar beet pectin. *Journal of Agricultural and Food Chemistry,* 56, 4164–4171.

Smith, J. M. (1986). *Ingeniería de la cinética química*. Capítulo 2, pp. 61. Compañía Editorial Continental, S.A. de C.V., México.

Tanhatan-Nasseri, A.; Crépeau, M. J.; Thibault, J. F., & Ralet, M. C. (2011). Isolation and characterization of model homogalacturonans of tailored methylesterification patterns. *Carbohydrate Polymers*, 86(3), 1236–1243.

Vincken, J. P.; Scholsm, H. A.; Oomen, R. J. F. J.; McCann, M. C.; Ulvskov, P.; Voragen, A. G. J., & Visser, R. G. F. (2003). If homogalacturonan were a side chain of rhamnogalacturonan I. Implications for cell wall architecture. *Plant Physiology*, 132, 1781–1789.

Vittadini, E.; Chinachoti, P. (2003). Effect of physico-chemical and molecular mobility parameters on *Staphylococcus aureus* growth. International *Journal of Food Science and Technology*, 38(8), 841–847.

Voragen, A. G. J.; Coenen, G. J.; Verhoef, R. P., & Schols, H. A. (2009). Pectin, a versatile polysaccharide present in plant cell walls. *Structural Chemistry,* 20, 263–275.

Willats, W. G.; Orfila, C.; Limberg, G.; Buchholt, H. C.; van Alebeek, G. J.; Voragen, A. G.; Marcus, S. E.; Christensen, T. M. I. E.; Mikkelsen, J. D.; Murray, B. S.; Knox, J. P. (2001). Modulation of the degree and pattern of methyl-esterification of pectic homogalacturonan in plant cell walls. Implications for pectin methyl esterase action, matrix properties, and cell adhesion. *Journal of Biological Chemistry,* 276, 19404–19413.

Willats, W. G. T.; Knox, J. P., & Mikkelsen, D. (2006). Pectin: new insights into an old polymer are starting to gel. Trends *in Food Science and Technology,* 17, 97–104.

Zhou, H.; Brock, J.; Liu, D.; Board, P. G., & Oakley, A. J. (2012). Structural insights into the dehydroascorbate reductase activity of human omega-class glutathione transferases. *Journal of Molecular Biology*, 420(3), 190–203.

Zsivanovits, G.; MacDougall, A. J.; Smith, A. C., & Ring, S. G. (2004). Material properties of concentrated pectin networks. *Carbohydrate Reasearch*, 339, 1317–1322.

In: Pectin: Chemical Properties, Uses and Health Benefits
Editor: Phillip L. Bush
ISBN: 978-1-63321-438-5

Chapter 9

Chemical Composition and Rheological Behaviour of Pectins from Unconventional Sources

Eliana Noemí Fissore[1,2], Florencia Basanta[1,3], Jhon Edinson Nieto Calvache[1,3], Ana María Rojas[1,2] and Lía Noemí Gerschenson*[*1,2]

[1]Industry Department, School of Natural and Exact Sciences (FCEN), Buenos Aires University (UBA), Ciudad Universitaria, Ciudad Autónoma de Buenos Aires, Argentina
[2]Member of the National Scientific and Technical Research Council, Argentina (CONICET)
[3]Fellow of the National Scientific and Technical Research Council, Argentina (CONICET)

Abstract

Pectins are soluble dietary fibres that constitute a family of complex polysaccharides present in the primary cell wall and middle lamella of plants. The major sources of commercial pectins are citrus peel and apple pomace. Most pectin is produced by the extraction of the raw material with hot aqueous mineral acid at pH~2. In the food industry, pectins are used as gelling agents, thickeners, and stabilizers. New applications are constantly developing and their use as emulsifiers is one of the latest new-comes.

Utilization of by-products of the fruit and vegetable industrialization as source of pectins may contribute to the efficiency of the processes and also to the sustainability of the environment.

In the present work, pectins were extracted through different procedures from three unconventional sources and were characterized in their chemical composition and rheological behaviour. Beetroot pectin isolated through cellulase digestion and alkaline pretreatment, presented 54 % (w/w) of uronic acids (UA) and showed low degree of

[*] Corresponding author: Lía Noemí Gerschenson, e-mail: lia@di.fcen.uba.ar. Phone: 54 – 11 – 4576-3366 / 3397, Fax number: 54 – 11 – 4576-3366.

methylation (DM) and acetylation. The aqueous solution of this pectin presented low viscosity and pseudoplastic behaviour in flow. Gels were formed by addition of Ca^{2+}. On the other hand, butternut pectin, isolated through cellulase digestion, also presented 54% (w/w) of UA and a high DM, and in the presence of high sugar concentrations and at low pH, produced viscous solutions with pseudoplastic behaviour. Pectins were also obtained from Japanese plums with water at different temperatures. Those extracted at room temperature contained 56% (w/w) of UA and low DM as well as pseudoplastic behaviour in water. Pectin fraction extracted with boiling water contained 50% of UA and showed high DM. Although the later procedure increased considerably the yield, the extracted pectin showed significant lower apparent viscosity in water, in spite of its high molecular weight.

The isolating procedures assayed permitted the extraction of pectin enriched fractions from non-conventional sources with interesting yields and diverse rheological characteristics.

Introduction

Pectins are a family of complex heteropolysaccharides that constitute a significant proportion (35%) of the primary cell wall of dicotyledonous plants and non-graminaceous plants, with important functions in their development, growth and maturation (Mohnen, 2008; Morris and Ralet, 2012). They are an abundant, ubiquitous and multifunctional component of the cell walls of all land plants and possess various pharmaceutical activities including healing wounds (Hokputsa et al., 2004), inhibiting lipase activity (Edashige et al., 2008; Kumar and Chauhan, 2010), inhibiting growth and metastasis and inducing apoptosis of human cancer cells (Jackson et al., 2007; Nangia-Makker et al., 2002). They also have immunostimulating activity, anti-ulcer activity and cholesterol decreasing effect (Yamada, 1996). Pectins are widely employed as functional ingredient in food applications, due to its valuable stabilizer, thickener and gelling properties.

According to Willats et al. (2006) 4–5 g of pectin are consumed every day in a western diet. Savary et al. (2003) reported an annual consumption of pectin of approximately 45,000 tonnes worldwide, representing a global market value of at least 400 million euros. The majority of commercially available pectins are extracted from citrus peels, as well as apple pomace (Kurita et al., 2008). Nowadays other novel sources, including sugar beets and sunflower heads (Joye and Luzio, 2000), are also being investigated but not at commercial scale.

Pectin consists mostly of polymers rich in D-galacturonic acid (GalA) and often contain significant amounts of L-rhamnose (Rha), D-arabinose (Ara) and D-galactose (Gal) as well as other 13 different monosaccharides (Vincken et al., 2003). The three major polysaccharide domains recognized in pectin macromolecules are homogalacturonan (HG), rhamnogalacturonan I (RG-I) and rhamnogalacturonan II (RG-II) (Pérez et al. 2003; Willats et al., 2006). The (HG) is the most abundant cell wall pectic polysaccharide, corresponding to approximately 50–90% of the total pectin (Mohnen, 2008; Yapo, 2011). HG is an α (1-4)-linked GalA-polymeric chain. The second most abundant type of pectic polysaccharide is rhamnogalacturonan I (RG-I), comprising between 20% and 35% of the total pectin (Santos et al., 2013). The RG-I domains are relatively short segments constituted by alternating α-(1-4)-linked GalA and α-(1-2)-linked rhamnopyranosyl monomers, where Rha can be more or less branched by lateral glycosidic chains of arabinogalactan I. The RG-II consists of very short segments of HG chains where GalA monomers are branched by lateral chains of alternating unusual monosaccharides (2-O-Me-xylose, 2-O-Me-fucose, 2-keto-3-deoxi-D-*lyxo*-heptulosaric acid or Dha; 2-keto-3-deoxi-D-*manno*-octulosonic acid or Kdo) and rhamnose (Rha). According to Vincken et al. (2003) the pectic macromolecules consist mostly of long "smooth" HG regions, which are side chains of the "hairy" RG-I core.

Depending on the degree of methylesterification (DM), which is the proportion of carboxyl groups present in the esterified form, pectins can be classified as low-methoxyl (LM) pectins when their DM is lower than 50% and high-methoxyl (HM) pectins when DM is equal to or greater than 50%. The formation of gels with HM pectins requires conditions of low pH (≈3.0) and high sugar content (≈ 65%). In contrast, LM pectins may form gels in the presence of calcium over a wide range of pH with or without sugar (Fu and Rao, 2001). Pectins are highly heterogeneous with respect to their molecular mass and chemical structure. The average molecular mass of pectins from various vegetable sources is typically in the range of 10^4 - 10^5Da (Izydorczyk et al., 2005).

Knowledge of the mechanism and kinetics of extraction process is generally needed for the design and optimization of industrial processes. The pectin network must be disrupted to enable extraction from vegetable tissues. This may involve extraction with calcium chelating agents, dilute alkali or dilute acid. Generally, commercial pectin is extracted with hot dilute mineral acid followed by recovery through precipitation in alcohol (Seggiani et al., 2009). However, the use of strong mineral acids (e.g., H_2SO_4 or HCl) can be harmful to the environment and require additional steps to remove toxic elements (Yapo, 2009). To produce pectin with superior yield and quality, it is of great importance to explore novel methods or modification of the existing methods to avoid using extreme extraction conditions (Yuting et al., 2014). Therefore researchers have proposed alternative processes for minimizing the use of detrimental chemicals during pectin extraction; these include thermal and/or mechanical treatments, ultrasound, autoclaving, extrusion, subcritical water extraction, catalysis with enzymes (Lim et al., 2012). These processes are influenced mainly by temperature, pH, and time.

The plant cell wall is mainly composed of cellulose, hemicellulose, pectin, and proteins. Therefore, the use of enzymes such as cellulases which are able to degrade the plant cell wall facilitates the isolation of pectins. The extracted materials will be heteropolymolecular and polydisperse having a diversity of chemical structures and molecular sizes (MacDougall and Ring, 2004) and its functional properties are influenced by the chemical structure (Willats et al., 2006). As a consequence, the composition, structure, and physiological properties of pectin might be affected by conditions of extraction as well as source, location and many other environmental factors (Fissore et al., 2013).

The objective of this research was to obtain pectins from three unconventional (beetroot, Japanese plum and butternut) sources. The extraction process of the pectins involved the use of non-pectolytic enzymes such as cellulases or an aqueous extraction. Products obtained were characterized in their chemical composition and rheological behaviour.

BUTTERNUT AND BEETROOT PECTINS

Extraction of butternut pectin. Butternut (*Cucumis moschata* Duch) mesocarp was dried (85 °C, 2 h), milled and sieved to obtain a powder enriched in cell wall material (CWM) which was submitted to digestion (30 °C, 20 h) into 0.5 M-sodium citrate buffer (pH 5.2) with cellulase (C9422, Sigma, St. Louis, USA). Insolubles obtained after digestion were separated through filtration with glass fibre filter (Schleicher and Schuell, Germany) under vacuum and cell wall polysaccharides were precipitated from the supernatant through 96% (v/v)-ethanol addition. These pectin enriched products were collected through filtration and they were, finally, freeze-dried. Chemical analyses were performed according to Fissore et al. (2007).

Extraction of beetroot pectin. *Beta vulgaris* L. var. conditiva roots were peeled, dried (85 °C, 2 h), milled and sieved for obtaining a powder enriched in cell wall material (CWM) which was treated with 2 M NaOH solution (25°C, 30 min) and filtrated. The residue obtained was submitted to the same procedure of cellulase digestion as for butternut pectin. Chemical analyses were performed according to Fissore et al. (2011).

Chemical Composition

Butternut and beetroot powders obtained were submitted to digestion with cellulase in sodium citrate buffer. Buffers act as cross-link cleaving agents of cell wall polysaccharides because of the salts they contain (Fry, 1986) and they are also useful for controlling pH which affects β-elimination (Kravtchenko et al., 1992). Non-covalent cross-links are disrupted by salt effect as well as by acid pH (5.2), mainly at temperatures higher than 18-20°C (Brett and Waldron, 1996; Parker and Waldron, 1995). Hydrolysis of cellulose is expected to facilitate pectin solubilization into buffer solution although some pectin could remain insolubilized due to non-disrupted covalent bonds.

Butternut pectin was obtained with a yield of 6 g / 100 g CWM powder. As can be observed in Table 1, this fraction had a total carbohydrate content of ≈ 66 g / 100 g and a GalA content of ≈ 54 g / 100 g. The degree of methylation was ≈ 83 %, indicating that the HG was highly methylated as it is naturally found into the cell wall. Neutral sugar composition presented Gal (4.07 g / 100g) and Ara (3.94 g / 100g), as the main monosaccharides, with Rha (1.68 g / 100g), glucose (Glc ; 0.92 g / 100g) and mannose (Man; 0.66 g / 100g) and lower proportions of Xylose (Xyl; 0.33 g / 100g) and fucose (Fuc; 0.12 g / 100g). This composition can be attributed to the RG-I core branched with arabinogalactans, galactans and arabinans. Degradation of cellulose may cause the dissociation of hydrogen bonded xyloglucan (XG) as well as removal of galactoglucomannan (GGM) (Schroeder et al., 2001). The Man proportion indicates certain degree of GGM removal after hydrolysis of the cellulose framework. It is presumed that XG and GGM persisted into the fraction obtained because they were not completely solubilized by the buffer solution. Presence of RG-II was confirmed through detectable amounts of the sugars 2-O-Me-fucose, 2-O-Me-xylose, Kdo and Dha (Table 1).

Table 1. Chemical[1] and sugar composition of pectins isolated from butternut[2] and beetroot[3]

	Butternut pectin	Beetroot pectin
Yield (g /100g CWM)[4]	6	21
Total Carbohydrates[5] (g/100g pectin fraction)	66 ± 1	88 ± 1
Uronic Acids (g/100g pectin fraction)	54 ± 1	54 ± 2
Neutral Sugars (g/100g pectin fraction)	12	34
Protein (g/100g pectin fraction)	7.5 ± 0.1	5.8 ± 0.4
DM[6] (mol /100 moles)	83	2
DA[7] (mol /100 moles)	16	1.8
Neutral sugar composition (g/100g pectin fraction)		
2-O-Me Fucose	0.13	tr
2-O-Me Xylose	0.12	0
Dha+Kdo[8]	0.28	0.01
Rhamnose	1.68	1.9
Fucose	0.12	0
Arabinose	3.94	29.8
Xylose	0.33	0
Mannose	0.66	0
Galactose	4.07	2.0
Glucose	0.92	0

[1] Mean and standard deviations are shown (n = 3).
[2] Fissore et al. (2007).
[3] Fissore et al. (2011).
[4] CWM: cell wall material.
[5] Total carbohydrates are expressed as g of galacturonic acid / 100g pectin fraction.
[6] DM: degree of methylation calculated as a percent molar ratio between methanol and uronic acids.
[7] DA: degree of acetylation calculated as a percent molar ratio between acetyl groups and total carbohydrates.
[8] Dha: 2-keto-3-deoxi-D-*lyxo*-heptulosaric acid; Kdo: 2-keto-3-deoxi-D-*manno*-octulosonic acid.

Beetroot pectin was isolated with high yield (≈ 21 %). When the same cellulase digestion was applied to the CWM but without the alkaline pre-treatment the yield of extraction was less than 1 % (Fissore et al., 2011). Ester-linked ferulic acid is present in pectin fractions isolated from certain plants like beetroots, conferring mechanical resistance to thermal processing (Waldron et al., 1997). Alkaline hydrolysis disrupted ferulic ester bonds allowing the removal of pectin from the cell wall.

This pectin was mostly constituted by water soluble polysaccharides with an uronic acid content of 54 g / 100 g and 34 g / 100 g of neutral sugars (Table 1). Approximately 94.0 % of uronic acid content was GalA and the remaining 6.0 % was glucuronic acid (GlcA). This fraction was characterized by an almost total demethylation (DM ≈ 2.0 %, molar ratio) as expectable, after saponification suffered by the methyl esters at the C6-carboxylic groups of GalA monomers. The low degree of acetylation (≈ 1.8 %, molar ratio) found was also attributed to the alkaline treatment applied, since beetroot cell wall analyzed in a previous work (Fissore et al., 2010) showed a 90%-degree of acetylation. It was also determined a protein content of 5.8 g / 100 g. Sugar composition of beetroot pectin showed that it was mainly constituted by HG side chains attached to the RG-I core. The high content of Ara (29.8 g / 100g) and the low content of Gal (2.0 g / 100 g) indicate that the hairy regions of RG-I are constituted mostly by arabinan chains and a very low proportion of arabinogalactans. Strasser and Amadò (2001) also found a high Ara content in beetroot pectin. The presence of some residual proportion of RG-II was revealed by a very low content of sugars like 2-O-Me-fucose, Dha and Kdo.

Rheological Behaviour

Thickening Capacity

For butternut and beetroot pectins, aqueous systems containing 2.00 % (w/v) of fraction were prepared and stored (25°C, 18 h) to attain swelling equilibria before measurement. Flow curves were determined at a constant temperature (25 °C) and data points were recorded after reaching steady-state. For fitting of the recorded data, Ostwald's law was considered (Fissore et al., 2012).

Swelling after hydration of polysaccharides involves polymer interactions with the water added to the system. The degree of thickening reached will depend on the level of polymer interaction with water. Figure 1 shows the flow curve for butternut pectin (2.00 % w/v) in water. At low shear rates, it can be observed a first Newtonian plateau which indicates that enough experimental time exists for the development of new interactions among the initially disrupted intermolecular entanglements (Ross-Murphy, 1994). At higher shear rates, it can be observed pseudoplastic behaviour which was confirmed by data fitting to Ostwald's law (n = 0.63) (Table 2). It was obtained a consistency index (k) of 0.86 Pa s^n and the apparent viscosity (η_a) calculated at a shear rate of 20 s^{-1} was 0.28 Pa s.

The aqueous solution of beetroot pectin (2.0 % w/v) showed pseudoplastic behaviour (n = 0.47) with a consistency index (k) of 0.12 Pa s^n (Table 2). The low apparent viscosity (≈ 0.025 Pa s) calculated at a shear rate of 20 s^{-1} indicates that these hydrated macromolecules interact poorly with each other through overlapping and entanglements (Fissore et al., 2012).

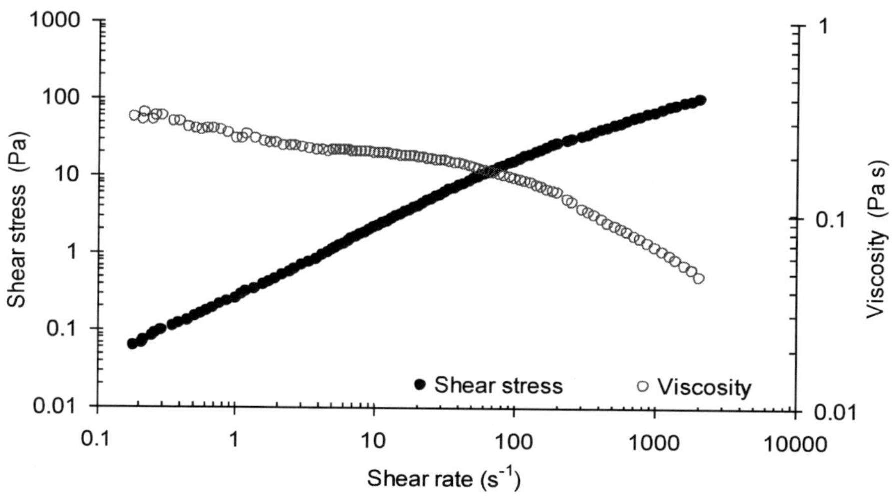

Figure 1. Flow behaviour (25 °C) of an aqueous butternut pectin (2.00 % w/v) system. (Fissore et al., 2009).

Table 2. Ostwald's law parameters[1] for aqueous butternut and beetroot[2] pectin (2.00 % w/v) systems

	k (Pa s^n)	n	R^2
Butternut pectin	0.86 ± 0.03	0.63 ± 0.01	0.997
Beetroot pectin	0.12 ± 0.01	0.47 ± 0.02	0.969

[1] Mean and standard errors are shown (n = 3).

[2] Fissore et al. (2012).

k: consistency index; n: pseudoplasticity index; R2: goodness of fit (α: 0.05).

Gelling Capacity

Butternut pectin had a high DM and its gelling capacity was assayed in the presence of sugar at low pH. Sample was prepared by dissolving pectin (1.00 % w/w) in water and heating (70 °C); 65.00 % (w/w) sucrose was added with continuous stirring and, when sugar was dissolved, pH was adjusted to 3.5 with tartaric acid (Fissore et al., 2013). Viscosity of the sample increased with sucrose addition but gelation did not occur. This could be ascribed to the fact that this fraction has a rather low GalA content and a high DA. The flow curve for the butternut pectin – sucrose system is shown in Figure 2 and it shows pseudoplastic behaviour, with a Newtonian plateau at low shear rates and shear thinning at higher shear rates.

Fitting of experimental data to Ostwald's law (Table 3) gave a consistency index (k) of 2.11 Pa s and a flow behaviour index (n) of 0.68. Apparent viscosity calculated at a shear rate of 20 s^{-1} was 0.81 Pa s.

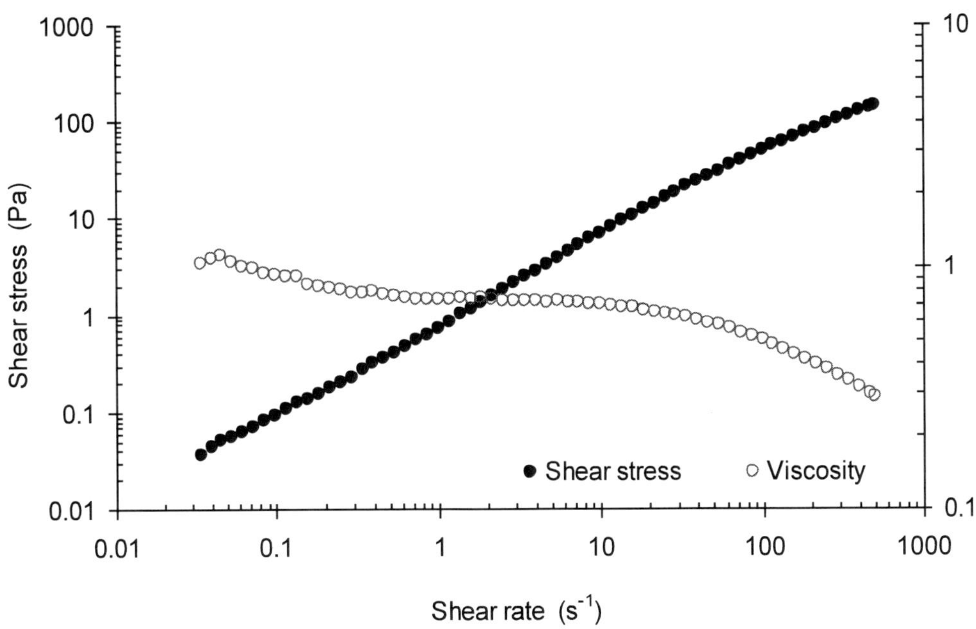

Figure 2. Flow behaviour (25 ºC) for butternut pectin (1.00% w/w) with sucrose (65.00% w/w) and pH 3.5 (Fissore et al., 2013).

Table 3. Ostwald's law parameters[1] for butternut[2] pectin (1.00% w/w) with sucrose (65.00% w/w) and pH 3.5

	k (Pa s^n)	n	R^2
Butternut pectin	2.11 ± 0.40	0.68 ± 0.03	0.996

[1] Mean and standard errors are shown (n = 3).

[2] Fissore et al. (2013).

k: consistency index; n: pseudoplasticity index; R^2: goodness of fit (α: 0.05).

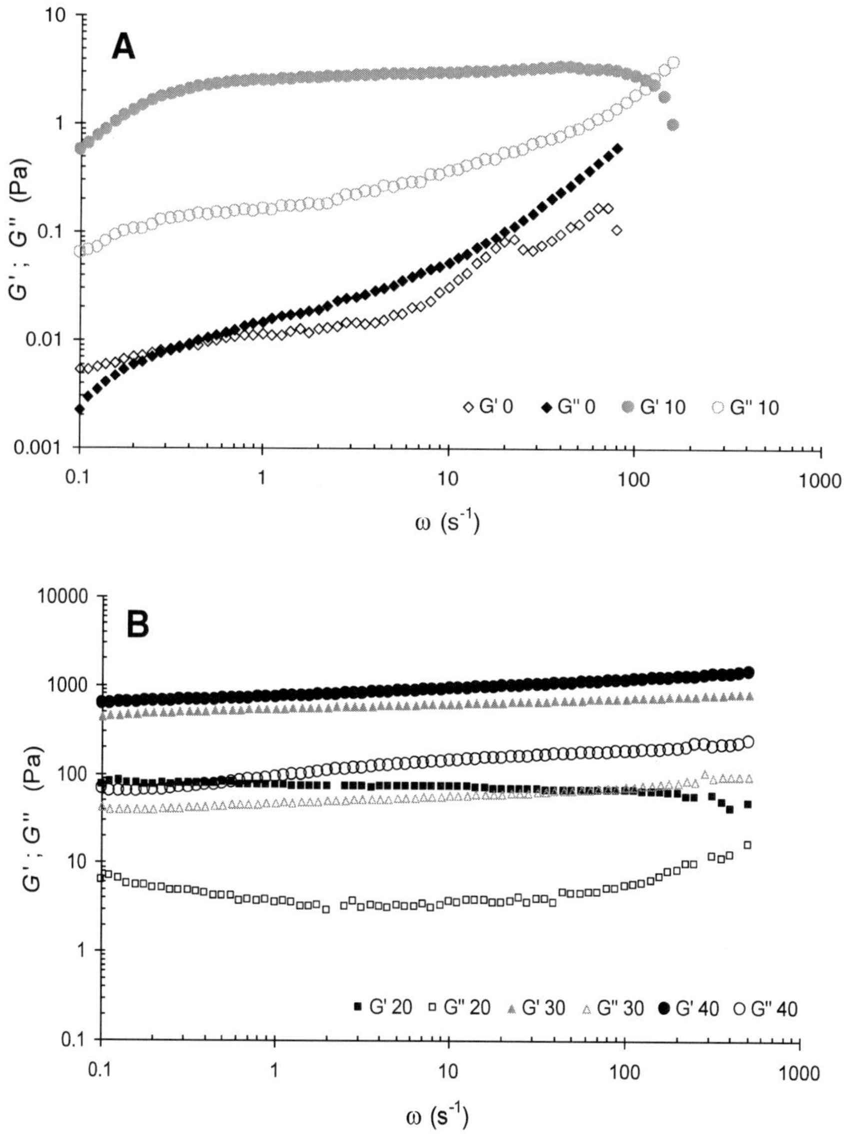

Figure 3. Storage (G') and loss (G") moduli profiles for beetroot pectin systems with different calcium concentrations: A) 0 and 10 mg Ca^{2+} / g pectin; B) 20, 30 and 40 mg Ca^{2+} / g pectin (Fissore et al., 2013).

Beetroot pectin had a low DM hence its gelling capacity was assayed in the presence of calcium ions. The 1.00 % (w/w) pectin systems were prepared by dissolving the pectin in water at 70 °C with continuous stirring and addition of $CaCl_2$ to give different Ca^{2+} concentrations (0, 10, 20, 30, 40 mg Ca^{2+} / g pectin) (Fissore et al., 2013). Calcium ions specifically bind to sites along the (1,4) linked α-D-galacturonic acid chains (Fry, 1986). As can be observed in Figure 3, in the absence of Ca^{2+} (0 mg / g pectin), *G''* was higher than *G'* and both moduli were frequency dependent (Figure 3A). This rheological behaviour corresponds to solutions systems, where the hydrocolloid concentration is below the critical one for coil overlap (Ross-Murphy, 1994). The system with a Ca^{2+} concentration of 10 mg / g pectin showed G' > G'' with some frequency dependence, which is the behaviour of weak gels. At Ca^{2+} concentrations higher than 10 mg Ca^{2+}/ g pectin strong gels were obtained. For the system containing 20 mg Ca^{2+}/ g pectin (Figure 3B), it can be observed a G' value of ~ 100 Pa with only a slight dependence on frequency, which is characteristic of the mechanical spectra of strong physical gels (Ross-Murphy, 1994; Doublier et al., 1992; Lapasin and Pricl, 1995). Systems containing 30 and 40 mg Ca^{2+}/ g pectin showed mechanical spectra different from that of 20 mg Ca^{2+}/ g pectin-system (Figure 3B). Their G' values were close to 1000 Pa, increasing slightly with the increase in frequency. The 40 mg Ca^{2+}/ g beetroot pectin-system showed a mechanical spectrum with G' values slightly more frequency dependent than that of 30 mg Ca^{2+}/ g pectin-system. Simultaneously, the G" profiles were considerably more dependent on frequency, showing a minimum near 0.1 s^{-1}. This suggests some instability of the gel networks developed from Ca^{2+} addition at 30 and 40 mg Ca^{2+}/ g pectin concentrations. It is known that the elastic modulus (G') is proportional to the number of cross links from which a network is developed (Doublier et al., 1992).

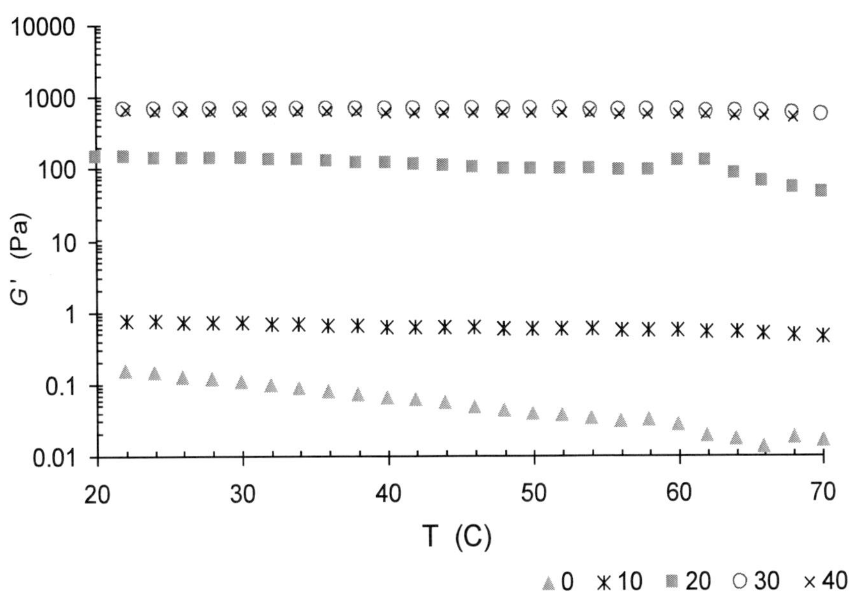

Figure 4. Storage (G') modulus profiles for a cooling ramp performed at 2 °C / min for beetroot pectin systems with different calcium concentrations (0, 10, 20,30 and 40 mg Ca^{2+} / g pectin) (Fissore et al., 2013).

Figure 4 shows the effect of temperature on the pectin - Ca^{2+} systems studied during a cooling ramp performed at 2 ºC / min. Gels developed with 30 and 40 mg Ca^{2+} showed a slight and continuous increase in G' with the decrease in temperature from 70 ºC to 60 and 53 ºC, respectively. After this, almost no influence was observed with the decrease in temperature down to 20 ºC. Storage modulus (G') also increased continuously and significantly from 70 to ~ 33 ºC for 10 and 20 mg Ca^{2+}-systems. On the other hand, the G' value increased continuously between 70 and 20 ºC, for the pectin aqueous solution without Ca^{2+}. The later phenomenon indicates the increasing development of interactions through hydrogen bonds between macromolecules as the temperature decreased to 20 ºC (Joesten and Schaad, 1974), in the absence of Ca^{2+} (Li, 2002). As calcium concentration increased, electrostatic junction zones increasingly developed between pectin chains, and their thermal stability prevails over the interaction through hydrogen bonding in the 70 - 20 ºC range, as inferred from the profiles observed in comparison to control (0 mg Ca^{2+}) system.

Japanese Plum Pectin

Extraction of plum pectin. After removal of the endocarp of *Prunus salicina* L var. Roysum, the fruit was put in ice-cold 80 % (v/v) ethanol and homogenized in a Warring blender and an Omni Mixer. The homogenate was boiled for 30 min, cooled, and filtered. The retentate was thoroughly washed with 96% (v/v) ethanol. The solids were then resuspended in chloroform:methanol (1:1 v/v), stirred for 15 min and filtered. The retentate was washed with the same solvent mixture. The insoluble material was washed with acetone, yielding the crude cell wall extract or alcohol insoluble material (AIR). Two pectin rich fractions were obtained from the Japanese plums AIR. The extractions were performed through: i) Stirring with water at room temperature for 2 h, ii) Stirring with boiling water for 2 h. The supernatants were recovered after centrifugation, dialyzed and liophilized, yielding fractions Plum-CW and Plum-HW, respectively (Basanta et al., 2012).

Chemical Composition

Yield and chemical composition of the fractions are reported in Table 4. A significantly lower yield was observed for the pectin fraction obtained at room temperature (≈ 12 %), while the yield of Plum-HW was ≈ 33 %. As it can be observed according to the contents of uronic acids (50 - 56 g / 100g) and neutral sugars (36 - 40 g / 100g), both fractions were constituted by pectic polysaccharides and contain a small amount of proteins (7 – 12 g / 100g). The neutral sugars were mainly constituted by arabinose and galactose (75% between both) in ratio 1:2, rhamnose (7 - 10 %) and, in lower proportion, by xylose, mannose and glucose. So, these pectins are constituted by homogalacturonan (HG) and rhamnogalacturonan-I (RG-I) branched with arabinogalactans chains (Carpita and Mc Cann, 2000). Taking into account (Ara+Gal)/Rha ratio (Plum-CW ≈ 9.6, Plum-HW ≈ 7.2), pectins extracted at room temperature presented more substituted RG-I. Also these pectins showed lower degree of methylation (DM ≈ 39 %), whereas pectin fraction extracted with boiling water showed a higher DM (54 %) and molecular weight (89.2 KDa).

Table 4. Yieds[1], composition[1] and molecular weights of pectin fractions extracted from Japanese plum AIR using water[2]

	Plum-CW[3]	Plum-HW[3]
Yield (g/100g pectin AIR)[4]	12.2 ± 0.1	33.6 ± 2.4
Protein (g/100g pectin fraction)	12	7
Neutral sugars (g/100g pectin fraction)	40 ± 1	36 ± 4
Uronic acids (g/100g pectin fraction)	56 ± 2	50 ± 4
DM[5] (mol/100 moles)	39	54
Molecular weight (kDa)	71.4	89.2
Neutral sugar composition (g/100 g pectin fraction)		
Rhamnose	3.1	3.8
Fucose	---	---
Arabinose	10.3	7.9
Xylose	1	1
Mannose	1.5	1.5
Galactose	19.5	19.5
Glucose	2.3	2.3

[1] Mean and standard deviations are shown (n = 2 - 4).
[2] Basanta et al. (2012).
[3] Plum-CW: Fraction obtained with water al room temperature. Plum-HW: Fraction obtained with boiling water.
[4] AIR: alcohol insoluble residue.
[5] DM: degree of methylation, calculated as a percent molar ratio between methanol and uronic acids.

Rheological Behaviour

For this characterization, 2.00 % (w/v) aqueous systems of the two pectin fractions were constituted and no insoluble material was observed showing that products isolated were constituted by "soluble" polysaccharides. The flow behaviour was determined through the recording of the viscosity curves over a range of shear rates (10^{-3}–2 s^{-1}).

As can be seen in Figure 5, both pectin systems exhibited a Newtonian plateau at the lowest shear rates followed by a pseudoplastic behaviour.

Table 5 shows results obtained when data were fitted to the Cross model. The Plum-CW solution showed an initial viscosity (η_0) of 238 Pa s and the Newtonian viscosity of the Plum-HW solution was lower ($\eta_0 = 5.5$ Pa s). A higher value of structural relaxation time (τ) along the shear thinning zone was observed for Plum-CW. Therefore, the macromolecular interactions in the Plum-HW aqueous solution produced more transient and less structured physical networks and Plum-CW generated a more structured aqueous system. The η_∞ values of the polymer solutions derived from fitting to the Cross model was higher than the water viscosity (0.001 Pa s). These trends are probably related to chemical composition, fundamental molecular properties and intramolecular and intermolecular interactions (Basanta et al., 2012).

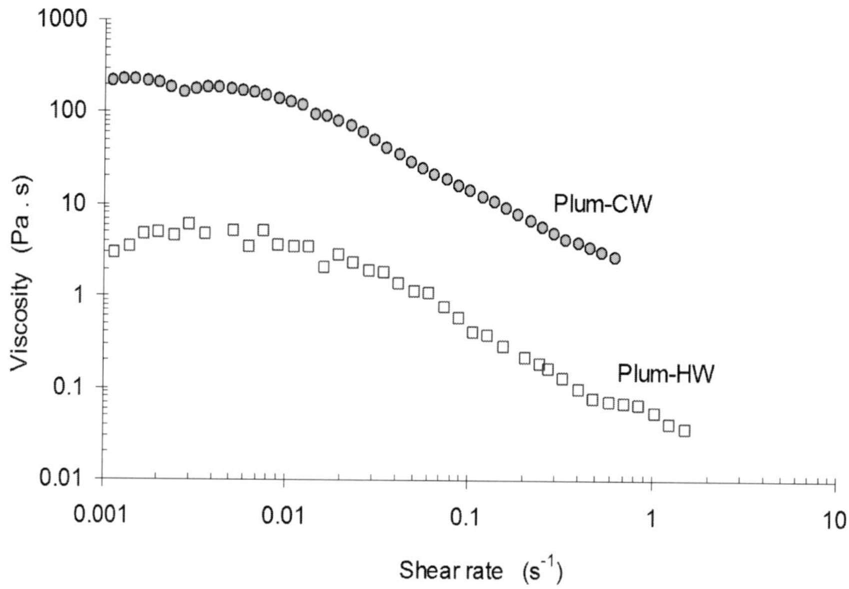

Figure 5. Flow behaviour (25°C) for Japanese plum pectin extracted by water at room temperature (Plum-CW) or boiling water (Plum-HW) Basanta et al. (2012).

Table 5. Cross model-parameters[1] calculated after fitting the flow experimental data (20 °C) for 2.00 % (w/v)-aqueous systems containing different Japanese plum pectin enriched products[2]

Sample	η_∞ (Pa s)	η_0 (Pa s)	τ (s)	m	R^2
Plum-CW	0.7 ± 0.2	238 ± 25	92 ± 6	1.3	0.998
Plum-HW	0.025 ± 0.05	5.5 ± 0.3	57 ± 8	1.3	0.976

[1] Mean and standard errors are shown (n = 3).

[2] Basanta et al. (2012).

η_0 Newtonian viscosity or viscosity at zero shear rate; η_∞: infinite or residual shear rate viscosity, τ: time constant, m:dimensionless parameter, R^2: goodness of fit.

Conclusion

According to results reported, butternut pectin can be used as a thickener and beetroot pectin as a gelling agent in the presence of calcium. The latter pectin showed a higher yield probably determined by the alkaline treatment that preceded the cellulase hydrolysis.

The yield of pectin obtained from Japanese plums was higher when extraction was performed with boiling water. Anyhow, the extraction with water at room temperature gave origin to a product which developed more structured physical networks in aqueous systems, showing a potential to be used as a thickener.

It can be concluded that butternut, beetroot and Japanese plum pectins, can be obtained as by-products of plant processing and feature potential for their use as food additives and dietary fibres.

The utilization of wastes of the industrialization of these tissues as source of pectins may contribute to the efficiency of the processes and also to the sustainability of the environment.

ACKNOWLEDGMENTS

This study was financially supported by University of Buenos Aires (UBACyT 20020100100726 and 20020100100700), National Agency of Scientific and Technical Research (PICT 38239) and CONICET (PIP 11220090100531 and 112-200801-03138).

REFERENCES

Basanta MF, Ponce NMA, Rojas AM and Stortz CA (2012). Effect of extraction time and temperature on the characteristics of loosely bound pectins from Japanese plum. *Carbohydrate Polymers* 89: 230–235.

Brett CT and Waldron KW (1996). *The physiology and biochemistry of plant cell walls,* 2nd edn. Chapman and Hall, London, pp. 26-32.

Carpita NC and Mc Cann MC (2000). The cell wall. In: Biochemistry & Molecular Biology of Plants. Buchanan BB, Gruissem W, Jones R (Eds.), *American Society of Plant Physiologists*, pp. 52-108.

Doublier JL, Launay B and Cuvelier G (1992). Viscoelastic properties of food gels. In: MA Rao & JF Steffe (Eds.), *Viscoelastic properties of food*, Elsevier Applied Science, New York.

Edashige Y, Murakami N and Tsujita T (2008). Inhibitory effect of pectin from the segment membrane of citrus fruits on lipase activity. *Journal of Nutritional Science and Vitaminology* 54(5): 409–415.

Fissore EN, Rojas AM, Gerschenson LN and Williams PA (2013). Butternut and beetroot pectins: Characterization and functional properties. *Food Hydrocolloids* 31: 172-182.

Fissore E, Rojas AM and Gerschenson LN (2012). Rheological performance of pectin enriched products isolated from red beet (Beta vulgaris L. var. conditiva) through alkaline and enzymatic treatments. *Food Hydrocolloids* 26(1): 249-260.

Fissore E, Ponce N, Matkovic L, Stortz C, Rojas AM and Gerschenson L (2011). Isolation of pectin enriched products from red beet (*Beta vulgaris* L. var. conditiva) wastes: composition and functional properties. *Food Science and Technology International* 17(6): 517-527.

Fissore E, Ponce N, de Escalada Pla M, Stortz C, Rojas AM and Gerschenson L (2010). Characterization of acid-extracted pectin-enriched products obtained from red beet (*Beta vulgaris* L. var. conditiva) and butternut (*Cucurbita moschata* Duch ex Poiret). *Journal of Agricultural and Food Chemistry* 58(6): 3793-3800.

Fissore EN, Matkovic L, Wider E, Rojas AM and Gerschenson LN (2009). Rheological properties of pectins isolated from butternut (*Cucurbita moschata* Duch ex Poiret). *LWT – Food Science and Technology* 42(8): 1413 – 1421.

Fissore EN, Ponce NM, Stortz CA, Rojas AM and Gerschenson LN (2007). Characterization of fibre obtained from pumpkin (*Cucumis moschata* Duch) mesocarp through enzymatic treatment. *Food Science and Technology International* 16(1): 1-7.

Fry SC (1986). Cross-linking of matrix polymers in the growing cell walls of angiosperms. *Annual Review of Plant Physiology* 37: 165-186.

Fu JT and Rao MA (2001). Rheology and structure development during gelation of low-methoxyl pectin gels: the effect of sucrose. *Food Hydrocolloids* 15: 93-100.

Hokputsa S, Gerddit W, Pongsamart S, Inngjerdingen K, Heinze T, Koschella A, Harding SE and Paulsen BS (2004). Water-soluble polysaccharides with pharmaceutical importance from Durian rinds (Durio zibethinus Murr.): isolation, fractionation, characterisation and bioactivity. *Carbohydrate Polymers* 56(4): 471–481.

Izydorczyk M, Cui SW and Wang Q (2005). Polysaccharide gums: structures, functional properties, and applications. In: SW Cui (Ed.), *Food carbohydrates: Chemistry, physical properties and applications.* London: Taylor & Francis, United Kingdom.

Jackson CL, Dreaden TM, Theobald LK, Tran NM, Beal TL, Eid M, Gao MY, Shirley RB, Stoffel MT, Kumar MV and Mohnen D (2007). Pectin induces apoptosis in human prostate cancer cells: correlation of apoptotic function with pectin structure. *Glycobiology* 17(8): 805–819.

Joesten MD and Schaad LJ (1974). *Hydrogen Bonding*. Marcel Dekker: New York; pp 182.

Joye DD and Luzio GA (2000). Process for selective extraction of pectins from plant material by different pH. *Carbohydrate Polymers* 43: 337-342.

Kravtchenko TP, Arnould I, Voragen AGJ and Pilnik W (1992). Improvement of the selective depolymerization of pectic substances by chemical β-elimination in aqueous solution. *Carbohydrate Polymers* 19(4): 237-242.

Kumar A and Chauhan GS (2010). Extraction and characterization of pectin from apple pomace and its evaluation as lipase (steapsin) inhibitor. *Carbohydrate Polymers* 82(2): 454–459.

Kurita O, Fujiwara T and Yamazaki E (2008). Characterization of the pectin extracted from citrus peel in the presence of citric acid. *Carbohydrate Polymers* 74: 725–730.

Lapasin R and Pricl S (1995). *Rheology of industrial polysaccharides. Theory and applications*. London: Chapman & Hall, Blackie Academic and Professional.

Li L (2002). Thermal gelation of methylcellulose in water: scaling and thermoreversibility. *Macromolecules* 35: 5990-5998.

Lim J, Yoo J, Ko S and Lee S (2012). Extraction and characterization of pectin from Yuza (Citrus junos) pomace: A comparison of conventional-chemical and combined physical-enzymatic extractions. *Food Hydrocolloids* 29: 160-165

MacDougall AJ and Ring SG (2004). Pectic polysaccharides. In: P Tomasik (Ed.), *Chemical and functional properties of food saccharides*. USA: CRC Press LLC.

Mohnen D (2008). Pectin structure and biosynthesis. *Current Opinion in Plant Biology* 11: 266–277.

Morris GA and Ralet MCA (2012). Copolymer analysis approach to estimate the neutral sugar distribution of sugar beet pectin using size exclusion chromatography. *Carbohydrate Polymers* 87(2): 1139–1143.

Nangia-Makker P, Hogan V, Honjo Y, Baccarini S, Tait L, Bresalier R and Raz A (2002). Inhibition of human cancer cell growth and metastasis in nude mice by oral intake of modified citrus pectin. *Journal of the National Cancer Institute* 94 (24): 1854–1862.

Parker ML and Waldron KW (1995). Texture of Chinese water chestnut: involvement of cell wall phenolics. *Journal of the Science of Food and Agriculture* 68: 337-346.

Pérez S, Rodríguez-Carvajal MA and Doco T (2003). A complex plant cell wall polysaccharide: rhamnogalacturonan II. A structure in quest of a function. *Biochimie* 85: 109-121.

Ross-Murphy SB (1994). Rheological Methods. In: SB Ross-Murphy (Ed.), *Physical Techniques for the Study of Food Biopolymers* London: Blackie Academic & Professional, Chapman & Hall, pp. 343-393.

Santos JDG, Espeleta AF, Branco A and De Assis SA (2013). Aqueous extraction of pectin from sisal waste. *Carbohydrate Polymers* 92: 1997– 2001

Savary BJ, Hotchkiss AT, Fishman ML, Cameron RG and Shatters R (2003). Development of a Valencia orange pectin methyl esterase for generating novel pectin products. In: F Voragen, H Schols & R Visser (Eds.), *Advances in pectin and pectinase research* The Netherland: Kluwer Academic Publishers, 345-361.

Schröder R, Nicolas P, Vincent SJF, Fischer M, Reymond S and Redgwell RJ (2001). Purification and characterisation of a galactoglucomannan from kiwifruit (Actinidia deliciosa). *Carbohydrate Research* 331: 291-306.

Seggiani M, Puccini M, Pierini M, Giovando S and Forneris C (2009). Effect of different extraction and precipitation methods on yield and quality of pectin. *International Journal of Food Science and Technology* 44: 574–580.

Strasser GR and Amadó R (2001). Pectic substances from red beet (Beta vulgaris conditiva). Part I. Structural analysis of rhamnogalacturonan I using enzymic degradation and methylation analysis. *Carbohydrate Polymers* 44(1): 63–70.

Vincken JP, Schols HA, Oomen RJFJ, McCann MC, Ulvskov P, Voragen AGJ and Visser RGF (2003) If homogalacturonan were a side chain of rhamnogalacturonan I. Implications for cell wall architecture. *Plant Physiology* 132: 1781-1789.

Waldron KW, Ng A, Parker ML and Parr AJ (1997). Ferulic acid dehydrodimers in the cell walls of Beta vulgaris and their possible role in texture. *Journal of the Science of Food and Agriculture* 74(2): 221-228.

Willats WGT, Knox JP and Mikkelsen JD (2006). Pectin: New into an Insights old polymer are starting to gel. *Trends in Food Science & Technology* 17: 97–104.

Yamada H (1996). Contribution of pectins on health care. *Progress in Biotechnology* 14: 173–190.

Yapo BM (2011). Pectic substances: From simple pectic polysaccharides to complex pectins A new hypothetical model. *Carbohydrate Polymers* 86: 373–385.

Yapo BM (2009). Biochemical characteristics and gelling capacity of pectin from yellow passion fruit rind as affected by acid extractant nature. *Carbohydrate Polymers* 98: 750–753.

Yuting X, Lifen Z, Yakufu B, Zhi G, Tian D, Xingqian Y and Donghong L (2014). Effects of ultrasound and/or heating on the extraction of pectin from grapefruit peel. *Journal of Food Engineering* 126: 72-81.

In: Pectin: Chemical Properties, Uses and Health Benefits ISBN: 978-1-63321-438-5
Editor: Phillip L. Bush

Chapter 10

PECTINS: FROM THE GELLING PROPERTIES TO THE BIOLOGICAL ACTIVITY

Mirian Angelene González-Ayón[1], Rosabel Vélez-de la Rocha[1], Mercedes Verdugo-Perales[12], José Luis Valenzuela-Lagarda[1], Raul Allende-Molar[1] and J. Adriana Sañudo-Barajas[1*]

[1]Centro de Investigación en Alimentación y Desarrollo, AC, Sinaloa, México
[2]Instituto Lightbourn, AC, Mexico

SUMMARY

Pectin is a heteropolysaccharide found in cell walls and middle lamellae of higher plants, its structure is linear or branched complex. It is composed of acidic and neutral sugars molecules and the molecular weight ranges from 50,000 to 180,000 daltons and are negatively charged in neutral pH. Pectins can be characterized by different molecular parameters such as the degree of methoxylation, the molecular weight, and the galacturonan content. The most common commercial sources for pectin are apple pomace and citrus peel, however other novel sources, including sugar beet, potato, sunflower heads and papaya, have been investigated. Commercial pectins have diverse composition, polymer size distribution, acylation pattern, esterification degree, neutral sugar substitution, and this variability can influence the optimal condition of extraction. In plant biology, this polysaccharide plays important roles in plant growth, development, morphogenesis, defense, cell-cell adhesion, wall structure, signaling, cell expansion, wall porosity, pollen tube growth, seed hydration, leaf abscission, and fruit development. Recently, remarkable progress has been made in elucidating pectin structure/function relationships at the molecular level, and this is leading to the design of pectins with specific functionalities in several applications. In human nutrition, pectin represents a prebiotic and soluble dietary fiber that has extended its use as a nutraceutical ingredient for its properties of hypoglycemic, hypolipidemic, immunostimulant and anticancer. Fermentative pectin is resistant to human digestion, but is degraded in the colon by species of *Aerobacillus*, *Lactobacillus*, *Micrococcus* and *Enterococcus*. The bacteria produce pectolytic and fermentative enzymes capable to hydrolyze pectin into short chain

* Corresponding author: adriana@ciad.edu.mx.

fatty acids (acetic acid, butyric acid, propionic acid) and carbon dioxide. In addition, many studies supports the benefit and protective effects of soluble fiber intake against debilitating diseases including obesity, diabetes, coronary heart disease and elevated cholesterol. In food technology, pectin is used as a gelling and stabilizing agent and is also widely used in industries of edible and biodegradable films, adhesives, paper substitutes, foams and plasticizers, medical devices, biomedical implants, and drug delivery. A recent application for modified-pectins have been explored as a specific targets, e.g., for binding to galectin-3 (Gal3), a multifaceted and prometastatic protein whose expression is up-regulated in several types of cancer. Studies suggest that specific pectin (arabinose and galactose ramified sugars) have the binding affinity to cancer cell receptors (Gal3), giving a promising application of pectin in the development of nanomaterials glycofunctionalized to direct drugs toward tumoral cells, increasing the efficiency of the anticancer therapy and reducing drug toxicity.

The multifaceted functions of native or modified pectins make them a good candidate for the nutrition of the future. In conclusion, pectin has evolved from a food ingredient to a potential healthy compound for highlighted applications.

1. PECTIN CHEMISTRY

Pectin is a complex group of diverse polysaccharides from plants, mostly consisting of galacturonic acid (GalA) forming a homopolymer or containing a diversity of sugars. This unit of GalA can be in either the free acid (salt) or methyl ester form (Carpita and McCann, 2000; Ridley et al., 2001; Coenen, 2007).

Pectins can be composed up to 17 different monosaccharides arranged with more than 20 linkages; however, its chain may be repetitive units or not, it means homopolysaccharides and heteropolysaccharides. The carboxyl group of certain number of GalA residues is methyl-esterified and some may be neutralized with cations, in different proportions depending on the plant source. Some hydroxyl groups might be acetylated or substituted with phenolic acids (Oakenfull, 1991; Ridley et al., 2001).

The polymerization degree of pectin is diverse and the molecular weight (MW) is also variable in function of its origin and method of extraction, for example the MW of pectin obtained from sugar beet is 20,000 to 25,000 Da; whereas the apple, pear and plum pectin range from 25,000 to 35,000 Da and citrus pectin is even larger with 40,000 to 50,000 Da (Datta, 1994).

Pectin is soluble in pure water but insoluble in alcohol and other polar solvents. This feature is used to obtain pectin from the aqueous slurry where they are usually extracted by hot acidic hydrolysis. Chemical properties of pectin correspond to a weak acid, in fact it is a polyacid and has no single and defined p*K,* although there are reports defining values from 3.5 to 4.5. Insoluble salts are formed from the reaction of pectin (polyanion) with di- and trivalents polycations; in addition, pectin reacts with positively charged polymers to form insoluble complexes.

In this Chapter to represent the structures of pectins, the extended form employed by the carbohydrate databank CarbBank is used. Long sequences are properly identified with their respective locants and anomeric descriptors. Through the document, common abbreviations for the main monosaccharides residues and derivatives are indicated in parentheses. Also a locant of the linkage (α/β) is given between the symbols (McNaught, 1996).

According to diversity of pectins, there are many structural elements rendering information about their composition, i.e., Galacturonans group comprise homogalacturonan (HG), the substituted xylogalacturonan (XG), rhamnogalacturonan type I (RG-I) and rhamnogalacturonan type II (RG-II) (Mohnen, 2008). Some authors have proposed alternative models of RG-I and its connection with chains of HG and RG-II (Coenen, 2007).

In terms of commercial pectin the composition is generally including in majority both, HG and the core of RG-I. Generally for HG, a chain of hundreds of (1→4)-α-D-GalA units are linked giving an important parameter in pectin application, the degree of esterification (DE). Nonetheless, in the RG-I backbone each GalA unit is attached to rhamnose, an important neutral sugar that has a torsion point in the linear chain of the polymer and provides the capacity of branching with other neutral sugars such as arabinose (Ara), galactose (Gal) and xylose (Oakenfull, 1991). The carboxyl group in the GalA moiety might be methyl esterified and according to the nature of the plant or developmental stage, the GalA can also be partially O-acetylated at C-3 and C-2 (Ishii, 1997).

Besides the identification of different pectin backbone, there is another important classification of pectins made according to the DE; this parameter is associated to the gelling properties and the behavior of pectins in solution but also in plant biology, has an important influence in cell-to-cell adhesion (Carpita and McCann, 2000). Pectins with DE greater than 50% are named high methoxyl (HM) pectins, and those with DE below 50% are low methoxyl (LM) pectins. There are some chemical modifications to produce pectins with specific application, e.g., low methyl-esterified amidated pectins (LMA) that are chemically amidated to obtain distinctive physical properties compared to HM and LM pectins. Calcium induced gelation is predominant in LM pectin and this occurs due to the formation of intermolecular junction zones between two linear HG regions of independent pectin polymers. Acetylation of GalA reduces the gelling effect of calcium or magnesium, divalent cations that in excess may produce pectin precipitation. Dissimilar behavior occurs with amidation of LM pectins because they become less susceptible to precipitation. The HM pectins produce gelling under conditions of high dissolved sugar and acidic conditions because produce a low availability of solvent-pectin interaction and create a dominance of pectin-pectin interactions where hydrophobic and hydrogen bonding attractions are induced (Schols and Voragen, 2002).

Among the main properties of pectin is its solubility in pure water. When water is added, dry powdered pectin tends to hydrate rapidly, so that clumps appear and the pectin forms a sheath coating the outer highly hydrated, hindering the solubility of the same (Rolin, 2002). This phenomenon can be prevented by pre-mixing the powdered pectin with other water-soluble material, thus improving the solubility thereof (DeVries, 1986). The ability to form gels is mostly an exploited property in industry; however, functionality is limited due to the particular structure of pectin. Unlike HM pectin, LM pectin may induce precipitation using aluminum or copper ions (Thakur et al., 1997).

Oakenfull (1991) points out that the main forces related to the aggregation of pectin molecules are hydrogen bonds and hydrophobic interactions. Gel formation occurs by hydrogen bonds between free carboxyl groups of the pectin molecules and the hydroxyl groups of some neighboring molecules. When pectin has a neutral or slightly acidic pH, most of the esterified carboxyl groups are present as partially ionized salts. When the pH of the solution is reduced, the carboxyl ions are converted into carboxylic acid groups in their ionized majority (Kohn, 1987). With the decrease of negative charges interactions between pectin and water molecules are reduced, also reducing the repulsion forces between the molecules of pectin. The presence of sugars significantly reduces the hydration of pectin, because these sugars compete for water. When pectin is exposed to these conditions reduces their ability to remain in a dispersed state (Grant et al., 1973).

Historically and by the large diversity, pectic substances were grouped in three categories, which are divided in basis to the size and number of esters groups in the molecule: Protopectin, pectinic acid and pectic acid. Protopectin is a high molecular weight methylated polymer of GalA found in immature fruit; it is insoluble in water and cannot form gels. As the fruit ripe, natural enzymatic hydrolysis of protopectin, lead the production of pectinic acids; this group of high molecular weight polymers and varying degree of methyl-esterification, are known as pectins. Pectins are soluble in water and can form gels in suitable conditions. Salts formed of pectinic acids are called pectinates (Nussinovitch, 1997).

The pectic acids do not contain methyl groups in its molecule, it found in overripe fruits, are formed by the action of the enzymes that cause depolymerization and demethylation of the pectinic acids, producing demethylated short-chains that cannot form gels, its salts are called pectates (Vaklavic and Christian, 2008).

Pectin consists of a set of polysaccharides in the cell wall of higher plants and can constitute a third of the dry matter. In the middle lamella of the cell wall are higher concentrations of pectin; this concentration is reduced while passing through the primary wall to the plasma membrane (Kertesz, 1993). Although pectin was discovered over 200 years ago, its specific structure and association with functions are still fields of continuous research and the main limitations are those related with changes due to the developmental process and isolation methods. In addition, some active fields of research are the determinations of changes as affected by storage and processing of plant material.

2. Pectin Biology Composition

2.1. Structural Diversity

Pectin structure is changed during growth, in function to species, tissues, developmental stages and presence of stress conditions. In optimal development, cell wall disassembly induce fruit softening and changes in structural components, like pectic polymers by high solubilisation of pectins, a loss of Gal and Ara, a reduction of the hemicellulosic content among others (Mohnen, 2008).

Residues of GalA comprise approximately 70% of pectin backbone. Pectic polysaccharides are covalently linked; HG is the most abundant with about 65% of the pectin, RG-I 20% to 35%, XG and RG-II only 10% each; while XG can be found in leaves and reproductive tissues.

Coenen (2007) revealed the existence of interconnections between pectin structural elements and their assembly within the cell wall, also discovered new points of junction between pectic subunits by using analytical techniques such as liquid chromatography-mass spectrometry (LC-MS) and characterized oligosaccharides obtained by controlled acid hydrolysis; this research showed that apple pectin obtained from HG and XG are covalently linked to RG-I, also degradation by beta-elimination showed that arabinans contain fragments

RG-I that are playing an important role with its complementary activity. A brief description of the majority groups of pectins is given bellow.

HG

Represents the most abundant pectic polysaccharide in cell wall, has a simple primary structure, it is a linear homopolymer formed by (1→4)-α-D-GalA residues. Contains about 200 GalA units and are about 100 nm long, the backbone residues, depending on the plant source, may be methyl-esterified at the C-6 carboxyl group or partially acetylated at the C-2 and C-3. HG is distinguished for its ability to form gels, conferring a functional application widely exploited in the food industry. DE allows to pectin the formation of gels by the way of binding at different intervals the carboxyl groups of neighboring chains thought ionic bridges with Ca^{2+} or Mg^{2+} (Carpita and McCann, 2000). It is assumed that methyl esterification has the major influence on the gelling properties of pectins, e.g., HG with a high degree of methyl esterification does not produce gel in the presence of Ca^{2+}, which could be due to more than 50% of the GalA are esterified with methyl group, however unmethylated may ionically interact with Ca^{2+} to form a calcium-pectate gel, if instead of adding calcium, the enzyme pectin esterase is added may slow the deesterification of pectins. The increase of calcium-pectate at the middle lamella in fruit like peach induces gel formation *in vivo*, whereas in mature cells, there is observed a decrement on methoxylation degree, this means, there is lower percentage of GalA units esterified by methanol (Zhou et al., 2000; Ridley et al., 2001). In terms of application at the level of food industry, there are necessary both type of pectins, LM pectin and HM pectin. There are two kinds of structurally modifies HGs, xyloglacturonano and RG-II, the latest described bellow.

RG-I

Is a rod-like heteropolymer of the repeating (1→2)-α-L-Rhamnose-(1→4)-α-D-GalA disaccharide units and nearly to 20 and 80% of the GalA residues may be O-acetylated on C-2 y C-3, backbone GalA residues may be O-acetylated, however its composition in plants depends on the source and isolation method. RG-I may contain arabinan, galactan and arabinogalactan of several configurations and sizes attached to the O-4 of many rhamnose residues (about half of the units). Its chains can be wrapped around the microfibrils, which enables anchoring of pectins and cellulose, keeping together, also act as union mechanisms between pectin and hemicellulose-cellulose networks (Caffall and Mohnen, 2009).

RG-II

Is a complex and highly conserved polysaccharide composed of a backbone of (1→4)-α-D-GalA residues substituted at C-2 or C-3, with side chains containing the richest diversity and linkage structures known. It is composed of at least 12 different glycosyl residues linked by at least 20 different glycosidic linkages, including some uncommon sugars like 2-*O*-methyl fucose, 2-*O*-methyl xylose, D-apiose, 3-C-carboxy-5-deoxy-L-xylose (aceric acid), 3-deoxy-D-lyxo-2-heptulosaric acid (Dha) and 3-deoxy-D-*manno*-2-octulosonic acid (Kdo) which is less abundant and appears to act as a cellular signal molecule in plants. However its complex structure is highly conserved in the plant kingdom and has an essential role in gymnosperms, monocotyledons and dicotyledons. It exists as dimers that are covalently linked by borate diesters, these dimers results in the cross-linking of the two homogalacturonan chains upon which the RG-II molecules are constructed and is essential for the formation of a three-dimensional pectic network *in muro*, interconnected by two diester bonds per boron atom, RG-II can be solubilized from the cell wall by treatment with endopolygalacturonase (Carpita and McCann, 2000; Mohnen, 2008).

2.2. Association with other Functional Groups

From the last two decades, great interest in studying the interactions between proteins and polysaccharides has aroused due to the influence of those interactions in various biological processes, such as the organization of the living cell, its importance in applications industry, such as micro-encapsulation, separation of proteins and processed foods. Several authors have been interested in studying the interactions of proteins with polysaccharides (Dubin et al., 1994; Schmitt et al., 1998; Doublier et al., 2000).

Proteins have molecules that bind to polysaccharide chains to form soluble complexes of protein-polysaccharide at the appropriated pH (Xia and Dubin, 1994). When the pH is under the isoelectric point of the protein (positive charged) the polysaccharide chain is found in its anionic form, therefore the protein-polysaccharide complex that was in dissolved form ultimately become a complex in its insoluble form, leading to the formation of clusters that increase its molecular weight and can precipitate. But, when the pH of the medium is reduced below the pK_a, carboxylic acid-based polysaccharides (e.g. pectin), can dissociate with insoluble complexes of proteins, even when these proteins are not found interacting with chain polysaccharides, this is due to the low load chain polysaccharides also the electrostatic repulsion between the protein under these conditions is positively charged (Dickinson, 1998). Interactions between pectins and proteins are usually carried out by electrostatic interaction, for this reason it is very important to control the physicochemical parameters that affect these interactions such as pH, ionic strength and polysaccharide-protein ratio (Schmitt et al., 1998).

The ability of pectins to interact with functional groups has been exploited in the food industry. Pectin interacts with whey proteins, improving the physicochemical and structural characteristics of the products made from them, because the pectin improves the interaction of these proteins with water, emulsification, foam formation and gelation (Cayot and Lorient, 1997).

In dairy products, the interaction of proteins with the dispersing medium has some instability, but when the generation of multilayer emulsions is favored, an effect of stabilizing might be reached. Pectins may interact with β-lactoglobulin by the hydrogen bond between the carboxyl groups of pectin and peptide linkage of protein, therefore their interaction might be affected by pH, ionic strength and the structural features of the pectin (Tolstoguzov, 1997). At the industrial level, pectin is used as an emulsifying agent, gelling, thickener and texturizer and due to the economic importance of these quality parameters, the control and manipulation of the macromolecular interactions are key factors in the sensorial quality (Tolstoguzov, 1997; Sawyer, 2003). The interaction of pectin with lipids is not included in this document.

On the other hand, pectins as a part of the dietary fiber can be naturally associated to phenolic compounds. Pectin polysaccharides react with phenolic acting as a trapping network or as a source of chemical interactions; this due to the hydrophilic hydroxyl groups and hydrophobic aromatic ring, respectively that can also bind to proteins giving a more complex matrix (Carpita and McCann, 2000). This nature allows the assembly of the cell wall components *in vivo*. The chemical associations responsible for the interaction are hydrogen bonds, hydrophobic interactions, and covalent bonds. These interactions may also be mediated by the specific characteristics of porosity and surface of contact (Saura-Calixto, 2010). In general, it is considered that GalA may interact with the phenolic compound.

2.3. Changes during Plant Growth and Development

Pectins are susceptible to numerous modifications that alter their conformations and connection to the cell wall. We discuss at the beginning of this chapter, the presence of numerous acidic residues in pectins that might be esterified with a methyl, acetyl or other unidentified functional group. This esterification occurs in Golgi, a complex organelle made of stacks of flattened vesicles containing proteins. Movement of pectins through Golgi toward the plasma membrane have been studied by immunocytochemical analysis using monoclonal antibodies (e.g. LM5 and LM6 that bind to side RG-I chains of arabinans and galactans), leading to the finding that the main traffic of pectins are identified in the cytoplasm. Also subcellular break of pectin biosynthetic enzymes reveals that pectins are synthesized in the Golgi followed by HG and RG-I transport through the endoplasmic reticulum in transition vesicles to the *cis* face of Golgi, continuing to the medial and *trans* where esterification occurs. Then pectic polysaccharides are packaged and assembled into vesicles in *trans*-Golgi network for their transport to vesicular compartments where are expelled to the space between cells and plasma membrane in secretory vesicles. Pectin is finally inserted into the wall (Carpita and McCann, 2000).

Biosynthesis of HG, RG-I and RG-II requires different activities and a large number of genes and enzymes. HG is deesterified to varying degrees by pectin methylesterases in the cell wall or at the cell plate. This action converts it to a more negatively charged molecule that is available to interact with ions, enzymes, proteins, and other HG molecules. Unesterified HG is mainly found in the spatial compartment of the middle lamella, so that it is believed that a deesterification occurs once pectin is delivered into the wall. General synthesis involve at least 67 transferase enzymes for the complete biosynthesis, modification and degradation of pectic polysaccharides among those are methyl-, glycosyl-, and acetyl-transferases, while esterases, lyases and various glycosil hydrolases degrade and modify polysaccharides. Each one needs specific nucleotide-sugar substrates and acceptors, UDP-glucosa and GDP-glucose formation are pathway for nucleotide sugar interconversion. Enzymes degrading pectic polysaccharides are very important as molecular tool for carbohydrates study, among the most important are endopolygalacturonase, pectin and pectate lyase, and pectin methyl esterase (Ridley et al., 2001; Mohnen et al., 2008).

Metabolism during fruit ripening is a key process in the reduction of fruit firmness. It involves the combined action of different proteins and hydrolytic enzymes on the cell wall, that additionally induces solubilization and enzymatic depolymerization at the level of pectin mainly but also hemicelluloses. Both solubilization and depolymerization process, causes a reduction and relaxation of the mechanical strength of polymer matrix producing dramatic effects in fruit firmness and resistance to the pathogen infection. Another characteristic derived of this metabolism during fruit ripening, is the loss of neutral sugars Gal and Ara from the cell wall but specially from pectins, this has been associated with a decrease in sugar concentration from the pectic polysaccharides branched structures (Mohnen et al., 2008).

Pectin solubilization is a central phenomenon to take into account when pectin distribution is considered to explain any cell wall-related process; this change cause a reduction of soluble pectin in chelating (connected by bridges of calcium) and alkali soluble pectins (linked pectins) that accumulates as a water soluble pectins. This in term of fruit ripening causes a dramatic reduction in cell wall integrity but for industrial purposes is desirable to increase pectin yields and to obtain the viscous properties depending of the specific industrialized product. Pectin depolymerization has another impact but now at the level of polymer interconnection and is represented by a characteristic downshift (reduction in molecular weight) from larger to smaller chains (Carpita and McCann, 2000). Therefore, considering that fruits and vegetables are the main source of commercial pectins, the developmental and ripening stage, as well as the tissue of extraction are key factors to take into account.

2.4. Sources of Commercial Pectins

Pectins are used in the food industry to provide texture, gelling properties and stabilization to foods. Most plant tissues have pectin in abundance, however there are few plants that can be used as a commercial source of pectin. This is because the ability of pectins to form a gel depends mainly on their molecular size and its DE, depending on the source of extraction capacity pectin gelation is different. Therefore, the fact that a plant containing a large amount of pectin is not a sufficient indicator of this plant is a good source of commercial pectin (Thakur et al., 1997).

Within the food industry, the main sources of pectin are apple pomace and citrus peel (Kertezs, 1993), both obtained from the residue of the processing industry. The apple pomace may contain from 10 to 15% of pectin (dry basis). While citrus peel contains from 20 to 30% (May, 1990). For purposes of the food industry, obtained pectins from apple and citrus have similar characteristics. Citrus pectins are light cream while apple pectins are generally darker. Alternative sources include sugar beet waste from sugar manufacturing, sunflower heads (seeds used for edible oil), and mango waste (Rolin, 2002). One of the main uses of pectins is as food additives, taking advantage of its gelling and stabilizing properties within the main products that are used are the jams, jellies, milk and sugar confectionery products (Sakai et al., 1993).

The commercial extraction of pectin is performed at low pH and high temperatures. For pectin extraction raw material is subjected to a pretreatment, generally blanching washing or drying, in order to inactivate not only lytic pectinases that can breakdown large pectins into small chains, but also pectin esterases that can modify the DE. Briefly, the extraction of pectins usually is made by treatment with acid (pH 1.5 to 3) and high temperatures (70-90 °C), within the main acids employed are hydrochloric acid, nitric acid and sulfuric acid. With this treatment the extraction and solidification of pectins found in the tissues of the plant source is allowed. Once a crude extract is obtained, pectin is separated from the pomace or peel residue by filtration or centrifugation. Subsequently, the pectin is separated from the purified extract by precipitation with alcohols or by salting out. After the removal of alcohol soluble impurities, the pectin is dried to remove moisture and the resulting product is powdered to a uniform size particle. Since pectins obtained by this extraction method are generally of high degree of esterification, in some instances if required, a deesterification is performed to produce pectin with a low DE (May, 1990; Sakai et al., 1993).

Under this common method of extraction, commercial pectins are essentially composed of HG and contain only small amounts of neutral sugars (Voragen et al., 1984).

3. Relation among Structure and Functionality

3.1. Ligand Affinity

Interactions between carbohydrates and proteins mediate numerous biological functions, such as signal transduction (Geijtenbeek et al., 2000), cell adhesion (Sacchettini et al., 2001), host–pathogen recognition (Karlsson et al., 1998), and the immune response (Henderson et al., 2009). Carbohydrate-recognizing proteins are involved in a great number of human disorders including biological responses and diseases, e.g., inflammation process and cancer. These key functional properties have stimulated significant efforts in drug design targeting carbohydrate-binding proteins (Ernst et al., 2009; Lepenies et al., 2010). Binding affinity is typically driven by a favourable enthalpic component that is partly offset by negative entropy. The nature of carbohydrate structure (lack of charges and hydrophobic surfaces) produces a low affinity and reduces the forming of strong interactions. In its place, the formation of carbohydrate–protein complexes involves relatively weak van der Waals interactions and hydrogen bonds to the carbohydrate hydroxyl groups, acetamides, and ring and glycosidic oxygens (Dam et al., 2002).

Some carbohydrates, like Gal, lactose, polylactosamine, and n-acetyl-lactosamine are considered as natural inhibitors, and have permitted the development of high-affinity carbohydrate-based inhibitors (Sörme et al., 2005) and Gal derivatives (Sörme et al., 2003). A degree of flexibility in the substitution at C-3 of the Gal has allowed the development of higher affinity ligands (Gunning et al., 2009).

Galectins are small soluble proteins that constitute a family of mammalian lectins, defined by a carbohydrate recognition domain (CRD) with a conserved sequence motif that confers affinity for β-galactoside-containing glycans (Leffler et al., 2004). At least 14 members of this family have been found in mammals and designated as galectin-1 through -14 (Visegrady et al., 2001). The structures of mammalian galectins can be identified as prototypes (galectins 1, 2, 5, 7, 10, and probably 11, 13, and 14) that exist as monomers or homodimers consisting of one carbohydrate recognition domain (CRD); chimera type (i.e. galectin-3) that contains a nonlectin N-terminal domain in addition to the CRD; and tandem-repeat type (galectins 4, 6, 8, and 9) composed of two different CRDs in a single polypeptide chain connected by a linker peptide (Barondes et al., 1994). Galectins have several important functions in both carbohydrate-dependent extracellular and carbohydrate-independent intracellular activities (Klyosov et al., 2012).

A specific example of carbohydrate-protein recognition is pectin type carbohydrate with protein galectin-3 (Gal3). The pectin is a Gal-based carbohydrate. Hydrogen bonding to O4 and O6 of the Gal and O3 of the glucose has been found to be important for stabilizing the lactose within the CRD (Morris et al., 2013). Studying the effects of pectin components on Gal3 at the molecular level, Gunning et al. (2009) showed that specific binding to Gal3 CRD involves neutral sugar side chains containing terminal Gal at the non-reducing end of the polysaccharide chain. The segments with higher capacity for binding to Gal3 were linear (1→4)-β-D-galactans. The simple linear structures might be expected to better facilitate binding within the Gal3 CRD. Potato RG-I, and enzymatically-modified RG-I, treated to remove Ara, showed that Gal3 specifically binds (1→4)-β-linear galactans. NMR data showed the existence of linear side chains linked to the RG-I backbone at an average length of 22 residues. Results showed that there is little evidence for specific binding of the polygalacturonic acid fragments with Gal3. This result is perhaps not surprising because the PGA is (1→4)-α-linked rather than (1→4)-β-linked, and the replacement of the secondary hydroxyl group with the carbonyl group will interfere with the hydrogen bonding to substituents on C-6 of the Gal within the primary Gal binding site in Gal3.

Yapo (2007) studied the removal of neutral sugar side chains from extracts of RG-I, enzymatically released from citrus pectin. He showed that a combination of the use of endo-α-L-arabinanase, endo-β-D-galactanase, α-L-arabinofuranosidase, and β-D-galactosidase can be used to debranch the RG-I simple. Treatment of RG-I to remove Ara significantly increased binding to Gal3, suggesting that the arabinans may sterically inhibit binding of Gal3 to galactans in the unmodified RG-I. RG-I regions can contain both (1→4)-β-linked and (1→3,6)-β-linked galactan side chains. The latter galactans are (1→3)-β-linked in their backbone, and the branching at C-6 would interfere with hydrogen bonding within the primary binding site on Gal3. The branching will sterically inhibit accommodation of terminal Gal residues within the binding site.

Thus, acid treatment of citrus pectin would enhance removal of entire chains of galactans and single units of Ara from RG-I regions and promote the release of both linear galactans and galactans lightly substituted with Ara. The stereochemistry of the linear (1→4)-β-D-arabinogalactan and (1→4)-β-D-galactan side chains suggest that they are the most likely candidates for binding to Gal3. This conclusion is reinforced by experimental evidence for the anticancer activity of arabinogalactans (Beuth et al., 1998).

HG-domains derived from citrus pectin have showed no specific interaction with Gal3. Interestingly, Fan et al. (2010) found that HG-rich ginseng pectin fractions significantly inhibited L-929 fibroblast cell migration, and this was enhanced when HG and RG-I-rich fractions were placed in contact. This observation led them to suggest that the HG domain might be an important functional element for the inhibition of cell migration, along with the RGI domain that could interact with Gal3. Vayssade et al. (2010) showed that an okra RGI reduced cell-cell contact and adhesion, increased apoptosis and decreased cell proliferation in a mouse melanoma cell line. As reported, okra pectin has a unique structure of almost pure RG-I with short galactan side chains (Gunning et al., 2009). Interaction of both Gal3 and okra RG-I has been observed on the cell membrane, however, knowing that pectin structures will vary significantly for different extracts, detailed structural information is required to investigate the specific mechanism of binding to Gal3.

3.2. Linking Capacity

The healthy effects of pectin are receiving increasing interest. It is generally accepted that a high fiber diet is beneficial to health and pectin is an important soluble fiber component of fruits and vegetables. Dietary supplementation of pectin showed decreased levels of blood cholesterol, serum glucose levels and may also have anti-cancer activities (Willats et al., 2006). Dietary fiber is a term used for a variety of plant substances that are resistant to digestion by human gastrointestinal enzymes; it can be classified depending on their solubility in water in insoluble fiber (lignins, cellulose, and some hemicelluloses) or soluble fiber (pectins, gums, mucilages, and the remainder of the hemicelluloses).

Soluble fibers from oats and psyllium or those in pure form likewise pectin and guar gum, decrease total and LDL cholesterol. In contrast, insoluble wheat fiber and cellulose have no effect on reduction of saturated fats and cholesterol. There is a discussion as to the degree of cholesterol reduction caused by soluble fibers. The range of effects on total cholesterol varies. The variations on the levels of cholesterol after consumption of fiber include small sample sizes, different dosages of fiber, different background diets, concurrent changes in body weight, varying dietary control, and different types of subjects, although is also possible that certain fibers reduce-lower cholesterol more effectively than others (Brown et al., 1999).

Some studies also suggest that the beneficial effects of pectin are closely related to its structural characteristics. For example, Briggs et al. (1997) suggested that pectin with less methoxyl content and lower molecular weight (<10000 kDa) is more efficient for cancer metastasis prevention, whereas pectin with greater methoxyl content and higher molecular weight is a better cholesterol-lowering agent. The correlation study of the structure-function relationships suggest that a higher inhibitory activity of pectin may be attained with higher methoxyl content, lower neutral sugar content, and lower proportion of low molecular weight pectin. Higher neutral sugar content of pectin does not reduce the cholesterol level (Liu et al., 2001).

On another hand, many studies have shown that pectins also reduce blood glucose levels (Giacco et al., 2002; Ou et al., 2001; Kalkwarf et al., 2001). Although the mechanism of this effect is unclear, there is evidence that several factors may influence glycemic index after an oral load of glucose, as rate of gastric emptying, rate of intestinal absorption, hormonal gastrointestinal response, hepatic glucose balance and cellular metabolism of glucose. The effect of various pectins (HM or LM pectins) on the intestinal absorption of glucose was investigated in gut-perfused rats finding that HM and LM pectins have effect on decrease glucose levels in rats (Kim et al., 2005). However, the inhibitive effect of HM pectin (56%) was greater than the observed in LM pectin (18%), probably due to the higher viscosity compared with LM pectin, this in consequence to the presence of soluble fibers (Flourie et al., 1984). The main effect of adding soluble fibers to the diet is a decrease in postprandial hyperglycemia. However, some early studies have demonstrated that dietary fibers have no effect on glucose absorption (Schwartz et al., 1980). Forster and Hoos (1977) observed that neither pectin nor cellulose impaired jejunal glucose absorption, but that pectin decreased serum glucose in response to an oral carbohydrate load.

Therefore we discuss here that the effect of soluble fibers on intestinal absorption of glucose is controversial. There are many contradictory data that may result from the differences in the type of soluble fiber (e.g. physicochemical property), experimental period, experimental technique, levels of soluble fiber and glucose, and species studied (Kim et al., 2005).

3.3. Water Binding Capacity

As we detailed before, pectin is a hydrophilic biopolymer due to the large number of polar hydroxyl and carboxyl groups in its molecule. The hydration properties or water binding capacity of pectin are crucial for their potential applications in different areas. When the pectin is dispersed in water, some of the acid groups are ionized and water binds both charged groups as the polar molecules. The negative charge of the pectin molecules, together with its attraction for water, can conduce to the formation of a stable gel. The interactions between this biopolymer and water depend on several interdependent intrinsic and extrinsic factors. Some important information is presented below.

3.3.1. Hydrophilic Groups Composition

The main hydrophilic groups in the major pectin component GalA are hydroxyl, amide and carboxyl groups. Each one of these groups differ in their water binding ability and, in case of carboxyl groups, the counter ions have an additional influence (Ping et al., 2001). The amount of bound water increased markedly with ionic groups introduced into the biopolymer structure. HM pectin shows almost twice the water-binding capacity of LM pectin. The inference is that hydrophobicity decreases with decreasing methylester content of pectins (Walter et al., 1991). However, Panchev et al. (2010) studied the absorption of water from hydrocolloids (as pectic substances) and concluded that is a complex process influenced from multiple factors as physico-chemical composition, size of the macromolecules, temperature, degree of crystallinity, and amorphous state, among others. Neutral sugars, mainly in the pectin side chains, contain hydroxyl groups and can influence the shape of sorption isotherms.

3.3.2. Availability of Polar Groups

Water molecules can form hydrogen bonds with many hydrophilic and charged groups of biopolymers, such as $-NH_2$, y -NH, -OH, -COOH, $-COO_2$. The water binding capacity may be based on weakening hydrogen bonds, dipole–dipole and intra-inter-macromolecular interactions due to the shielding of these mainly attractive forces by water molecules. Materials with strong intra- and intermolecular interactions, in case of pectins mainly hydrogen bonds and hydrophobic interactions, require a certain amount of water for breaking the hydrogen bonds and starting the water binding (Matveev et al., 2000). The availability can additionally depend on molecule conformation and resulting position of the polar groups; equatorial hydrophilic groups are known to be more accessible for water binding than axial hydrophilic groups. In HM pectin, the methoxyl group might shield the polar groups from hydrogen bonding (Bettelheim et al., 1956; Lewicki et al., 2004).

3.3.3. State

The state is crucial for softening, swelling and any water binding depends strongly on sample pretreatment and thermal history, such as drying or storage (Mathlouthi et al., 2001; Tsami et al., 1999). In crystalline and glassy materials water is bound at first to polar groups at the surface until -after certain swelling, softening and plasticization into a rubber state - internal groups are better available for hydration. Amorphous materials can bind water rapidly also to the more internal polar groups (Panchev et al., 2010). In general, a high resistance of the material to swelling inhibits the total water absorption (Ping et al., 2001).

About particle and bulk powder properties, a smooth and partly horned surface can delay swelling and inhibit water binding considerably; a rough inhomogeneous surface can promote it. The porosity of the particles as well as the bulk powder is highly important, too. More voids between as well as within particles accelerate water uptake and binding (Tsami et al., 1999).

4. Pectin As a Dietary Fiber

Pectin is the part of plant derived products and together with gums, contributes with the major portion of soluble fibers in foods. As a hydrosoluble compound, pectin constitutes the fraction of soluble dietary fiber and modifies the properties of water mainly producing thickness. As pectin cannot be digested or broken down by the digestive machinery, adds almost no energy to the diet but its function occurs after it passes through the stomach to intestines. Evidences of the health integral benefits of soluble fiber are widely reported around the world; however, despite progressive advances in the understanding about fiber interaction with human and microbial populations, the recommended intake is not well established and is dependent of genre, age, and origin of consumer, among other factors.

Pectins as a portion of the dietary soluble fiber can have immediate physiological effects after consumption and so far, this has been the most popularized functional property. Pectin is used in short-term to prevent constipation and to stabilize the levels of glucose uptake; or, in long-term the prolonged intake of soluble dietary fiber has been used to increase glucose tolerance, increase insulin sensitivity, and reduce serum cholesterol levels. The mode of action in the gastrointestinal tract are explained in part by the effects of changes in the timing of digestion and nutrient absorption as a consequence of properties as water retention, increment of viscosity, gelling forming matrices and specific ion affinity among others. In addition, its fermentation in the large intestine produces short chain fatty acids that influence cell function, especially of the endocrine system. The physicochemical properties and the biological properties of dietary fiber are dependent on several factors, including the plant source, tissue, particle size and type of processing. This is due the chemical-structural features such as polymer molecular weight, composition, surface area, solubility and pH (Wong et al., 2006). Is important to highlight that fermentation of soluble fiber is a function of the resident microbiota, having important variations among individuals.

The moderate consumption of fiber is necessary to promote an effect in the control of certain types of diabetes and to reduce the reabsorption of bile salts in the small intestine that benefit the cholesterol metabolism and the development of heart disease, may also limit the absorption of important ions such as calcium and iron (Sriamornsak, 2003).

The enzymes required for the degradation of complex carbohydrates are absent in the human genome but present and expressed from intestinal microorganisms, in consequence the degradation and further fermentation of pectins is dependent of the diversity of the microbiota at the digestive tract. Some common fermentative species are *Aerobacillus*, *Lactobacillus*, *Micrococcus* and *Enterococcus*. This bacteria produce pectolytic and fermentative enzymes capable to hydrolyze pectin into short chain fatty acids (acetic acid, butyric acid, propionic acid) and carbon dioxide. Is important to highlight that short chain fatty acids are widely studied by its benefit and protective effects against disorders and diseases including obesity, diabetes, coronary heart disease and elevated cholesterol (Rideout et al., 2008; Cantarel et al., 2012; Huttenhower et al., 2012).

In deduction, the functional properties of the soluble fiber can vary depending on the diet itself and consumption amount, period of use and type of fiber. For example, dextrans and peptidoglycan are mainly degraded in the mouth while the gastrointestinal tract has a greater potential for degradation of other soluble gums. Finally, the fiber properties are also influenced by the genetic basis of the consumer and the microbiome present in the gastrointestinal tract. Detailed functional benefits derived from evidence *in vitro* and *in vivo* is described in the following sections.

5. Other Functional Nutraceutical Properties of Pectins

5.1. Immunological and Prebiotic

A wide range of polysaccharides, extending from homopolymers to highly complex heteropolymers, has been reported to exhibit immunomodulating activities. So far, no clear information has been obtained on how a polysaccharide has to be structurally designed in order to have an optimal inducing effect on specific immune cells. However, for both glucans and pectic polymers, high molecular weight appears to increase the bioactivities, and branching of side chains are needed for optimal activity (Lin et al., 2005).

Previous studies have shown that pectic polysaccharides from several Malian medicinal plants like Entada africana, Trichilia emetica, Vernonia kostchyana, Cochlospermum tinctorium, Biophytum petersianum, and Glinus oppositifolius have immunomodulatory activities in biological screening assays, with effects varying from complement fixation to activation of macrophages and dendritic cells. The ability of pectic polysaccharides to modulate components of the immune system responses, may in part explain some of the beneficial effects of medicinal plants. Previously published data on the structure of bioactive pectin indicates that their observed biological activities are due to rhamnogalacturonan regions rich in neutral sugar side chains such as arabinan, galactan and arabinogalactan (Inngjerdingen et al., 2010).

On another hand, although specific nutrients are known to be important in the development and function of the immune system, less is known about the potential of dietary fibers to impact on immune function. However, studies demonstrating a lower incidence of bacterial translocation across the gut barrier with the administration of dietary fiber suggest that this dietary nutrient modulates immunity (Xu et al., 1998). The mechanism for the effect of fermentable dietary fibers on immune function in the gut has not been established. A prebiotic fiber is neither hydrolyzed nor absorbed in the upper part of the gastrointestinal tract, and becomes a selective substrate for one or a limited number of beneficial colonic bacteria thereby altering the microflora of the gut (Gibson et al., 1995). There is strong evidence indicating that consumption of prebiotic fibers increases the proportion of beneficial lactic acid bacteria in the human colon (Gibson et al., 1995). Also other oligomers that may be prebiotics are lactulose, and oligosaccharides containing xylose, mannose, and Gal (Gibson et al., 1998). The studies conducted with recognized prebiotic fibers have shown increased lymphocyte and/or leucocyte numbers in the gut-associated lymphoid tissues (Pierre et al., 1997) and peripheral blood (Kaufhold et al., 2000).

5.2. Antioxidant

The development of many human diseases is accompanied by activation of free radical peroxide oxidation of lipids, considered as a mechanism of the destruction of biological membranes. The induction of peroxide oxidation of lipids occurs during inflammatory processes, intoxications, cardiovascular problems, cancer, allergic reactions, and many other diseases. The protection of cells and tissues against the aggressive effects of free radicals is provided by the endogenous antioxidant system. However, the endogenous antioxidants cannot always prevent the development of oxidant stress. For this reason, it is of interest to search for some exogenous substances with antioxidant properties that might be used as drugs for prophylactics or therapy of diseases accompanied with increased reactions of free radical oxidation (Khasina et al., 2003). Antioxidants are substances used in order to both overcome the effects of free radicals to human health and to improve the quality of foods. Their efficiency is due to their capability to bind free radicals thus, preventing their destructive oxidant activity. There are natural and synthetic antioxidants, with the natural ones being safer. Fruits, vegetables, grains and herbs are some sources of natural antioxidants. Tocopherols, flavonoids and phenolic acids are the most important groups of natural antioxidants and can be found in microorganisms and plants (mainly tea and herbs).

The function of natural antioxidants and dietary fiber in foods and biological systems has received much attention. Fruits and vegetables play a significant role in the human diet providing protection against cellular damage caused by exposure to high levels of free radicals, while also aiding digestion (Dillard et al., 2000). This is attributed to the fact that these foods provide an optimal mix of antioxidants such as vitamin C and E, polyphenols, carotenoids (Eastwood et al., 1999) and complex carbohydrates.

Although traditionally pectin is used as a thickening agent in the food industry, such as in the production of jams and jellies, it has been shown to have ability to inhibit free radicals. Recent studies also suggest that the beneficial effects of pectin are closely related to its structural characteristics (Liu et al., 2001) and the ability to bind phenolic compounds to its structure as we discuss earlier. Hydroxycinnamic acids, especially ferulic acid and p-coumaric acid are linked by ester-linkage to pectic polysaccharides in spinach and other dicotyledonous. Unlike monocots, dicots ferulic acid is linked to pectic polysaccharides through ester at C-2 hydroxyl group of arabinofuranose or the C-6 hydroxyl group linked galactopyranose residues (50-55% or 45-50%, respectively) (Fry, 1982; Ralet et al., 1994).

The crude pectin hydrolyzates derived by pectinase cleavage exhibited significant antioxidant activity in the serum of hyperlipidemic mice. The results revealed great prospects for the application of haw pectin oligosaccharides as antioxidants in the development of functional food in human subjects (Li et al., 2014). This is an interesting field of research with potential application in the production of antioxidant pectin oligosaccharides.

5.3. Apoptosis

Cancer therapy is aimed at either the primary tumor or metastatic cells. Because of the differences in the response of primary and metastatic cancers, most therapies do not address both cancer types. Pectin is a compound that appears to be able to inhibit cancer metastasis and primary tumor growth in multiple types of cancer in animals. Although pectins were initially recognized as compounds capable of inhibiting metastatic lesions, more recently, pectins have been shown to reduce primary tumor growth (Jackson et al., 2007).

Recently, great progress has been made in elucidating structure/function relationships of pectin at a molecular level, and this is leading to the design of pectins with specific functionalities. Studies suggest that small molecular weight pectin fragments, rich in galactans, can bind to the carbohydrate recognition domain (CRD) on the pro- metastatic protein Gal3. Galectins facilitate cell–cell interactions by binding to Gal-containing molecules on neighboring cancer cells. In human colon, stomach and thyroid cancers, the amount of galectin increased with the progression of cancer. Blocking Gal3 expression in highly malignant human breast, papillary, and tongue carcinoma cells led to reversion of the transformed phenotype and suppression of tumor growth in nude mice (Nangia-Makker et al., 2002). It has been proposed that the pH-modified citric pectin blocks binding of galectins, and thus, inhibits tumor cell–cell interactions. The potential impact of blocking galectin action includes inhibition of the aggregation of cancer cells to each other and to normal cells, thus inhibiting metastatic lesions.

Furthermore, it has been reported the proapoptotic effects of modified citrus pectin in combination with cytotoxic drugs (such as cisplatin, staurosporine, etoposide, bortezomib, dexamethasone, and doxorubicin) on apoptosis and cell proliferation for different types of cancer cells. One example is the synthesis of pectin nanoparticles for antineoplastic drugs delivery, where pharmacokinetics studies in healthy mice confirmed that the drug possessed a much longer half-life that the free drug in the circulation system. However, recognition studies to Gal3 for these nanoparticles are still in progress.

This raises the possibility that modified pectins may be utilized in a potentially safe, nontoxic approach for preventing or reducing different kind of cancer, as well as, prostate cancer (Glinskii et al., 2012), colon cancer (Nangia-Makker et al., 2002), breast cancer (Nangia-Makker et al., 2002), melanoma (Inohara et al., 1994), multiple myeloma (Chauhan et al., 2005) and hemangiosarcoma (Johnson et al., 2007).

Although the usefulness of pectin in cancer therapy is beginning to be appreciated, the mechanism of induction of apoptosis by pectin is not known. The elucidation of the mechanism(s) of action of pectin is complicated by (i) the structural complexity of this plant-derived cell wall polysaccharide, (ii) the modifications in pectin structure resulting from the process of its extraction from plants, and (iii) the additional modifications of pectin structure that result from the diverse fragmentation techniques used to produce specialized pectin (Glinsky et al., 2009).

ACKNOWLEDGMENTS

Authors express our gratitude to Conacyt (National Council of Science and Technology, México) for financial support under project CB–2010–158342 and doctoral scholarships. We thank the careful review and comments of Dra. Marcela Vergara-Jiménez and Ing. Nancy Varela-Bojórquez.

REFERENCES

Barondes, S. H.; Cooper, D. N. W.; Gitt, M. A. & Leffler, H. (1994). Galectins: structure and function of a large family of animal lectins. *J. Biol. Chem., 269*, 20807–20810.

Bettelheim, F. A.; Sterling, C. & Volman, D. H. (1956). Pectin substancesewater. I. Structural changes in the polygalacturonide chains during water adsorption. *J. Polym. Sci., 22*, 303–314.

Beuth, J.; Ko, H. L.; Schirrmacher, V.; Uhlenbruck, G. & Pulverer, G. (1998). Inhibition of liver-tumor cell colonisation in 2 animal tumor-models by lectin blocking with D-galactose or arabinogalactan. *Clin. Exper. Metastasis, 6*, 115–120.

Briggs, S. (1997). Modified citrus pectin may halt metastasis. *Nutr. Sci. News, 2*, 216–218.

Brown, L.; Rosner, B.; Willett, W. W. & Sacks, F. M. (1999). Cholesterol-lowering effects of dietary fiber: a meta-analysis. *Am. J. Clin. Nutr., 69*, 30–42.

Caffall, K. H. & Mohnen, D. (2009). The structure, function, and biosynthesis of plant cell wall pectic polysaccharides. *Carbohydr. Res., 344,* 1879–1900.

Cantarel, B. L.; Lombard, V. & Henrissat, B. (2012). Complex carbohydrate utilization by the healthy human microbiome. *PLoS ONE* 7, e28742.

Carpita, N. & McCann, M. (2000). The cell wall: Biochemistry & molecular biology of plants (First Edition). Rockville, Maryland, USA: American Society of Plant Physiologists.

Cayot, P. & Lorient, D. (1997). Structure-function relationship of whey proteins. In: Damodaran, S., Paraf, A. (eds): Food Proteins and Their Applications. Marcel Dekker Inc., New York. 1997. 225–291.

Chauhan, D. L. (2005). A novel carbohydrate-based therapeutic GCS-100 overcomes bortezomib resistance and enhances dexamethasone-induced apoptosis in multiple myeloma cells. *Cancer Res.*, 65, 8350–8358.

Coenen, G. J. (2007). Structural characterization of native pectins. Ph.D. thesis, Wageningen University, the Netherlands.

Dam, T. K. & Brewer, C. F. (2002). Thermodynamic studies of lectin-carbohydrate interactions by isothermal titration calorimetry. *Chem. Rev.*, 102, 387–429.

Datta, S.C. (1994). Plant physiology. New Age Intl. Ltd. Delhi.

DeVries, J. A.; Hansen, M.; Søderberg, J.; Glahn, P. E. & Pedersen, J. K. (1986). Distribution of methoxyl groups in pectins. *Carbohyd. Polym.*, 6, 165–176.

Dickinson, E. (1998). Stability and rheological implications of electrostatic milk protein-polysaccharide interactions. *Trends Food Sci. Technol.*, 9, 347–354.

Doublier, J. L.; Garnier, C.; Renard, D. & Sanchez, C. (2000). Protein–polysaccharide interactions. *Curr. Opin. Colloid Interface Sci.*, 5, 202–214.

Dubin, P. L.; Gao, J. & Mattison, K. W. (1994). Protein purification by selective phase separation with polyelectrolytes. *Sep. Purif. Method.*, 23, 1–16.

Eastwood, M. A. (1999). Interaction of dietary antioxidans *in vivo*: how fruit and vegetables prevent disease? *QJM.*, 93, 527–530.

Ernst, B., & Magnani, J. L. (2009). From carbohydrate leads to glycomimetic drugs. *Nat. Rev. Drug Discov., 8*, 661–677.

Fan, Y.; Cheng, H.; Li, S.; Wang, J.; Liu, D. & Hao, M. (2010). Relationship of the inhibition of cell migration with the structure of ginseng pectic polysaccharides. *Carbohydr. Polym., 81*, 340–347.

Flourie, B.; Vidon, N.; Florent, C. H. & Bernier, J. J. (1984). Effect of pectin on jejunal glucose absorption and unstirred layer thickness in normal man. *Gut, 25*, 936–941.

Forster, H. & Hoos, I. (1977). Influence of gums on intestinal absorption. *Nutr. Metab.*, 21, 262–264.

Fry SC. (1982). Phenolic components of the primary cell wall: feruloylated disaccharides of D-galactose and L-arabinose from spinach polysaccharide. *Biochem. J.* 203, 493–504.

Geijtenbeek, T. B.; Torensma, R.; Van Vliet, S. J.; Van Duijnhoven, G. C, Adema, G. J., Van Kooyk, Y.; Figdor, C. G. & Geijtenbeek, T. B. (2000). Identification of DC-SIGN, a novel dendritic cell-specific ICAM-3 receptor that supports primary immune responses. *Cell, 100*, 575–585.

Giacco, R.; Clemente, G. & Riccardi, G. (2002). Dietary fiber in treatment of diabetes: myth or reality? *Dig. Liver. Dis., 34*, S140–144.

Gibson, G. R. & Roberfroid, M. B. (1995). Dietary modulation of the human colonic microbiota: introducing the concept of prebiotics. *J. Nutr., 125*, 1401–1412.

Gibson, G. R. (1998). Dietary modulation of the human gut microflora using prebiotics. *Br. J. Nutr., 80*, 209–212.

Glinskii, O. V. (2012). Inhibition of prostate cancer bone metastasis by synthetic TF antigen mimic/ galectin-3 inhibitor lactulose-L-leucine. *Neoplasia, 14*, 65–73.

Glinsky, V. V. (2009). Modified citrus pectin anti-metastatic properties: one bullet, multiple targets. *Carbohydr. Res., 344*, 1788–1791.

Grant, G. T.; Morris, E. R.; Rees, D. A.; Smith, P. J. C. & Thom, D. (1973). Biological interactions between polysaccharides and divalent cations: The egg-box model. *FEBS Lett.*, 32, 195–198.

Henderson, N. C. & Sethi, T. (2009). The regulation of inflammation by galectin-3. *Immunol. Rev., 230*, 160–171.

Huttenhower, C.; Abubucker, S.; Segata, N.; Goll, J.; Schubert, A. M.; Izard, J.; Cantarel, B. L.; Rodriguez-Muller, B.; Zucker, J.; Thiagarajan, M.; Henrissat, B.; White, O.; Kelley, S. T.; Methé, B.; Shloss, P. D.; Gevers, D. & Mitreva, M. (2012). Metabolic reconstruction for metagenomic data and its application to the human microbiome. *Plos Comput. Biol., 8*, e1002358.

Inngjerdingen, M.; Ghildyal, P.; Gronhaug, T. E.; Rusten, A.; Skogsrud, M.; Rolstad, B.; Diallo, D.; Michaelen, T. E. & Paulsen, B. S. (2010). Chemical composition and immunological activities of polysaccharides isolated from the malian medicinal plant Syzygium guineense. *J. Pharmacogn. Phytother., 2*, 76–85.

Inohara, H. A. (1994). Effects of natural complex carbohydrate (Citrus pectin) on murine melanoma cell properties related to galectin-3 functions. *Glycoconjugate J., 11*, 527–532.

Ishii, T. (1997). O-acetylated oligosaccharides from pectins of potato tuber cell walls. *Plant Physiol.* 113, 1265–1272.

Jackson, C. L.; Dreaden, T. M.; Theobald, L. K.; Tran, N. M.; Beal, T. L.; Eid, M.; Gao, M. Y.; Shirley, R. B.; Stoffel, M. T.; Kumar, M. V. & Mohnen, D. (2007). Pectin induces apoptosis in human prostate cancer cells: correlation of apoptotic function with pectin structure. *Glycobiology, 17*, 805–819.

Johnson, K. D. (2007). Galectin-3 as a potential therapeutic target in tumors arising from malignant endothelia. *Neoplasia, 9*, 662–670.

Kalkwarf, H. J.; Bell, R. C.; Khoury, J. C.; Gouge, A. L. & Miodovnik, M. (2001). Dietary fiber intakes and insulin requirements in pregnant women with type 1 diabetes. *J. Am. Diet. Assoc., 101*, 305–310.

Karlsson, A.; Follin, P.; Leffler, H. & Dahlgren, C. (1998). Galectin-3 activates the NADPH-Oxidase in exudated but not peripheral. *Blood, 91*, 3430–3438.

Kaufhold, J.; Hammon, H. M. & Blum, J. W. (2000). Fructo-oligosaccharide supplementation: effects on metabolic, endocrine and hematological traits in veal calves. *J. Vet. Med. A, 47*, 17–29.

Kertesz, Z. I. (1993). The pectic substances. Rolin, C. Pectin. In: Industrial gums. Interscience Publishers, Whistler R. L. & BeMiller J. N. (eds), New York. (3rd edn), New York.

Khasina, E. I.; Kolenchenko, E. A.; Sgrebneva, M. N.; Kovalev, V. V. & Khotimchenko, Y. S. (2003). Antioxidant activities of a low etherified pectin from the seagrass *Zostera marina*. *Russ. J. Mar. Biol., 29*, 259–261.

Kim, M. (2005). High-methoxyl pectin has greater enhancing effect on glucose uptake in intestinal perfused rats. *Nutrition, 21*, 372–377.

Klyosov, A. A. & Traber, P. G. (2012). Galectins in disease and potential therapeutic approaches. Washingtong, DC.: ACS Symposium Series.

Kohn, R. (1987). Binding of divalent cations to oligomeric fragment of pectin. *Carbohydr. Res.* 160, 343–353.

Komalavilas, P. & Mort, A. J. (1989). The acetylation at O-3 of galacturonic acid in the rhamnose-rich region of pectins. *Carbohydr. Res.*, 189, 261–272.

Leffler, H.; Carlsson, S.; Hedlund, M.; Qian, Y. & Poirier, F. (2004). Introduction to galectins. *Glycoconjugate J., 19*, 433–440.

Lepenies, B.; Yin, J. & Seeberger, P. H. (2010). Applications of synthetic carbohydrates to chemical biology. *Curr. Opin. Chem. Biol, 14*, 404–411.

Lewicki, P. P. (2004). Water as the determinant of food engineering properties. A review. *J. Food Eng., 61*, 483–495.

Li, P.; Li, S.; Dong, Y.; Zhu, R. & Liu, Y. (2014). Antioxidant activity of penta-oligogalacturonide, isolated from haw pectin, suppresses triglyceride synthesis in mice fed with a high-fat diet. *Food Chem.*, 145, 335–341.

Lin, Z. B. (2005). Cellular and molecular mechanisms of immuno-modulation by Ganoderma lucidum. *J. Pharmacol. Sci.*, 99, 144–153.

Liu, Y.; Ahmad, H.; Luo, Y.; Gardiner, D. T. & Gunasekera, R. S. (2001). Citrus pectin: Characterization and inhibitory effect on fibroblast growth factor-receptor interaction. *J. Agric. Food Chem.*, 49, 3051–3057.

Mathlouthi, M. (2001). Water content, water activity, water structure and the stability of foodstuffs. *Food Control*, 12, 409–417.

Matveev, Y. I.; Grinberg, V. Y. & Tolstoguzov, V. B. (2000). The plasticizing effect of water on proteins, polysaccharides and their mixtures. Glassy state of biopolymers, food and seeds. *Food Hydrocolloid,* 14, 425–437.

May, C. D. (1990). Industrial pectins: Sources, production and applications. *Carbohyd. Polym.*, 12, 79–99.

McNaught, A. D. (1996). Nomenclature of carbohydrates. International Union of Pure and Applied Chemistry and International Union of Biochemistry and Molecular Biology, 68, 1919–2008.

Mohnen, D. (2008). Pectin structure and biosynthesis. *Plant Biology*, 11, 266–277.

Mohnen, D.; Bar-Peled, M. & Somerville, C. (2008). Biosynthesis of plant cell walls. In Biomass Recalcitrance; Himmel, M., Ed.; Blackwell Publishing: Oxford.

Morris, V. J.; Belshaw, N. J.; Waldron, K. W. & Maxwel, E. G.Y (2013). The bioactivity of modified pectin fragments. *Bioac. Carbohydr. Diet. Fibre,* 1, 21–37.

Mukhiddinov, Z. K. (2000). Isolation and structural characterization of a pectin homo and rhamnogalacturonan. *Talanta*, 53, 171–176.

Nangia-Makker, P.; Hogan, V.; Honjo, Y.; Baccarini, S. & Tait, L. (2002). Inhibition of Human Cancer Cell Growth and Metastasis in Nude Mice by Oral Intake of Modified Citrus Pectin. *J Natl Cancer Inst, 24*, 1854–1862.

Nussinovitch, A. (1997). Hydrocolloid Applications. Gum Technology in The Food and Other Industries. Blackie Academic & Professional. Chapman & Hall, 82–103.

Oakenfull, D. G. (1991). The Chemistry of High Methoxyl Pectins. In: Walter, R. The Chemistry and Technology of Pectins. San Diego CA, USA. Academic Press.

O'Neill, M. A.; Eberhard, S.; Albersheim, P.; Darvill, A. G. (2001). Requirement of Borate Cross-Linking of Cell Wall Rhamnogalacturonan II for Arabidopsis Growth. Science, 294, 846–849.

Ou, S.; Kwok, K.; Li, Y. & Fu, L. (2001). *In vitro* study of possible role of dietary fiber in lowering postprandial serum glucose. *J. Agric. Food Chem.*, 49, 1026–1029.

Panchev, I. N.; Slavov, A.; Nikolova, K. & Kovacheva, D. (2010). On the water-sorption properties of pectin. *Food Hydrocolloid,* 24, 763–769.

Gunning, A. P.; Bongaerts, R. J. & Morris, V. J. (2009). Recognition of galactan components of pectin by galectin-3. *FASEB J.*, 23, 415–424.

Pierre, F.; Perrin, P.; Champ, M.; Bornet, F.; Meflah, K. & Menanteau, J. (1997). Short-chain fructo-oligosaccharides reduce the occurrence of colon tumors and develop gut-associated lymphoid tissue in min mice. *Cancer Res.*, 57, 225–228.

Ping, Z. H.; Nguyen, Q. T.; Chen, S. M.; Zhou, J. Q. & Ding, Y. D. (2001). States of water in different hydrophilic polymers-DSC and FTIR studies. *Polymer,* 42, 8461–8467.

Ralet, M. C.; Thibault, J. F.; Faulds, C. B. & Williamson, G. (1994). Isolation and purification of feruloylated oligosaccharides from cell walls of sugar-beet pulp. *Carbohydr. Res.* 263, 227–241.

Rideout, T. C.; Harding, S. V.; Jones, P. H. & Fan, M. Z. (2008). Guar gum and similar soluble fibers in the regulation of cholesterol metabolism: Current understandings and future research. *Vasc. Health Risk Manag.*, 4, 1023–1033.

Ridley, B. L.; O'Neill, M. A. & Mohnen, D. (2001). Pectins: Structure, biosynthesis, and oligogalacturonide-related signaling. *Phytochemistry*, 57, 929–967.

Rolin, C. (2002). Commercial pectin preparation. In: Pectins and their manipulation. Blackwell, CRC Press.

Sacchettini, J. C.; Baum, L. G. & Brewer, C. F. (2001). Multivalent protein-carbohydrate interactions. A new paradigm for supermolecular assembly and signal transduction. *Biochemistry,* 40, 3009–3015.

Sakai, T.; Sakamoto, T.; Hallaert, J. & Vandamme, E. J. (1993). Pectin, pectinase, and protopectinase: Production, properties, and applications. *Adv. Appl. Microbiol.*, 39, 213–29.

Saura-Calixto, F. (2010). Dietary fiber as a carrier of dietary antioxidants: An essential physiological function. *J. Agr. Food Chem.*, 59, 43–49.

Sawyer, L. (2003). β-lactoglobulin. In: Fox, P. F. & McSweeney, P. L. H. (eds): Advanced Dairy Chemistry. Volume 1: Proteins. Kluwer Academic/Plenum Publishers, New York.

Schmitt, C.; Sanchez, C.; Desobry-banon, S. & Hardy, J. (1998). Structure and technofunctional properties of protein–polysaccharide complexes: a Review. *Crit. Rev. Food Sci. Nutr.*, 38, 689–753.

Schols, H. A. & Voragen, A. G. J. (2002). The chemical structure of pectins. In: Pectins and their Manipulation. Blackwell, CRC Press.

Schwartz, S. E. & Levine, G. D. (1980). Effects of dietary fiber on intestinal glucose absorption and glucose tolerance in rats. *Gastroenterology,* 79, 833–836.

Sörme, P.; Arnoux, P.; Kahl-Knutsson, B.; Leffler, H. & Rini, J. M. (2005). Structural and thermodynamic studies on cation-II interactions in lectin-ligand complexes: high-affinity galectin-3 inhibitors through fine-tuning of an arginine-arene interaction. *J. Am. Chem. Soc.,* 127, 1737–1743.

Sörme, P.; Kahl-Knutsson, B.; Carlsson, S. & Nilsson, U. J. (2003). The design of monovalent galectin inhibitors based on 3-OH-modified galactosides. *FASEB J.,* 17, A162.

Sriamornsak, P. (2003) Chemistry of pectin and its pharmaceutical uses: A review. *Silpakorn Univ. Int. J.* 3, 206–228.

Thakur, B. R.; Singh, R. K. & Handa, A. K. (1997). Chemistry and uses of pectin — A review. *Crit. Rev. Food Sci. Nutr.*, 37, 47–73.

Tolstoguzov, V. B. (1997). Protein-Polysaccharide Interactions. In: Damodaran S., Paraf A. (eds): Food Proteins and Their Applications. Marcel Dekker Inc., New York.

Tsami, E.; Krokida, M. K. & Drouzas A. E. (1999). Effect of drying method on the sorption characteristics of Model Fruit Powders. *J. Food Eng.,* 38, 381–392.

Vaclavick, V. A. & Christian, E. W. (2008). Pectins and gums. Essential of Food Science. Foods Science Text Series. Springer, NY.

Vayssade, M.; El-Kirat, K. & Bernier, M. C. (2010). Anatase titanium dioxide nanoparticles induce the secretion of the cytokine IL6 by pre-osteoblast cells. *Toxicol. Lett.,* 196, S276–279.

Visegrady, B.; Than, N.G.; Kilar, F.; Sumegi, B. & Bohn, H. (2001). Homology modelling and molecular dynamics studies of human placental tissue protein 13 (galectin-13). *Protein Eng.,* 14, 875–880.

Visser, J.; Voragen, A. G. J. (1997). Progress in biotechnology 14: Pectins and pectinases. Elsevier, Amsterdam.

Voragen, A. G. J.; Schols, H. A.; Clement, A. J. & Pilnik, W. (1984). Enzymatic analysis of pectins. In: Philips, G.O.; Wedlock, D.J.; Williams, P.A. Eds, Gums and Stabilizers for the Food Industry 2. Applications of Hydrocolloids; Elsevier London.

Walter, R. H. & Taylor, S. (1991). *The Chemistry and Technology of Pectin.* united states of america: Academic Pres Inc.

Willats, W.G.; Knox, P. & Mikkelsen, J. D. (2006). Pectin: new insights into an old polymer are starting to gel. *Trends Food Sci. Tech.,* 17, 97–104.

Wong, J. M.; de Souza, R.; Kendall, C. W.; Emam, A. & Jenkins, D. J. (2006). Colonic health: fermentation and short chain fatty acids. *J Clin Gastroenterol.,* 40, 235–243.

Xia, J. & Dubin, P. L. (1994). Protein–polyelectrolyte complexes. In Macromolecular Complexes in Chemistry and Biology. Springer-Verlag: Berlin.

Xu, D. L.; Lu, Q. & Deitch, E. A. (1998). Elemental diet-induce bacterial translocation associated with systemic and intestinal immune suppression. *JPEN-Parent. Enter.,* 22, 37–41.

Yapo, B. M.; Lerouge, P.; Thibault, J. F. & Ralet, M. C. (2007). Pectins from citrus peel cell walls contain homogalacturonans homogeneous with respect to molar mass, rhamnogalacturonan I and rhamnogalacturonan II. *Carbohydr. Polym.,* 69, 426–435.

Zhou, H.; Ben-Arie, R. & Lurie, S. (2000). Pectin esterase, polygalacturonase and gel formation in peach pectin fractions. *Phytochemistry,* 55, 191–195.

In: Pectin: Chemical Properties, Uses and Health Benefits
Editor: Phillip L. Bush
ISBN: 978-1-63321-438-5

Chapter 11

PECTIN FILMS FOR APPLICATION IN FOOD PACKAGING: REVIEW

Aleksandra R. Nesic
Vinca Institute of Nuclear Science, University of Belgrade,
Belgrade, Serbia

ABSTRACT

Pectin represents a family of complex polysaccharides and commercially can be easily obtained by extraction from citrus peel, apple pomace, sugar beets, mango and other plants. Pectin is hydrophilic and has tendency to form gel at acidic conditions or in presence of divalent cations. Due to its high capacity to form gel, pectin is widely used as a gelling agent and stabilizer, but can also act as a water binder or thickener. Pectin has in recent years been getting a great attention for potential application in food packaging due to its nontoxicity, biodegradability, edibility, biocompatibility and selective gas permeability. However, the main disadvantages of pure pectin films are poor water vapor barrier and low mechanical properties. In order to overcome these problems, pectin has been investigated in combination with other polysaccharides such as chitosan, alginate or cellulose. This chapter will present an introduction of new concept of biodegradable food packaging materials, criteria for food packaging materials and overview of latest developments of pectin films intended for application in food packaging. Finally, the physical-mechanical and antibacterial properties of these films will also be discussed.

ABBREVIATIONS

Acetyl cellulose	AcC
Brain Heart Infusion	BHI
Carbon dioxide transmission rate	CO_2TR
Contact angle	θ
Degree of blockiness	DB
Degree of esterification	DE
Degree of methylation	DM

Elastic modulus	E
Endo-polygalacturonase	endo-PG
Essential oil from cinnamon leaves	CLO
Fish skin gelatin	FSG
Galacturonic acid	GalA
High density polyethylene	HDPE
High methylated pectin	HM pectin
Homogalacturonan	HG
Liquid vapor interfacial energy	γ_{LV}
Low density polyethylene	LDPE
Low methylated pectin	LM pectin
Nylon 11	NY11
Oxygen permeability coefficient	OPC
Oxygen transmission rate	OTR
Pectatelyase	PL
Pectin lyase	PNL
Pectinmethylesterases	PME
Pectinmethylesterase inhibitor	PMEI
Polyamide	PA
Poly(butylene succinate)	PBS
Polycaprolactone	PCL
Polyethylene	PE
Polyethyleneterephthalate	PET
Polygalacturonase	PG
Poly(hydroxybutyrate)	PHB
Poly(lactide)	PLA
Polypropylene	PP
Polystyrene	PS
Polyvinylchloride	PVC
Proteinphaseolin	Ph
Ready-to-eat	RTE
Rhamnogalacturonan-I	RG-I
Rhamnogalacturonan-II	RG-II
Solid liquid interfacial energy	γ_{SL}
Solid vapor interfacial energy	γ_{SV}
Soy protein isolate	SPI
Soybean flour protein	SFP
Surface density	ρ_s
Transglutaminase	TG
Water vapor permeability	WVP
Water vapor transmission	WVT
Water vapor transmission rate	WVTR
2-keto-3-deoxy-D-lyxo heptulosaric acid	Dha
2-keto-3-deoxy-D-manno octulosonic acid	Kdo

1. INTRODUCTION

The packaging process is important part of the food manufacturing processes, due to maintenance of quality and safety of food products during storage and transportation and extension of shelf-life. Packaging protects food from environmental influences such as heat, light, presence or absence of moisture, oxygen, enzymes, dust particles, gas emission etc. Although food packaging has evolved in its various functions, every package still has to meet the basic requirements. Good food packaging systems should possess good mechanical, optical and thermal properties, hinder gain or loss of moisture, prevent microbial contamination and act as a barrier against permeation of water vapor, oxygen, carbon dioxide and other volatile compounds [1]. Also, the reduction of food waste and spoilage during distribution, cost of preservation and recovery of food packaging materials are very important tasks. Recovery of food packaging materials includes recycling, reuse, energy recovery and composting [2].

Petroleum-based plastic materials such as polyethylene terephthalate (PET), polyvinylchloride (PVC), polyethylene(PE), polypropylene (PP), polystyrene (PS) and polyamide(PA) have been mostly used for the packaging of various types of food due to their good transport properties, acceptable mechanical thermal, chemical and optical properties and large availability at low prices. The use of plastic materials has increasingly replaced metal and glass for food and beverage packaging in recent years because they are lighter in weight, flexible and less susceptible to breakage. Although plastic materials possess good criteria for food packaging, they cause a serious environmental problem since they are not easily degraded in the environment after use. Also, the recycling of these materials is economically inconvenient due to often contamination by food or microorganism [3, 4]. Therefore, in the last decade there has been an increased interest in development of biodegradable and bio based packaging materials from renewable sources as an alternative replacement for synthetic plastic packaging materials. Biodegradable packaging materials have been especially investigated for use in short-term packaging such as disposable plate, cups and utensils, trash bags, cutlery, beverage containers etc. Bio based packaging materials obtained by naturally renewable resources such as polysaccharides, proteins, and lipids or combinations of those components may offer favorable environmental advantages due to their recyclability and reutilization compared to conventional petroleum-based synthetic polymers [5]. Bio based films and coatings may serve too as good barriers against permeation of gas, or good inhibitors of the migration of moisture and complement other types of packaging by minimizing food quality deterioration and extending the shelf life of foods [6, 7].

This chapter gives the concept and criteria for food packaging and short overview of the main bio based packaging materials. Furthermore, the main focus will be given to pectins-natural polymers obtained by extraction from fruits, as a challenge for new food packaging materials. The detailed information about pectin structure, characterization and review of most common limitations or possible solutions for potential use in food packaging will be given.

1.1. Food Packaging Definition

Packaging materials protect food from surrounding challenges. Innovations in food packaging have created an array of new terms associated with the role of packaging in the improvement of safety, shelf life, and convenience of the food product. Food packaging can be divided into three large categories: primary, secondary and tertiary packaging. The primary package is the package which is in direct contact with food. The functions of primary packaging are protection, convenience, safety and storage the foods. A can of tuna, a bag of peanuts, a jar of jam or the wrap around a chocolate candy are examples of primary packages. A number of primary packages are usually contained in an outer or secondary package for transportation, storage and delivery. Secondary packaging is often used for protection of primary packaging from mechanical damage. Carton box of tuna containing, 24 cans of tuna, or a "six-pack" of beer containing six bottles of beer, are examples of secondary packaging. Finally, the tertiary packaging is perceived as whatever specific form of packaging is used for the purpose of wide spread distribution, such as a pallet system for example, where a large number of cases of cans of soda are wrapped in shrink wrap and carried on large wooden pallets to their destination.

The food industry has developed new techniques of packaging due to rising demand for higher quality, better storage of foods and improvement of the existent packaging technology. These packaging includes edible films and coatings, active packaging, modified atmosphere packaging or combination of these types.

An edible film is defined as a thin layer, which can be consumed or coated on a food by casting and drying the liquid film-forming solution or by extrusion. Edible films can be used as a food wrap without changing the original ingredients or the processing method. Edible coatings are formed on the food by spraying, dipping or fluidizing the liquid film-forming solution or molten compounds. Edible films and coatings are produced mainly from edible biopolymers and food-grade additives including proteins, polysaccharides (carbohydrates and gums), and lipids. In order to improve mechanical properties, plasticizers and other additives are often added during the production of these packaging materials. Edible films and coatings have been used to improve the gas, oil and moisture barriers, sensory perceptions, convenience, and microbial protection and to reduce amount of primary synthetic packaging materials [8, 9].

Active packaging employs a packaging material that interacts with the internal gas environment to extend the shelf life of a food. Such new technologies continuously modify the gas environment (and may interact with the surface of the food) by removing gases from or adding gases to the headspace inside a package. Recent technological innovations for control of specific gases within a package involve the use of chemical scavengers (oxygen-, carbon dioxide-, moisture-scavengers) to absorb a gas or alternatively other chemicals that may release a specific gas as required. Also, new active packaging system provides intelligent functions such as antimicrobial activity, edibility, biodegradability. Oxygen-scavenging systems have been commercialized in the form of a sachet that removes oxygen. An oxygen-free environment can prevent food oxidation, rancidity and growth of aerobic bacteria and mold. An active packaging system that scavenges carbon dioxide has been used for products that require fermentation or aging, such as pickles, sauces, cheese, coffee, or products that produce gas after packaging. Moisture-scavenging systems have been used for packaging of dried foods, moisture-sensitive foods, pharmaceuticals and electronic devices and usually may be found in the form of a sachet which includes desiccant material. Antimicrobial packaging is a novel development that incorporates antimicrobial agent into packaging material to suppress the activities of targeted microorganisms. Controlled migration of the active compound from the packaging material into the food enables not only the initial inhibition of undesirable microorganisms present in food, but also create a residual activity over time, during transport, storage and distribution of food [10, 11, 12].

Modified atmosphere packaging presents packaging materials with an optimal amount of pure oxygen, carbon dioxide and nitrogen within a high barrier or permeable package. This system is developed to meet the specific respiration needs for each packaged food product. Modified atmosphere enables fresh and minimally processed packaged food products to maintain visual, textural and nutritional appeal. The controlled modified atmosphere packaging environment enables food packaging to provide an extended shelf life without requiring the addition of chemical preservatives or stabilizers. The use of modified atmosphere packaging systems is attractive to the food industry due to fast-growing market for minimally processed fruits and vegetables, non-frozen chilled meats, ready-to-eat meals, and semi-processed bulk foods. Despite the advantages of modified atmosphere packaging, this type of packaging is not convenient where the development of anaerobic pathogens is possible. In such cases, the concentration of oxygen should be left intentionally in the package to avoid extreme growth of anaerobic pathogens [13, 14, 15].

1.2. Food Packaging Criteria

Today's packaging applications demand a great deal from their materials of construction, including: barrier to flavor, oxygen, moisture and water vapor, mechanical strength, high aesthetics, including (in some cases) transparency and gloss, high-temperature resistance, food-contact approval, flexibility or rigidity, biodegradability etc. In most cases, these demanding and often conflicting properties cannot be provided by a single material; hence multilayer structures are widely used. The broad array of polymers used in multilayer films used in these concepts are often incompatible and therefore need a "tie-layer" to join them together to provide packaging material that provide all the desired properties. Next section presents the main properties that food packaging materials should satisfy.

1.2.1. Barrier Properties

One of the functions of packaging is to act as a barrier that separates and protects the product from exposure to the environment. The determination of the barrier properties of packaging materials is crucial to estimate and predict the product-package shelf-life. Regarding the barrier properties of packaging materials, the critical compounds that can penetrate the packaging materials and degrade food quality are water vapor and oxygen of the surrounding atmosphere. Other important components to which food packaging should be less permeable are carbon dioxide, nitrogen moisture, fats and oils [16]. The extent of the barrier provided by a package depends on the chemical properties of the used material. However, environmental conditions, such as temperature, relative humidity and the stress and handling of the product by consumers also influence the performance of the package.

1.2.1.1. Oxygen Transmission Rate

The oxygen transmission rate (OTR) of a packaging material plays an important role in influencing the shelf-life of oxygen sensitive products such as fresh produce (fruits, vegetables and salads) and high fat (e.g. nuts, donuts) products. OTR is the quantity of oxygen gas passing through a unit area of the parallel surfaces of a packaging material per unit time under the same environmental temperature and relative humidity conditions of the test (mol/(m^2s)). Generally the biodegradable and bio based packaging materials present a value one or more order of magnitude below the synthetic-based packaging materials used in the same field. The oxygen permeability coefficient (OPC)(kg m/m^2sPa) can be calculated by average OTR multiplied by the average film thickness and then divided by the pressure differences between the two sides of films:

$$\mathrm{OPC} = \mathrm{OTR} * \frac{l}{\Delta \mathrm{P}} \quad (1)$$

where l is the thickness of the film (m), ΔP= partial oxygen pressure difference between the two sides of the film (Pa) [17]. $\Delta P = p_1 - p_2$, where p_1 is the oxygen partial pressure at the temperature test on the test side and p_2 is equal to zero on the detector side.

1.2.1.2. Water Vapor Transmission Rate (WVTR)

The water vapor barrier properties for the packaged product are of great importance for maintaining or extending its shelf-life. The water vapor transmission (WVT) of a packaging material can be determined according to the standard ASTM E96-92 method. In this method, salt solutions or any ingredient with a known relative humidity is placed inside a WVT cup and sealed by the packaging material. The WVT cup is placed into a humidity-controlled chamber. Cup is weighed at scheduled times until the steady state is reached and the amount of water vapor transmission rate through the packaging material is estimated by the linear portion of the diagram obtained by plotting the weight increment of the cup as a function of time. The difference in the relative humidity inside the cup and that of the environment is the direction of driving force of the water vapor movement through the packaging material. Weight changes in the cup indicate the driving force of the environment (dehydration or absorption). The weight of the ingredient inside the WVT cup decreases in a dehydrating condition, whereas the weight increased occurs in an adsorbing condition. For fresh food products it is important to avoid dehydration while for bakery or delicatessen is important to avoid water permeation. The permeability of packaging material to water vapor is usually expressed as the water vapor transmission rate, WVTR, which is the quantity of water vapor transmitted per unit area and unit time, by a film of unit thickness, under specified conditions of vapor pressure difference and temperature. The WVTR (g·m/m^2hPa) can be calculated by the following equation:

$$\mathrm{WVTR} = \frac{w*l}{A*t*(P_2-P_1)} \quad (2)$$

where w is the weight gain of the cup (g), l is the thickness of packaging material (m), A is the area of exposed packaging material (m^2), t is the time of gain (s) and (P_2 – P_1) is the difference in the vapor pressure across the film (Pa).

1.2.1.3. Carbon Dioxide Transmission Rate (CO_2TR)

The carbon dioxide transmission rate can be measured using a technique similar to the oxygen permeability method. This method is used to determine the amount of carbon dioxide that is transmitted through a packaging material for a certain period of time, exposed to a specific area, at a given temperature and CO_2 partial pressure. Carbon dioxide transmission rate is important in cases when the respiration rate of a product affects the quality of the food. For the modified atmosphere packaging of fruits (e.g. banana, pineapple, papaya) the rate of carbon dioxide loss from the package is one of the keys to prolonging the shelf life of these products.

1.2.2. Wettability

Wetting properties are essential surface features of packaging materials. Almost all packaging materials are known as hydrophobic materials with low surface energy. In order to increase surface energy and improve adhesion and wettability of packaging materials, polar groups are introduced on the surface of packaging material. The wettability of a solid surface can be determined by measuring the contact angle formed between the test liquid and the packaging material. Water contact angle determines surface hydrophilicity by measuring how much a droplet of water spreads on a surface. The lower the contact angle, the greater is the tendency for the liquid to wet the solid. When a surface has more polar groups introduced to it, these groups form strong hydrogen bonds with water and the droplet spreads along the hydrophilic surface, resulting in a lower contact angle [18]. However, this method is limited by its inability to distinguish between different hydrophilic functional groups and by the measurement errors including difference in operator measurement and changes in environmental temperature and humidity [19]. The wetting of solid surface present equilibrium relationship between contact angle θ and the liquid vapor interfacial energy (γ_{LV}), the solid vapor interfacial energy (γ_{SV}) and the solid liquid interfacial energy (γ_{SL}) (Figure 1.) and can be expressed by equation of Thomas Young:

$$\gamma_{LV} cos\theta = \gamma_{SV} - \gamma_{SL} \tag{3}$$

This equation can be applied for determination of contact angle on homogenous and smooth solid surface, while for solid surfaces with mechanical roughness or asperities and chemical heterogeneity correction factor r must be include in equation.

The surface energy can be estimated by measuring the contact angles of two or three test liquids including polar and non- polar liquids using Zisman approach [20].

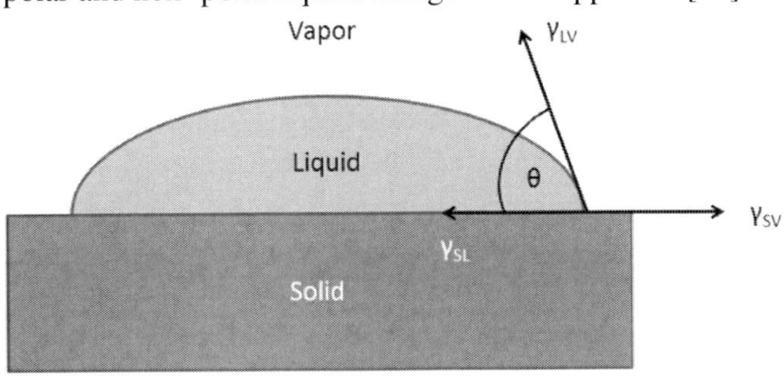

Figure 1. Schematic diagram of the contact angle and its surface free energy (tension) components.

The solid surface energy is the sum of polar and non-polar (dispersive) contributions. It permits to take into account the effect of these two contributions on the surface properties and interaction processes [21, 22].

1.2.3. Mechanical Properties

Protection of package contents against external forces depends on its mechanical properties. In packaging technology, mechanical properties should be considered and evaluated at the level of the packaging material, the formed empty package, the product package assembly and the outer packages. The values of tensile strength (MPa), the percent elongation at yield (%), the percent elongation at break (%) and the elastic modulus (GPa) of the packaging material are important to get mechanical information of used food packaging material. Standard test method for determination of mechanical properties of packaging materials can be obtained by ASTM D822-02. In this test method, the sample must be conditioned at 23°C, and 50% relative humidity for over 48 hours and then be cut to a specific length and width. Tensile strength is calculated by dividing the maximum load for breaking the packaging material by the original cross-sectional area. The percent elongation is calculated by dividing the elongation of packaging material at rupture by the initial length. Elongation is calculated by dividing the change in the dimension by the original dimension. The stress/strain curves obtained from mechanical tests provide information about the flexibility, toughness, and elongation that can be used for prediction of performance of packaging material during handling. Compressive resistance of containers, components and unit loads can be determined using the ASTM D642, standard compression test method. The compression strength depends on type of material and design (shape and size).

1.2.4. Optical Properties

Optical properties of packaging materials are very important because they can affect the protective function of the package. Many deteriorative reactions like lipid oxidation, off-flavor generation, discoloration, and destruction of nutritionally important components are catalyzed by visible and ultraviolet light [23]. On the other side, haze, transparency or clarity and transmittance are also significant indicators in packaging applications, since these parameters allow the consumer to see what is contained within the package. Hence the optical properties are important features in the marketability of a particular product and control of these properties is therefore required whether the film is utilized for food packaging, pallet wrap, or other related applications.

1.2.5. Chemical Reactivity

The most packaging materials incorporated with chemical compounds may react with the food components during processing or storage and migrate into the food. Hence, the migration of chemical compounds into the food affects seriously safety and food quality. Possible chemical compound-migrants that are incorporated within packaging materials to improve functionality include plasticizers, antioxidants, thermal stabilizers, slip compounds, and monomers [24]. Chemical migration from food packaging is affected by various parameters such as chemical properties of food, storage condition and temperature of the system, the type of packaging material and the properties of the migrants [25]. Against the accepted food safety risk, two types of migration are regulated, namely ‘specific migration’ and 'global migration'. Specific migration refers to the migration of compounds that are considered to present health hazards even when present in quite small quantities. In such cases, their migration into foods needs to be controlled rigorously. The concept of global migration, by contrast, has been adopted to regulate the ingress of compounds which, though they do not present the same degree of risk, are nevertheless undesirable. The advantage of considering global migration is that it avoids having to analyze individual migrating compounds separately.

1.3. Biobased Food Packaging Materials

The most packaging production derives from fossil fuel and the difficulty in recycling these materials presents the most important drawbacks for their use in food packaging. Replacing petroleum-based plastics with plastics made from renewable raw materials, such as plants, reduces dependence on fossil fuels, fosters the development of more sustainable products and increases the diversion of food waste from landfills [3, 4]. Bioplastics such as bio-PET, bio-PE or bio-PP are chemically identical to their petroleum-based materials, but are synthesized from biomass, mostly from bioethanol. However, bioplastic polymers do not solve the problem of public waste management, as they do not necessary have to be biodegradable or compostable. For example, polycaprolactone (PCL), and poly (butylene succinate) (PBS) are petroleum based, but they can be degraded by microorganisms, while poly (hydroxybutyrate) (PHB) and poly(lactide) (PLA) are produced from biomass or renewable resources, and are thus biodegradable. On the other side, despite the fact that polyethylene (PE) and Nylon 11 (NY11) can be produced from biomass or renewable resources, they are non-biodegradable. The biodegradation of acetyl cellulose (AcC) depends on the degree of acetylation.

Biobased polymers may be broadly divided into three large groups: polymers extracted from renewable resources (polysaccharides, proteins), polymers obtained from renewable biobased monomers (poly lactic acid and other polyesters) and polymers produced by microorganism or genetically modified bacteria (polyhydroxyalkonoates) as shown in Figure 2. [26]. The compostability is very important parameter for biobased materials because while recycling is energy expensive, composting allows disposal of the packages in the soil. By biological degradation it produced only water, carbon dioxide and inorganic compounds without toxic residues. Polymers extracted from renewable resources are the most commonly available and may be obtained from marine and agricultural animals and plants. The representatives of this group are polysaccharides such as cellulose, starch, alginate and chitin and proteins such as casein whey, collagen and soy.

Polysaccharide and protein-based films demonstrate adequate gas barrier properties, but due to their hydrophilic nature exhibit poor water vapor barrier properties [27, 28]. These films possess a relatively high degree of crystallinity causing processing and performance problems. Therefore, food packages made from polysaccharide or protein demonstrate high moisture sensitivity, poor water barrier and poor mechanical properties compared to those made from synthetic polymers [29]. Though many research efforts focused on improving the film properties of biobased packaging films indicated a significant improvement in film properties, their physical, thermal, and mechanical properties are still not satisfactory and find difficulties in industrial applications. In order to improve physical and mechanical properties of these films, bio based materials are usually modified by incorporation of nanofillers or blending with compatible plasticizers [30]. Addition of plasticizers takes part in an increase of film flexibility and decrease the glass transition temperature, while incorporation of nanofillers into polymer matrix improves the mechanical, thermal and barrier properties.

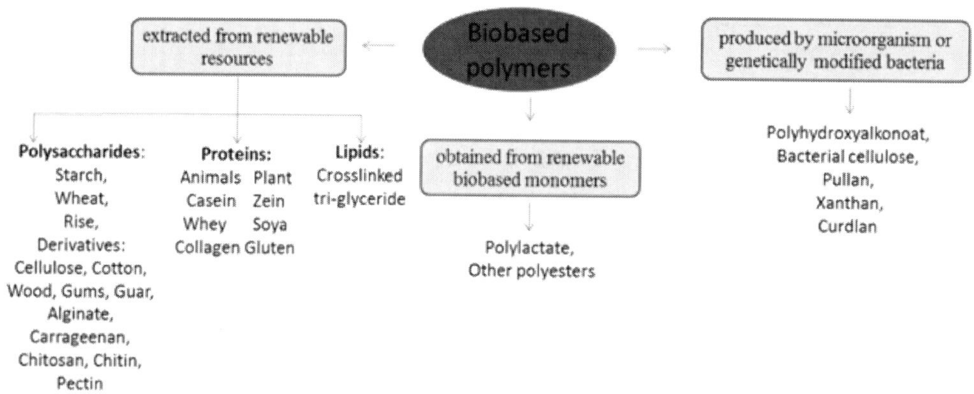

Figure 2. Scheme of bio based polymers.

Recently, pectin which belong to the group of polysaccharides, has gained more attention as a promising polymer for variety of very important uses in more than just the food industry. As a cheap and versatile raw material obtained from renewable sources, pectin is poised for potential application in food packaging. Further part of this chapter will be focused on chemistry of pectin, its main properties and an overview of food packaging films based on pectin and its derivative reported in literature.

2. PECTIN

Pectin is a family of polysaccharides that together with cellulose and hemicellulose participate in building the cell wall of plants. The largest natural source of pectin is fruits. The amount, structure and chemical composition of pectin differs of extraction fruit source. Pectin has been widely used in a food and beverage industry as a thickening agent, a gelling agent and a colloidal stabilizer, due to capability to form gel in acidic conditions. In the large intestine and colon, pectin can be digested under the influence of enzyme pectinase which is present only in the colon, and therefore can be remained undigested in the GI tract which contains enzymes like protease and amylase [31]. During the digestion of pectin, it can liberate the short chain fatty acids that have favorable influence on health. These properties make pectin suitable as a matrix for drug delivery and as dietary fiber [32, 33]. Pectins are also used to reduce blood cholesterol levels and gastrointestinal disorders. Pectin can form a gel in the presence of divalent and trivalent cations and therefore found use in preparation of membranes for ultracentrifugation, electrodialysis and wastewater treatments. The benefits of pectin such as nontoxicity, biodegradability and biocompatibility have enabled it to be investigated in biomedical application and as edible films for food packaging [34, 35].

2.1. Chemical Structure of Pectin

The composition and structure of pectin are still not completely understood although pectin was discovered over 200 years ago. The structure of pectin is very difficult to determine because pectin can change during isolation from plants, storage, and processing of plant material.

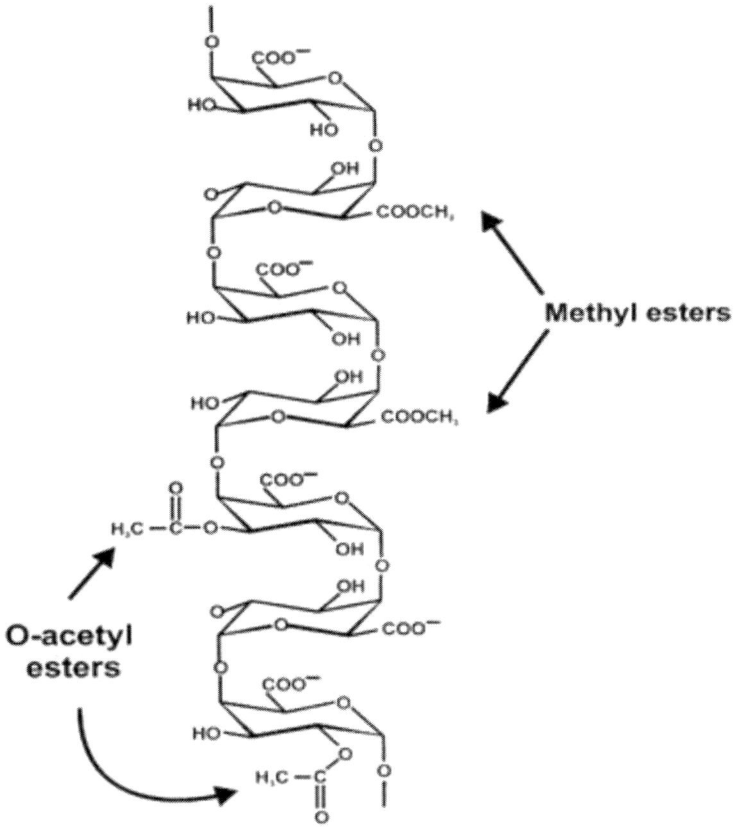

Reproduced with permission from Ridley, B.L. et al. *Phytochem.* 2001, *57*, 929–967. Copyright © 2001, Elsevier B.V.

Figure 3. The primary structure of homogalacturonan [40].

Pectin is a family of polysaccharides present in all plant primary cell walls. It is composed of 1,4-linked α-D-galacturonic acid (GalA) residues and at least 17 different monosaccharides containing more than 20 linkages. Three major pectic polysaccharides groups have been identified and characterized within the pectin structure: homogalacturonan (HG), rhamnogalacturonan-I (RG-I) and rhamnogalacturonan-II (RG-II) [36]. All of these pectic polysaccharides contain D-galacturonic acid linked at the O1 and the O4 position [37]. Homogalacturonan is the most abundant pectic polysaccharide (65% of pectin structure) formed by 1,4-linked α-D-galacturonic acid residues which can be partially methyl esterified at the C6 atom of carboxyl group and can be acetylated at O2 or O3 atom (Figure 3.) [38]. The pattern and degree of methyl esterification and acetylation varies from source to source. The unmethylated C-6 of HG GalA residues is negatively charged and may ionically interact with Ca^{2+} ions to form a stable gel with other pectic molecules if >10 consecutive unmethyl esterified GalA residues are coordinated [39]. The HG-calcium complex can be explained by "egg box" model (Section 2.3).

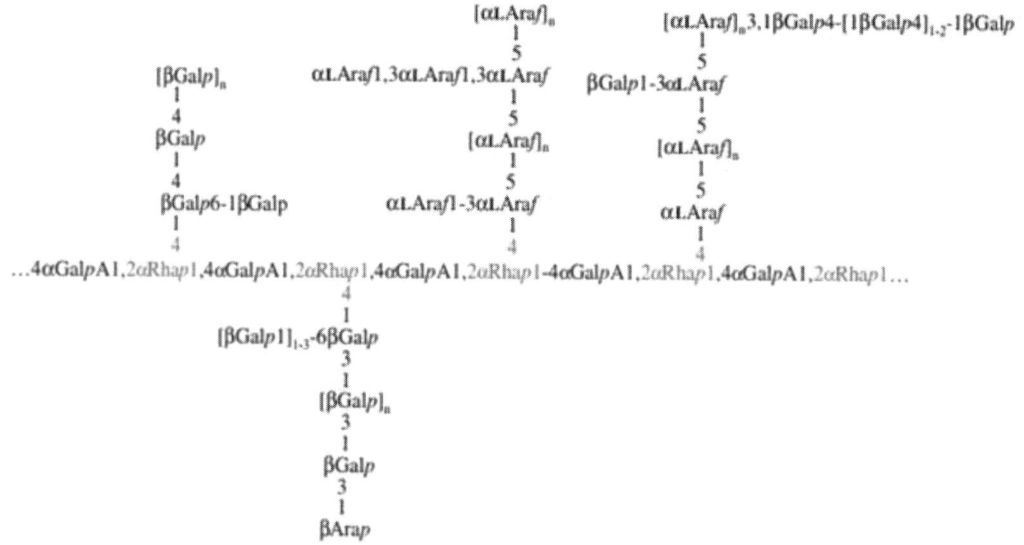

Reproduced with permission from Caffal, K.H. et al. *Carbohydr. Res.* 2009, *344*, 1879–1900. Copyright © 2009, Elsevier B.V.

Figure 4. The structure of RG-I backbone and representative side chains of:α-(1,5)-L-arabinan, β-(1,4)-galactan, and Type-I arabinogalactan. The α-1,5-L-arabinan chains that originate from the RG-I backbone may be branched with long chains of 3-linked branches of mono- or di-meric L-arabinanor mono-, di-, or oligomeric branches of β-(1,3)-linked Gal. Type-II arabinogalactan may have branches of 6-linked or 3,6-linked galactose residues [42].

RG-I represents 25-30% of pectin and contains a backbone of the repeating rhamnose-galacturonic acid disaccharide $[\text{-}\alpha\text{-D-GalA-1,2-a-L-Rha-1-4-}]_n$. RG-I exhibits a high degree of cell type and develop-dependent expression in the type and number of sugars as galactose, arabinose and xylose, oligosaccharides, and branched oligosaccharides attached to its backbone (Figure 4.) [41]. The backbone residues can be acetylated on C2 or C3 atom, while there are no proofs about methyl esterified backbone residues.

Rhamnogalacturonan II makes up only 10% of pectin, but presents the most structurally complex pectic polysaccharide. It consists of an HG backbone of at least 8 1,4-linked α-D-GalA residues with side branches that consist 12 different types of glycosyl residues including the rare sugar species 2-O-methyl xylose, 2-O-methyl fucose, aceric acid, 2-keto-3-deoxy-D-lyxo heptulosaric acid (Dha), and 2-keto-3-deoxy-D-manno octulosonic acid (Kdo) [43, 44, 45, 46]. RG-II usually exists inplant walls as RG-II dimers that are crosslinked by a 1:2 boratediol ester between OH-2 and OH-3 of the apiosyl residues in each monomeric RG-II unit (Figure 5.) [47].

A.

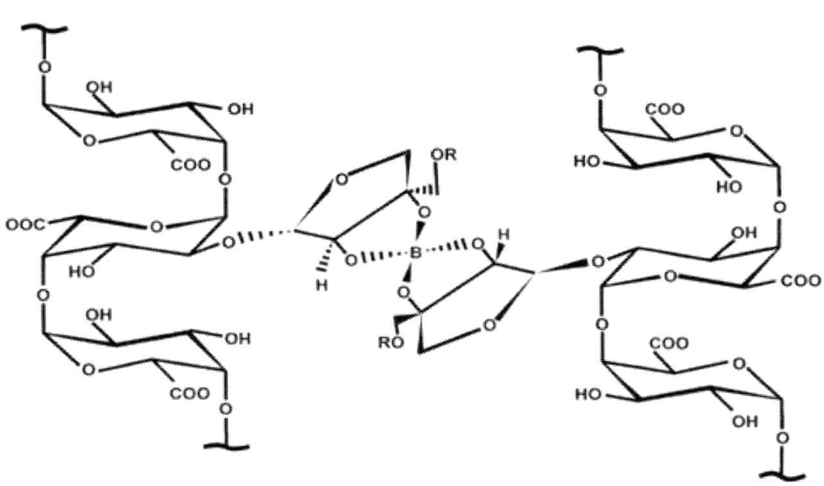

B.

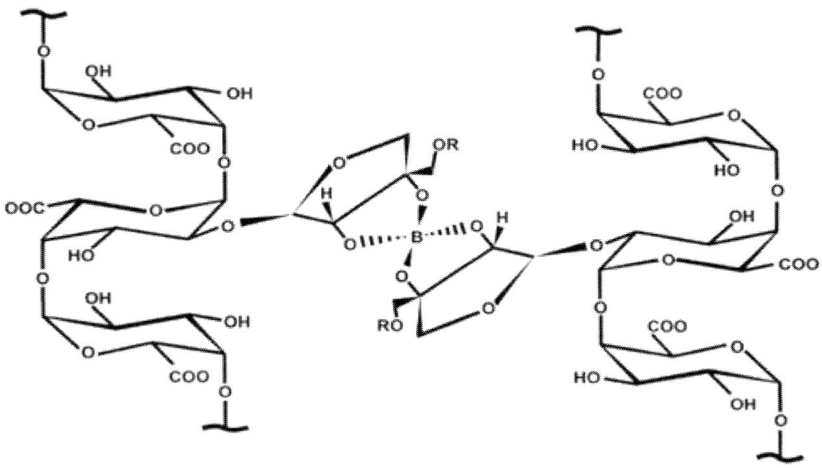

Reproduced with permission from Ridley, B.L. et al. *Phytochem.* 2001, *57*, 929-967. Copyright © 2009, Elsevier B.V.

Figure 5. The borate 1:2 diolester that cross links two monomeric units of RG-II. The borate ester is formed between OH-2 and OH-3 of the 30 linked apiosyl residues. ''R'' represents the oligoglycose that is linked to 30 of the apiosyl residues. The ester can exist in either of two diastereomeric forms. In A one ''R'' group is pointing upward and the other is pointing downward. In B both ''R'' groups are pointing upward. [40].

2.2. Physicalproperties of Pectins and Its Characterization

The main properties of pectins are the content of galacturonic acid and neutral sugar, degree of esterification (DE), degree of methylation (DM), degree of blockiness (DB) and their molecular weight distribution. However, two pectins with same chemical properties might differ significantly in their physical properties indicating that distribution of methyl esters over the molecule is more important than absolute amount of ester groups.

The degree of esterification presents the ratio of esterified GalA groups per total amount of GalA in pectin structure. Pectins can be divided in to majors groups on the basis of their degree of esterification: high methylated (HM) pectin and low methylated (LM) pectin. In HM pectin, a high portion (>50%) of carboxyl groups exists as a methyl ester, while the remaining carboxyl groups exist as the free acid, ammonium, sodium, calcium, or other rarer salts. LM pectin has less than 50% of methyl esterified groups. LM pectin is usually obtained from mild alkali or acidic treatment of HM pectin. DE value for commercial HM pectin typically ranges from 60 to 75% and for LM pectin ranges from 20 to 40%. Amidated LM pectin can be obtained from HM pectin by treatment with ammonia in the alkaline deesterification process. During this process, some of the carboxyl groups are converted to the acid amide. Small quantities of amidated pectin can be found naturally in sugar beets and certain other sources. DE is commonly measured by titrimetric where free carboxyl groups are protonated via washing the pectin with acidic alcohol and then drying. Subsequent dissolution in water and titration with a standard base, DE of the pectin sample is determined.

Degree of methylation describes the extent to which carboxyl groups in pectin molecules exist as the methyl ester in ratio to all esterified groups. Methoxyl content can be determined through enzymatic or alkaline demethylation. Analysis of methanol is achieved through various techniques such as spectrophotometric technique involving reaction with potassium permanganate or alcohol oxidase, subsequent condensation with pentane-2,4-dione to yield a colored product. Total acid hydrolysis of a pectin sample is typically used to determine total neutral monosaccharide content. This modern standard of analysis involves methanolysis with 2M HCl prior to trifluoroacetic acid hydrolysis. Subsequent monosaccharides are converted to alditol acetates and analyzed via GLC or HPLC with refractive index detection. In order to determine the galacturonic acid content, pectin must also be hydrolyzed. Acid hydrolysis may cause the degradation of galacturonic acid, therefore the enzymatic hydrolysis is more suitable for determination of galacturonic acid content. Several methods have been used for quantitative analysis of galacturonic acid such as colorimetric method, thin-layer chromatography, gas chromatography, high-performance liquid chromatography or capillary electrophoresis [48, 49, 50, 51].

One way to characterize differences in distribution of substituent groups in pectin, for example, methoxy and amide groups, is establishing its degree of blockiness. Pectin should be digested with an endo-polygalacturonase (endo-PG) in order to determine the value of DB. The degradation products can be quantified using high-performance anion exchange chromatography and pulsed amperometric detection [52]. Average molar mass of pectin is usually determined by high performance size-exclusion chromatography.

2.3. Gelling Properties of Pectins

The most unique property of pectin is its ability to form gel in the presence of divalent and trivalent cations or in acidic solutions. The gelation mechanism of pectin is governed mainly by its degree of esterification. HM-pectin can form gels only in the presence of sugar and at acidic condition. The formation of gel is caused by suppression of carboxylate group ionization at low pHs which leads to reduction in hydration of carboxylic acid groups. High concentration of sugar leads to a decrease of water activity and dehydration of pectin. Decrease in hydration of carboxylic acid groups lowers the attraction between pectin and water molecules and the repulsive forces between pectin chains, resulting in association of pectin molecules and formation of gel [53]. Therefore, the gelling mechanism of HM pectins is explained by hydrophobic interactions and hydrogen bonds between undissociated carboxyl groups [54]. A higher DE (i.e. pectin with a DE of above 72%) causes more rapid gelling than pectin with low values of DE. HM pectin does not contain sufficient acid groups to gel or precipitate in the presence of calcium ions, although other ions such as aluminium or copper may cause precipitation under certain conditions. LM and amidated pectins can form gel relatively independent from soluble solids content and pH-value in the presence of calcium ions forming intermolecular ionic junction zones between smooth regions of neighbored chains.

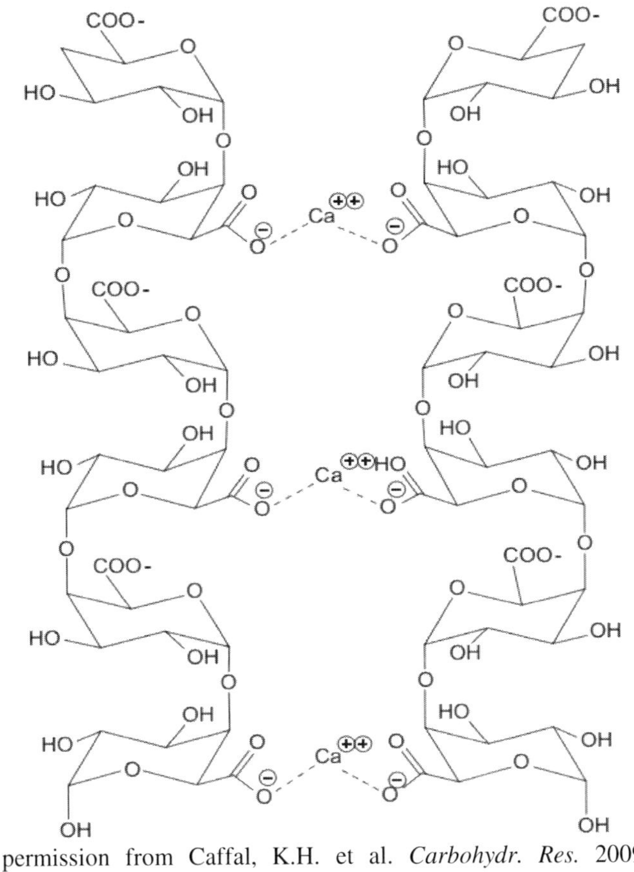

Reproduced with permission from Caffal, K.H. et al. *Carbohydr. Res.* 2009, *344*, 1879–1900. Copyright © 2009, Elsevier B.V.

Figure 6. Shematic representation of "egg-box" mechanism of calcium ions binding to pectin [42].

The mechanism of gelation of LM pectin is explained by "egg-box" model originally suggested for the alginates (Figure 6.) [55, 56]. According to this model, cross-links are formed by calcium ions occupying electronegative cavities in the twofold buckled ribbon structure of the carboxylic groups. The mechanism of binding calcium ions involve formation of three possible types of junction zones: hydrophobic interactions between methoxyl ester groups, hydrophilic interactions between undissociated carboxyl groups and/or hydroxyl groups via hydrogen bonds as well as ionic interactions between dissociated carboxyl groups via Ca-bridges. There is a requirement of a minimal length of subsequent carboxyl units, estimated to be between 6 and 20, in order to form typical egg box structure [57].

The gel forming ability of LM-pectin increases with decreasing the value of DM, while the high content of sugar interfere with gel formation because the dehydration of the sugar favors hydrogen bonding and decreases cross-linking by divalent ion forces. Amidation improves the gelling ability of LM pectins.

The more amid groups are present, that means the more bonding zones are affected, the stronger the resulting gels are. Monovalent ions such as sodium, which can also react with free carboxyl groups, can affect gel formation because they decrease the cross-linking reaction of calcium and improve the solubility of LM-pectin in the presence of calcium.

2.4. Source of Pectin

The highest concentrations of pectin are found in the middle lamella of cell wall and gradually decrease in pass through the primary wall toward the plasma membrane. Commercial pectins are almost exclusively derived from citrus peel or apple pomace, both by-products from juice (or cider) manufacturing.

The content of pectin on a dry matter basis in apple pomace and citrus peel is 10-15% and 20-30%, respectively. Alternative sources of pectin are sugar beet waste from sugar manufacturing, sunflower heads (seeds used for edible oil) and mango waste.

Apples, quince, plums, gooseberries, oranges and other citrus fruits contain higher content of pectin compared to soft fruits like cherries, grapes, strawberries etc. Typical levels of pectin in plants are: apples 1-1.5%, apricot 1%, cherries 0.4%, oranges 0.5-3.5%, carrots 1.4% and citrus peel 30% [58, 59, 60].

2.5. Extraction of Pectin

Pectin extraction is a multiple stage physicochemical process which involves hydrolysis and extraction of pectin macromolecules from plant tissue, purification of the liquid extract and isolation of the extracted pectin from the liquid. These processes are influenced by various factors, mainly temperature, pH and time. The most commonly used methods for extraction of pectin are direct boiling and microwave heating extraction. The direct boiling extraction of pectin includes treatment of raw material with hot dilute mineral acid (sulfuric, hydrochloride or phosphoric acid) at pH about 2 [61]. The operated temperature is in the range of 60-100 °C, while duration of process may vary between 2-10h. Thereafter, pectin is precipitated using ethanol or isopropyl alcohol. The yield and degree of esterification of pectin depends on extraction temperature and operating time. This process produces HM pectin with approximately value of degree esterification 70%. To produce other types of pectin, some of the ester groups must be hydrolyzed by addition of acid, either before or during a prolonged extraction, in the concentrated liquid, or in alcoholic slurry before separation and drying. This process can produce a range of low methoxyl pectin, while hydrolysis using ammonia results in the conversion of some of the ester groups into amide groups, producing amidated low methoxyl pectin.

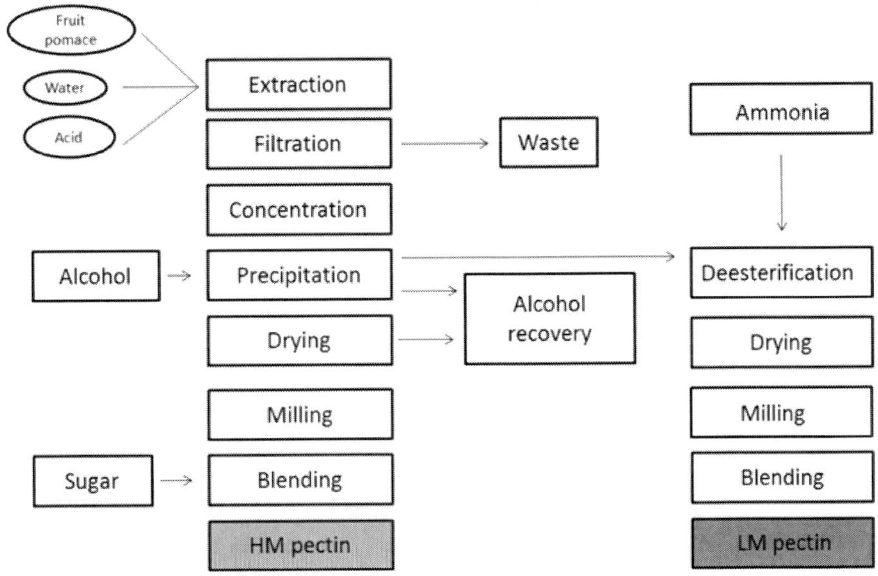

Figure 7. Industrial procedure for production of pectin [61].

Microwave heating extraction is generally more efficient method due to higher yield of produced pectin. Commonly used solvents in microwave extraction procedure are ethanol, 0.05M ethylenediamine tetra acetic acid, 1M sodium hydroxide. Only 15 minutes of microwave heating extraction is enough to produce similar amount of pectin as one obtained by direct boiling extraction which last 3 hours [62]. The industrial procedure for pectin production is presented in Figure 7.

2.6. Degradation of Pectin

Chemical (non-enzymatic) degradation of pectin may occur spontaneously when pectin is heated at neutral or even weakly acidic conditions. This situation leads to the split of glycosidic linkages between GalA residues via ß-elimination reaction and this reaction is competing with the demethoxylation reaction. The ß-elimination reaction occurs on uronic acid, which possesses a glycosidic linkage on C4 in the ß-position of the carboxyl group at C5 atom. A requirement is the presence of a methyl-ester group at C6, rendering H5 sufficiently acidic to be removed by an alkali. This results in formation of double bonds between C4 and C5 at the non-reducing end. Alkali treatment promotes demethoxylation rather than the competitive ß-elimination reaction, even at room temperature [63, 64]. Methyl-esters at C6 of GalA residues are hydrolysed under alkaline or mild-acidic conditions (approx. pH > 5) through saponification.

Pectin degradation may also occur during thermal treatment at acidic conditions (pH<3) trough mechanism of acidic hydrolysis. The reaction proceeds through an initial protonation of the glycosidic oxygen, followed by a rate-limiting unimolecular heterolysis of the conjugated acid. In acidic conditions, pectin with low DM hydrolyzes faster and the reaction is boosted by the simultaneous chemical demethoxylation of the polymer under such conditions [65]. The rate of hydrolysis increases with the decrease of pH, addition of NaCl in system or with increase of temperature. In general, maximum stability of pectin is found at pH 4, while at pH in a range of 5-6, HM pectin is stable at room temperature only. A further increase in temperature or pH leads to β- elimination reaction which results in chain cleavage and very rapid loss of viscosity and gelling properties. LM pectin shows a somewhat better stability at these conditions. The possible scheme of chemical degradation of pectin occurring at the HG chains is presented in Figure 8.

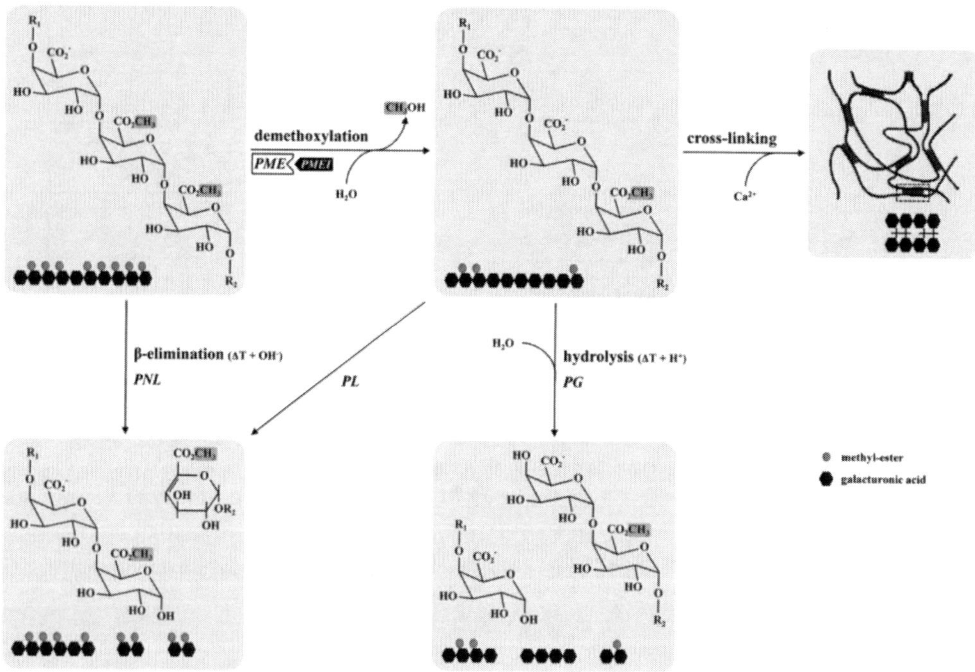

Reproduced with permission from Jolie, R.P. et al. *Trends Food Sci. Technol.* 2012, *24*, 103-118. Copyright © 2009, Elsevier B.V.

Figure 8. Schematic presentation of chemical and enzymatic degradation of pectin. Names of enzymes and inhibitors are formatted italic: PME = pectinmethylesterase, PMEI= PME inhibitor; PNL= pectin lyase; PG = polygalacturonase, PL = pectatelyase, *R1*/*R2* = initial/terminal fragment of the pectin polymer, ΔT= heating [66].

Next to the chemical degradation methods, enzymes have been widely used to hydrolyze complex carbohydrates, in order to reveal structural characteristics [67]. A wide range of endogenous and exogenous enzymes can synergistically modify and degrade pectin smooth and hairy regions. Pectin methylesterases (PME) catalyze the demethoxylation of HG at O-6 of GalA within plant cell walls, releasing methanol and protons (and creating negatively charged carboxyl groups). The demethoxylated HG may form gel via crosslinking with divalent ions (such as Ca^{2+}, Mg^{2+}) and may form a substrate for pectin depolymerizing enzymes [68]. The pectin-depolymerizing enzymes consist of hydrolases and lyases. Hydrolases split the α-(1→4)-D linkages between two neighboring GalA residues in HG chains by acid/base assisted hydrolysis, while the lyases fragment the polymers via a *β*-elimination reaction mechanism resulting in the formation of a double bond between C-4 and C-5 at the newly formed nonreducing end. Lyases can be endo or exo-acting and require calcium ions for catalysis. Endo-hydrolases hydrolyses pectin randomly which causes the fast decrease in molar mass of pectin, while exo-hydrolases the are confined to cleave off GalA monomers (EC 3.2.1.67) or dimers (EC 3.2.1.82) from the non-reducing end of a chain [69, 70]. When pectin structures are not degradable by the available enzymes, a combination of chemical and enzymatic approaches can be applied. For example, the enzyme activity of endo-hydrolases is improved after removal of acetyl groups and/or methyl esters [71].

3. Application of Pectin in Food Packaging

Giancone et al. investigated the influence of pectin surface density (ρ_s) on the physical-mechanical properties of high methoxyl (HM) pectin-based edible films [72]. The results showed that an increase of ρ_s from 2.5 to 5.8 mg cm^{-2}, did not affect changes in structure of HM pectin films. HM pectin film had a tensile strength of 20±7 MPa and an elastic modulus (E)equal to 2,400±200 MPa. Although the film structure was unaffected by ρ_s, water vapor permeability increased with the rise in ρ_s while oxygen and carbon dioxide permeability decreased. Water vapor permeability (WVP) of HM pectin films increased from 18 to 54 pg $m^{-1}s^{-1}Pa^{-1}$ as ρ_s increased from 2.5 to 5.8 mg cm^{-2}. This effect could be attributed to the higher number of free hydroxyl groups, enhancing interaction with water and favoring water vapor transmission through the films [73]. The WVP of HM pectin edible films at low ρ_s revealed the same order of magnitude as biodegradable commercial films Mater-Bi which have a WVP equal to 22.6 pg m^{-1} $s^{-1}Pa^{-1}$. HM pectin edible films showed approximately the same water vapor barrier properties as sodium caseinate–starch film [74], chitosan-based edible films [75, 76, 77] and caseinate-based edible films [78] and almost two orders of magnitude lower than the WVP of starch-based edible films [79, 80], but about two orders of magnitude greater than that (0.55 or 0.23 pg $m^{-1}s^{-1}Pa^{-1}$) of commercially used packaging materials low- or high-density polyethylene (LDPE or HDPE) films, respectively. The oxygen and carbon dioxide barrier properties of HM pectin films at high ρ_s are of the same orders as magnitude of LDPE (21 and 94 pg $m^{-1}s^{-1}Pa^{-1}$, respectively, for O_2 and CO_2) films, but one order of magnitude higher than that of Mater-bi biodegradable films (for O_2 =8 pg $m^{-1}s^{-1}$ Pa^{-1}) and about two orders of magnitude greater than that of HDPE films (0.7 and 1.9 pg $m^{-1}s^{-1}$ Pa^{-1}, respectively, for O_2 and CO_2).

Alves et al. investigated barrier properties of biodegradable films for food packaging based on commercial pectin and k-carrageenan [81]. The results showed that water vapor permeability increased with an increase of k-carrageenan in film, but reached a plateau at about 67% (dry basis) of kappa-carrageenan. A further increase in the k-carrageenan content had not a significant influence on the films water vapor permeability. In order to improve water vapor barrier properties of these films, different amounts of organically modified nanoclay (1, 5 and 10%) were incorporated into polymer matrix [82]. The value of water vapor permeability was reduced about 35% for composite films with nanoclay content of 10 %. A significant barrier improvement was also observed for CO_2 permeability (50%).

Composite films from pectin and fish skin gelatin (FSG) or pectin and soybean flour protein (SFP) showed an increase in stiffness and strength and a decrease in water solubility and water vapor transmission rate, in comparison with pure pectin films [83]. The crosslinking of the composite films with glutaraldehyde/methanol reduced the interstitial spaces among the macromolecules and, consequently, improved their mechanical properties and water resistance. Pure pectin films crosslinked with glutaraldehyde/methanol also improved the Young's modulus and tensile strength, but showed little effect on the water resistance, due to dehydration of pectin during the crosslinking treatment.

Park and Zhao investigated edible films based on cranberry pomace extract and effects of type and concentration of added pectins and plasticizers on the basic physical, mechanical, and moisture barrier properties of films [84]. Obtained films had bright red color and strong cranberry flavor. Films plasticized with sorbitol were denser in matrix structure and had higher color intensity than those of glycerol plasticized films. Films produced with LM pectin and sorbitol as plasticizer showed the highest tensile strength and the lowest elongation at break and water vapor permeability.

Treatments of pectin and agar by autoclaving at a ratio of 0.1%, prior to their being added to calcium caseinate and whey protein solutions, generated films with improved water vapor barrier properties of the films by 24% with respect to the formulation without the polysaccharides: from 85 x 10^{-2} to 64 x 10^{-2} g mm/(m^2day mmHg) [85]. γ-irradiation significantly reduced the barrier property of protein solution to moisture from 70 x $10^{-2}$for noniridiated solutions to 59 x 10^{-2} g mm/(m^2day mmHg) for irradiated solutions. Moreover, the addition of autoclaved pectin and agar to the treated protein solution further enhances the moisture barrier by 18%: from 59 x 10^{-2} to 48 x 10^{-2} g mm/(m^2 day mm Hg), due to the enhancement of aggregation caused by the interactions between the proteins and the polysaccharides.

Maftoonazad and Ramaswamy reported the effect of HM pectin-based coating on the kinetics of quality change in stored avocados [86]. Avocados were immersed in coating solution containing pectin, sorbitol and bees wax for 1 min at 20 ◦C and afterwards dried 3h and stored at three different temperatures (5, 10 and 20℃) for various times. Results revealed that avocados became softer and darker during the time and higher temperatures resulted in more rapid changes in the different quality parameters. Whereas the rate of CO_2 evolution at the climacteric peak in control samples was 287, 253 and 186 mL/(kg·h) at 20, 15 and 10 ◦C, respectively, reached after 6, 12 and 16 days, in coated samples, the peak values of 232, 210 and 152 mL/(kg·h) were observed after 8,14 and 22 days, respectively. Control fruits lost 5.2, 6.8 and 11.5% of their original weight during 7 days of storage at 10, 15 and 20 ◦C, respectively, while coated fruits lost 3.8, 4.5 and 9.1% under the same conditions. These results indicated that pectin-coating extended the shelf life of avocado.

It was observed that addition of a lipid component such as beeswax, glycerol monostearate or palmitic acid significantly enhanced effectiveness of polysaccharide coatings, indicating their regulation of the hydrophilic–hydrophobic balance, which would in turn, restrict the water loss. Moalemiyan et al. investigated the effect of pectin-based coating on the shelf-life and quality of mangoes [87]. The coating was consisted of HM pectin, sorbitol as plasticizer, beeswax and monoglyceride as emulsifiers. Mangoes were then dipped in the coating emulsion for 1 min at 20 ◦C and dried in a cold air draft for 4 h. After 6 days of storage, control mangoes lost 6.3% of their original weight, whereas coated mangoes lost only 4.4%. The control fruits could only be stored for 7 days before becoming unacceptable, whereas the coated fruits remained good up to 13 days of storage at room temperature, thereby improving the shelf-life of mangoes at room temperature.

The inclusion of nanoreinforcements emerges as a suitable approach for enhancing the physical properties of pectin-based films. Therefore, Moreira et al. investigated HM and LM pectin-based films modified by magnesium hydroxide ($Mg(OH)_2$) nanoplates [88]. The pectin–$Mg(OH)_2$ films were intended as bioactive food packaging materials. The inclusion of $Mg(OH)_2$could render these hybrids good fortifiers to foods, given the importance of Mg to human health. Migration studies using arugula leaves revealed an increase in content of Mg in the leaves after contact with the pectin–$Mg(OH)_2$film. The greatest increase of Mg content was detected in the leaves after 2.5 days contact with the HM pectin–$Mg(OH)_2$film (490 ±60 mg 100 g^{-1}). This investigation showed that the Mg content in arugula leaves was significantly incremented by 45%after contact with the films demonstrating good potential of pectin–$Mg(OH)_2$films as fortifiers of the magnesium content in foods.

The production of an edible hydrocolloid film based on citrus pectin and the protein phaseolin (Ph) without and with crosslink enzyme microbial transglutaminase (TG) was reported by Giosafatto et al. [89]. The effect of storage temperature on tensile strength and elongation to break was investigated. In particular, the materials were kept at room temperature 25◦C, 4◦C and −20◦C for 8 days and both tensile strength and elongation to break of Pectin/Ph films were found to remain fairly constant as the temperature changed. Films prepared in the presence of TG were more resistant at low temperatures (mostly at 4◦C) indicating that these films can also be used for the protection of food products stored at low temperatures. Both films prepared with and without TG exhibited a goodCO_2barrier property being almost 5 times lower than that exhibited by commercially available biodegradable Mater-Bi, while the film crosslinked by TG showed permeability to $O_2$240-fold and 620-fold lower than Mater-Bi and the synthetic LDPE, respectively. The permeability characteristics and water-solubility of edible pectin–soy flour films obtained in the absence or presence of the enzyme transglutaminase were also investigated and compared with the commonly used packaging films: high density polyethylene film and the biodegradable Mater-Bi® film [90]. The soy protein transglutaminase-catalyzed crosslinking was found to determine a significant decrease in the solubility of the pectin–soy flour films both at different pH and in different denaturing conditions compared to the films obtained in the absence of the enzyme, even though their solubility remained higher than that of the commercially available polyethylene and Mater-Bi® films. The films obtained in the presence of the TG exhibited permeability to oxygen and carbon dioxide lower than that possessed by polyethylene films.

Pectin-based films have been incorporated with several antimicrobial substances from natural sources in order to obtain antimicrobial active packaging that contributes to extend product shelf life and reduce the risk of pathogen growth on food surfaces. Pectin played an important role in embedding nisin-antimicrobial peptide into the film. Nisin has been used as a biopreservative in a variety of foods including dairy, eggs, vegetables, meat, fish, beverages and cereal-based products to inhibit growths of food borne pathogens including Listeria monocytogenes due to its non-toxicity, thermal stability and non-contribution of off-flavors. The antimicrobial activity of pectin films incorporated with 0.025% of nisin and its combination with treatment of ionizing radiation was prepared to control Listeria monocytogenes on a ready-to-eat (RTE) turkey meat [91]. Results showed that treatment of pectin films with nisin without irradiation reduced L. monocytogenes by 1.76 log CFU/cm^2, while the irradiation resulted in a 3.95 log CFU/cm^2 reduction at1 kGy and a 5.35 log CFU/cm^2 reduction at 2 kGy; indicating asynergistic effect on Listeria viability on the surface of RTE turkey meat. Pectin-nisin films used did not prevent but did significantly slow the proliferation of the L. monocytogenes cells that survived irradiation during 8 weeks of storage at 10◦C. The Jin et al. investigated the mechanical properties of pectin-polylactic acids (PLAs) films incorporated with nisin and influence of these films on inhibition of Listeria monocytogenes [92]. The pectin /PLA film had a similar stiffness as pure PLA film, but reduced tensile strength and fracture energy for 19% and 40%, respectively. The incorporation of nisin in pectin/PLA film didn't affect the mechanical properties. The results showed that cells of L. monocytogenes were reduced by 2.1, 4.5 and 3.7 log units m/L by the pectin /PLA composite film containing nisin (1000 IU m/L of tested liquid) in Brain Heart Infusion (BHI) broth, liquid egg white and orange juice, respectively, after 48 h at 24 ◦C. These experiments proved that nisin incorporated into the pectin /PLA film was an effective approach to reducing L. monocytogenes in a typical growth medium (e.g. BHI broth) as well as in foods (e.g. orange juice and liquid egg). Hence, the use of pectin and PLA in combination with nisin has shown a great potential in antimicrobial food packaging to reduce post-process growth of food pathogens.

In order to increase the antioxidant status and reduce bacterial growth of fresh-cut peach, Ayala-Zavala et al. developed pectin edible films with enriched with the essential oil from cinnamon leaves (CLO) [93]. Among essential oils, cinnamon leaf oil is a potent antibacterial and antioxidant agent. The addition of different concentrations of CLO had no significant effect on the water vapor permeability. WVP values ranged from 115 to 180 g/m^2/day. The coating treatment of fresh-cut peach using pectin/CLO films increased the antioxidant and antibacterial capacity with an increase of CLO concentration and reached the highest efficiency at a CLO concentration of 36.1 g/l.

The effects of cinnamon oil and ratio of soy protein isolate (SPI) to pectin on microbial growth in dry tofu during storage at 37 and 28 ◦C for 120 h were investigated by Liu et al. [94]. SPI-pectin films showed the antibacterial activity after the incorporation of cinnamon oil in concentrations above 0.2%. As the concentration of cinnamon oil increased, the zone of inhibition for both bacteria and mold/fungi increased significantly. The microbial growth in dry tofu was highly dependent on the medium, temperature and package of dry tofu. The microbial growth in control sample was 11.61cfu/g after 5 days of storage at 37 ◦C, while the number of microbial flora in the SPI-pectin film treatment increased to10.44 cfu/g. For fungal growth, the number of microbial flora without SPI-pectin film increased to 7.54 cfu/g in dry tofu after 5 days of storage at 28 ◦C, the number of microbial flora in the SPI-pectin film treatment reached value of 4.11 cfu/g. Compared to the control sample, it could be concluded that SPI-pectin film could provide microbiological safety of food products.

The effects of *Laurusnobilis* L. oil and oleic acid on the physical properties of apple pectin/cassava starch films were investigated by Taqi et al. [95]. The mechanical property of 1.5% (v/w) *L. nobilis* L. oil in edible film was significantly greater than the films with oleic acid and also with the control film. The films with different concentrations of L. nobilis L. oil had the lowest value of oxygen permeability rate (0.40, 0.37 and 0.26 cc/m^2/day) compared to the films with addition of different concentrations of oleic acid(0.43, 0.36 and 0.31 cc/m2/day) and pure apple pectin/cassava starch film (0.4833 cc/m^2/day), respectively. The similar dependence was observed for the water vapor permeability values. The WVP values of the film with different concentrations of L. nobilis L. oil vary between 4.49and 1.56 g·mm/m^2·day·Pa.

Edible films based on pectin and apple puree incorporated with oregano, cinnamon or lemongrass essential oils were investigated by Rojas-Grau et al. [96]. The antimicrobial activity of oregano oil in these films against E. coli O157:H7 was significantly greater than the activities of cinnamon oil and lemongrass oil. Whereas water vapor barrier properties decreased with an increase of concentration of incorporated essential oils, the oxygen permeability values increased. This effect was more prominent with oregano oil, reaching oxygen permeability value of 38.12 $cm^3 \mu m/m^2 d$ kPa compared to the control value of 22.64 $cm^3 \mu m/m^2 d$ kPa. The tensile strength of the films containing certain levels of antimicrobials did not differ significantly from control films without added antimicrobials.

Ravishankar et al. tested the influence of cinnamaldehyde or carvacrolin corporation in pectin-apple puree films on antimicrobial activity on chicken breasts and ham during storage at 23℃ and 4 ◦C for 72 h [97]. Results showed that films containing carvacrol exhibited stronger antimicrobial activity against E. coli on chicken breast surface at4 ◦C and 23 ◦C than those films with cinnamaldehyde, while the reduction of Salmonella enteritidis was similar for both carvacrol and cinnamaldehyde at the same temperature. For L. monocytogeneson ham, carvacrol films induced greater reductions than did cinnamaldehyde films at all concentrations tested. Cinnamaldehyde films had better barrier properties (water vapor release or absorption) than carvacrol films, and as the concentration of cinnamaldehyde increased in pectin-apple puree films, WVP values also increased. Carvacrol incorporated in pectin-apple puree films can also inactivated antibiotic-resistant Campylobacter jejuni in a phosphate buffer [98].

CONCLUSION

Consumer demand for minimizing the food waste has led to the development of new packaging materials. Pectin-based films are an environmental friendly alternative to petroleum-based polymers due to its low cost, availability, biodegradability, and its feasibility to produce films alone or in combination with other polymeric matrixes. When incorporated with natural bioactive compounds, pectin films showed antimicrobial properties against food borne pathogens. Although pectins fulfill the environmental concerns, they show some limitations in terms of performance like barrier and mechanical properties compared to the petroleum-based packaging materials. Therefore, this kind of packaging materials needs more research and introduction of smart and intelligent molecules able to make efficient replacement for petroleum-based packaging materials.

ACKNOWLEDGMENTS

The author acknowledges funding from the Ministry of Education, Science and Technological Development of the Republic of Serbia, Science Projects No. 43009.

REFERENCES

[1] Yokoyama, Y. In: *Package Design in Japan, Materials in packaging.;* Editor, Hashimoto, S., Rikuyo-sha Publishing, Tokyo, Japan, 1985, 1,113-115.

[2] Restuccia, D.; Gianfranco Spizzirri, U.; Parisi, O.I.; Cirillo, G.; Curcio, M.; Iemma, *F. Food Control.* 2010, 21 (11), 1425-1435.

[3] Sorrentino, A.; Gorrasi, G.; Vittoria, V. *Trends Food Sci. Techn.* 2007, 18, 84-95.

[4] Kirwan, M. J.; Strawbridge, J. W. *Food Packaging Technology* 2003, 174-240.
[5] Imam, S.H.; Glenn, G.M. *Emerging Food Packaging Technologies Principles and Practice* 2012, 435–468.
[6] Chiumarelli, M.; Hubinger, M.D. *Food Hydrocolloid*. 2014, 38, 20-27.
[7] Zhong, Y.; Cavender, G.; Zhao, Y. LWT - *Food Sci. Technol.* 2014, 56, 1-8.
[8] Mikkonen, K.S.; Rita, H.; Hele, H.; Talja, R.A.; Hyvonen, L.; Tenkanen, M. *Biomacromolecules*. 2007, 8, 3198-3205.
[9] Campos, C.A.; Gerschenson, L.N.; Flores, S.K. Food *Bioprocess Technol.* 2011, 4, 849–875.
[10] Lim, L.T.; 2011. In: *Comprehensive Biotechnology. Active and intelligent packaging materials.*; Editors, Cooney, C.L.; Humphrey, A.E. second ed. Pergamon Press, Oxford, U.K, 2011, 4, 629-644.
[11] Han, J.H. *Int. Rev. Food Sci. Technol.* 2003, 11, 106-109.
[12] Suppakul, P.; Miltz, J.; Sonneveld, K.; Bigger, S.W. *J. Food Sci.* 2003, 68 (2), 408-420.
[13] Stiles, M.E. In: *Modified Atmosphere Packaging of Food. Modified atmosphere packaging of meat, poultry and their products.*; Editors, Ooraikul, B.; Stiles, M.E. Ellis Horwood, Chichester, UK, 1991.
[14] Speranza, B.; Corbo, M.R.; Conte, A.; Sinigaglia, M.; Del Nobile, M.A. *J. Food Sci.* 2009, 74 (9), M473-M478.
[15] Skura, B.J. In: *Modified Atmosphere Packaging of Foods. Modified atmosphere packaging of fish and fish products.*; Editors, Ooraikul, B.; Stiles, M.E. Ellis Horwood, Chichester, UK, 1991.
[16] Germain, Y. *Industrie´eAlimentaireet Agricules* 1997, 137-140.
[17] Siracusa, V.; Rocculi, P.; Romani, S.; Rosa, M.D. *Trends Food Sci. Tech.* 2008, 19, 634-643.
[18] Lee, K.T.; Goddard, J.M.; Hotchkiss, J.H.; *Packag. Technol. Sci.* 2009, 22, 139-150.
[19] Kiery, L.J.; Olson, N.F. *Food Microbiol.* 2000, 17, 277-291.
[20] Fox, H.W.; Zisman, W.A. *J. Colloid Sci.* 1950, 5, 520.
[21] Sanchis, M.R.; Blanes, V.; Blanes, M.; Garcia, D.; Balart, R. *Eur. Polym. J.* 2006, 42, 1558-1568.
[22] Abbas, A.; Vivien, C.; Bocquet, B.; Guillochon, D.; Supiot, P. Plasma *Processes Polym.* 2009, 6, 593-604.
[23] Bosset, J.Q.; Sieber, R.; Gallman, P.U. In: *Food Packaging and Preservation. Influence of light transmittance of packaging materials on the shelf life of milk and milk products -a review.*; Editor, Mathlouthi, M. Blackie Academic and Professional, Glasgow, UK,1994.
[24] Ferrara, G.; Bertolodo, M.; Scoponi, M.; Ciardelli, F. *Polym. Degrad. Stab.* 2001, 73, 411–416.
[25] Auras, R.; Singh, S. P.; Singh, J. J. *Packag. Techn. Sci.* 2005, 18, 207-216.
[26] Guilbert, S.; Gontard, N.; Gorris, L.G.M. *Lebensm. Wiss. Technol.* 1996, 29, 10-17.
[27] Avena-Bustillos, R.J.; Krochta, J.M. *J. Food Sci.*1993, 58, 904–907.
[28] Hernandez-Izquierdo, V.M.; Krochta, J.M. *J. Food Sci.* 2008, 73(2), R30–39.
[29] Weber, G.H.; Haugard, V.; Festersen, R. *Food Addit. Contamin* 2002, 19, 172–177.
[30] Zhang, Y.; Liu, Z. In: *Environmentally compatible food packaging. Starch-based edible films.;* Editor, Chiellini, E. Cambridge, UK: Wood head Publishing Ltd. 2009, 108–136.
[31] Madziva, H.; Kailasapathy, K.; Phillips, M. J. *Microencapsul.* 2005, 22, 343-351.

[32] Bernabé, P.; Peniche, C.; Monal, W.A. *Polym. Bull.* 2005, 55, 367-375.
[33] Emaga, T.H.; Robert, C.; Ronkart, S.N.; Wathelet, B.; Paquot, M. *Biores. Technol.* 2008, 99, 4346–4354.
[34] Munarin, F.; Tanzi, M.C.; Petrini, P. *Inter. J. Biolog. Macromolecules.* 2012, 51, 681–689.
[35] Espitia, P.J.P.; Du, W.X.; Avena-Bustillos, R.; Ferreira Soares, N.; McHugh, T.H. *Food Hydrocolloid.* 2014, 35, 287-296.
[36] Visser, J.; *Pectins and pectinases, progress in biotechnology.*; Editor, Voragen, A.G.J. Elsevier, Amsterdam, 1996, 14.
[37] Mohnen, D. *Curr. Opin. Plant. Biol.* 2008, 11, 266–277.
[38] O'Neill, M.; Albersheim, P.; Darvill, A. In: *Methods in Plant Biochemistry. The pectic polysaccharides of primary cell walls.;* Editor, Dey, P.M. London: Academic Press, 1990, 415-441.
[39] Jarvis, M. C.; Apperley, D. C. *Carbohydr. Res.* 1995, 275, 131–145.
[40] Ridley, B.L.; O'Neill, M.A.; Mohnen, D. *Phytochem.* 2001, 57, 929–967.
[41] Willats, W.G.T.; Knox, P.; Mikkelsen, J.D. *Trends Food Sci. Technol.* 2006; 17, 97-104.
[42] Caffall, K.H.; Mohnen, D. *Carbohydr. Res.* 2009, 344, 1879–1900.
[43] Darvill, A.; McNeil, M.; Albersheim, P. *Plant Physiol.* 1987, 62, 418–422.
[44] Spellman, M. W.; McNeil, M.; Darvill, A. G.; Albersheim, P. *Carbohydr. Res.* 1983, 122, 115-129.
[45] Stevenson, T. T.; Darvill, A. G.; Albersheim, P. *Carbohydr. Res.* 1988, 179, 269–288.
[46] York, W. S.; Darvill, A. G.; McNeil, M.; Albersheim, P. *Carbohydr. Res.* 1985, 138, 109–126.
[47] Kobayashi, M.; Matoh, T.; Azuma, J.; *Plant. Physiol.* 1996, 110, 1017–1020.
[48] Brummer, Y.; Cui, S.W. In: *Food carbohydrates, chemistry physical properties, and applications. Understanding carbohydrate analysis.;* Editor, Cui, S.W. New York: CRS Press, 2005, 67–104.
[49] Chaplin, M. F. In: *Carbohydrate analysis. Monosaccharides.*; Editors, Chaplin, M.F.; Kennedy, J.F. Oxford: Oxford University Press, 1994, 1-41.
[50] Yamada, H.; Sun, X.-B.; Matsumoto, T.; Ra, K.-S.; Hirano, M.; & Kiyohara, H. *Planta Medica.* 1991, 57, 555–559.
[51] Lv, Y.; Yang, X.; Zhao, Y.; Ruan, Y.; Yang, Y.; Wang, Z. *Food Chem.* 2009, 112, 742–746.
[52] Daas, P.J.H.; Voragen, A.G.J.; Schols, H.A. *Biopolymers.* 2001, 58,195–203.
[53] Sharma, B. R.; Naresh, L.; Dhuldhoya, N. C.; Merchant, S. U.; Merchant, U. C. *Times Food Process J.* 2006, 4, 44-51.
[54] Oakenfull, D.; Scott, A. *J. Food Sci.* 1984, 49, 1093-1098.
[55] Grant, G. T.; Morris, E. R.; Rees, D. A.; Smith, P. J. C.; Thom, D. *FEBS Letters.* 1973, 32(1), 195–198.
[56] Morris, E. R.; Powell, D. A.; Gidley, M. J.;Rees, D. A. *J. Molec. Biol.* 1982, 155(4), 507–516.
[57] Fraeye, I.; Duvetter, T.; Doungla, E.; van Loey, A.; Hendrickx, M. *Trends Food Sci. Technol.* 2010, 21, 219-228.
[58] Thakur, B.R.; Singh, R.K.; Handa, A.K. *Crit. Rev. Food Sci. Nutr.* 1997, 37, 47-73.
[59] May, C.D. *Carbohyd. Polym.* 1990, 12, 79-99.

[60] Rolin, C. In: *Industrial gums. Pectin.;* Editors, Whistler, R.L.; BeMiller, J.N. New York: Academic Press, 1993.

[61] May, C. D. *Carbohydr. Polym.* 1990, 12, (1), 79-99.

[62] *Extraction of pectin by microwave heating under pressure*, United States patent 6143337, 2007.

[63] Kravtchenko, T. P.; Arnould, I.; Voragen, A. G. J.; Pilnik, W. *Carbohydr. Polym.* 1992, 19 (4), 237.

[64] Renard, C.M.G.C.; Thibault, J.F. *Progr. Biotechnol.* 1996, 14, 603-608.

[65] Krall,.S; McFeeters R. *J. Agric. Food Chem.* 1998, 46,1311–1315.

[66] Jolie, R.P.; Christiaens, S.; De Roeck, A.; Fraeye,I.; Houben,K.; Buggenhout, S.V.; Van Loey, A.M; Hendrickx M.E. Trends *Food Sci. Technol.* 2012, 24, 103-118.

[67] Schols, H.A.; Gereads, C.C.J.M.; Searle-van Leeuwen, M.F.; Kormelink, F.J. M.; Voragen, A.G.J. *Carbohydr. Res.* 1990, 206, 105-115.

[68] Jolie, R.P.; Duvetter, T.; Van Loey, A.M.; Hendrickx, M. E. *Carbohydr. Res.* 2010, 345, 2583-2595.

[69] Benen, J.A.E.;Visser, J. In: *Handbook of food enzymology. Pectate and pectin lyases.;* Editors, Whitaker, J.R.; Voragen, A.G.J.;Wong, D.W.S. New York: Marcel Dekker Inc 2003a, 1029-1041.

[70] Benen, J.A.E.; Visser, J. In: *Handbook of food enzymology. Polygalacturonases.;* Editors: Whitaker, J.R.; Voragen, A.G.J.; Wong, D.W.S. New York: Marcel Dekker Inc 2003b.

[71] Searle-van Leeuwen, M. J. F.; Vincken, J.P.; Schipper, D.; Voragen, A.G.J.; Beldman, G. In: *Pectins and Pectinases. Acetyl esterases of Aspergillusniger: Purification and mode of action on pectins.;* Editors, Visser, J.; Voragen, A.G.J. Amsterdam: Elsevier Science 1996, 793-798.

[72] Giancone, T.; Torrieri, E.; Di Pierro, P.; Cavella, S.;Giosafatto, C.V.L.; Masi, P. *Food Bioprocess Technol.* 2011,4, 1228–1236.

[73] Miller, K.S.; Krochta, J.M. *Trends Food Sci. Technol.* 1997, 8, 226–237.

[74] Arvanitoyannis, I.;Psomiadou, E.; Nakayama, N. *Carbohydr. Polym.* 1996, 31, 179–192.

[75] Chillo, S.; Flores, S.; Mastromatteo, M.; Conte, A.; Gerschenson, L.; Del Nobile, M.A. *J. Food Eng.* 2008,88, 159–168.

[76] Di Pierro, P., Chico, B., Villalonga, R., Mariniello, L., Masi, P., & Porta, R. *Enzyme Microb. Tech.* 2007, 40, 437–441.

[77] Di Pierro, P.; Chico, B.; Villalonga, R.; Mariniello, L.; Damiao, A.E., Masi, P. *Biomacromolecules.* 2006,7, 744–749.

[78] Bruno, M.; Giancone, T.; Torrieri, E.; Masi, P.; Moresi, M. *Food Bioprocess Technol.* 2008,1, 393–404.

[79] Bertuzzi, M. A.; Castro Vidaurre, E. F.; Armada, M.; Gottifredi, J.C. Water vapour permeability of edible starch based films. *J. Food Eng.* 2007, 80, 972–978.

[80] Flores, S.; Fama, L.; Rojas, A. M.; Goyanes, S.; Gerschenson, L. *Food Res. Inter.* 2007,40, 257–265.

[81] Alves, V.; Costa, N.; Hilliou, L.; Larotonda, F.; Gonçalves, M.; Sereno, A.; Coelhoso, I. M. *Desalination.* 2006,199, 331–333.

[82] Alves, V.D.; Castelló, R.; Ferreira, A.; Costa, N.; Fonseca, I.M.; Coelhoso, I.M. *Procedia Food Sci.* 2011, 1, 240 – 245.

[83] Liu, L.; Liu, Ch.; Fishman,M.L.; Hicks, K.B. *J. Agric. Food Chem.* 2007, 55, 2349-2355.

[84] Park, S.; Zhao, Y. *Journal of food science* 2006, 71(2), E95-E101.

[85] Letendre, M.; D'Aprano, G.; Lacroix, M.; Salmieri, S.; St-Gelais, D.J *Agric. Food Chem.* 2002, 50, 6017-6022.

[86] Maftoonazad, N.; Ramaswamy, H.S. *J. Food Process Pres.* 2008, 32, 621–643.

[87] Moalemiyan, M.; Ramaswamy, H.S.; Maftoonazad, N. *J. Food Process Eng.* 2012, 35, 572–600.

[88] Moreira, F. K.V.; De Camargo, L.A.; Marconcini, J.M.; Mattoso, L.H.C. *J. Agric. Food Chem.* 2013, 61, 7110−7119.

[89] Valeria, C.;Giosafatto, L.; Di Pierro, P.; Gunning, P.; Mackie,A.; Porta, R.; Mariniello, L. *Carbohydr. Polym.* 2014, 106, 200–208.

[90] Di Pierro, P.; Mariniello, L.; Giosafatto, C. V. L.; Masi, P.; Porta, R. *Food Biotechnol.* 2005, 19, 37-49.

[91] Jin, T.; Liu, L., Sommers, C. H.; Boyd, G.; Zhang, H. *J. Food Protect.* 2009, 72(3), 644-649.

[92] Jin, T.; Liu, L.; Zhang, H.; Hicks, K. *Inter. J. Food Sci. Technol.* 2009, 44, 322–329.

[93] Ayala-Zavala, J.F.; Silva-Espinoza, B.A.; Cruz-Valenzuela, M.R.; Leyva, J.M.; Ortega-Ramírez, L.A.; Carrazco-Lugo, D.K.; Pérez-Carlón, J.J.; Melgarejo-Flores, B.G.; González-Aguilara G.A.; Miranda, M.R.A. *Flavour Fragr. J. 2013*, 28, 39–45.

[94] Liu, H.; Li, J.; Zhu, D.; Wang, Y.; Zhao, Z.; Li, J. *J. Food Process Pres*, 2013, ISSN 1745-4549, DOI: 10.1111/jfpp.12117.

[95] Taqi, A.; Mutihac, L.; Stamatin, I. *J. Food Process Pres*, 2013, ISSN 1745-4549, DOI: 10.1111/jfpp.12174.

[96] Rojas-Graü, M. A.; Avena-Bustillos, R. J.; Friedman, M.; Henika, P. R.; Martín-Belloso, O.; McHugh, T. H. *J. Agricult. Food Chem.* 2006, 54(24), 9262-9267.

[97] Ravishankar, S.; Zhu, L.; Olsen, C. W.; McHugh, T. H.; Friedman, M. *J. Food Sci.* 2009, 74(8), M440-M445.

[98] Ravishankar, S.; Zhu, L.; Law, B.; Joens, L.; Friedman, M. *J. Food Prot.* 2008, 71(6), 1145–1149.

INDEX

D

E

G

H

J

K

L

M

Q

R

S

T

U

V